Operator Theory: Advances and
Applications
Vol. 154

Spectral Methods for Operators of Mathematical Physics

J. Janas
P. Kurasov
S. Naboko
Editors

Springer Basel AG

Editors:

Jan Janas
Institute of Mathematics PAN
ul. Sw. Tomasz 30
31-027 Cracow
Poland
e-mail: najanas@cyf-kr.edu.pl

Pavel Kurasov
Department of Mathematics
Lund Institute of Technology
Box 118
22100 Lund
Sweden
e-mail: kurasov@maths.lth.se

Sergei Naboko
Department of Mathematical Physics
Institute of Physics
St. Petersburg University
Ulianovskaia 1
198904 St. Petersburg
Russia

2000 Mathematics Subject Classification 47-06; 47A10, 47A45, 47A75, 47B25, 47B39

A CIP catalogue record for this book is available from the
Library of Congress, Washington D.C., USA

Bibliographic information published by Die Deutsche Bibliothek
Die Deutsche Bibliothek lists this publication in the Deutsche Nationalbibliografie; detailed bibliographic data is available in the Internet at <http://dnb.ddb.de>.

ISBN 978-3-0348-9632-0 ISBN 978-3-0348-7947-7 (eBook)
DOI 10.1007/978-3-0348-7947-7

Originally published by Birkhäuser Verlag , Basel, Switzerland in 2004
Softcover reprint of the hardcover 1st edition 2004
Printed on acid-free paper produced from chlorine-free pulp. TCF ∞
Cover design: Heinz Hiltbrunner, Basel

ISBN 978-3-0348-9632-0

9 8 7 6 5 4 3 2 1

www.birkhauser-science.com

Contents

Introduction

This volume contains proceedings of the International Conference: Operator Theory and its Applications in Mathematical Physics – OTAMP 2002 held at Mathematical Research and Conference Center in Bedlewo near Poznan.

The aim of the conference was to gather researches working in close areas of operator theory, analysis and mathematical physics, which is reflected by the titles of scientific sessions

- Random and quasiperiodic Schrödinger operators
  (P. Stollmann and G. Stolz);
- Jacobi matrices and orthogonal polynomials
  (J.S. Geronimo and W. Van Assche);
- Singular perturbations of self-adjoint operators
  (W. Karwowski and P. Kurasov).

The current volume contains in addition to materials of the lectures given at the conference original research articles, several ones initiated during the conference. Two main entirely connected themes dominate the volume

- spectral properties of 1-dimensional Schrödinger operators
  and infinite Jacobi matrices,
- theory of self-adjoint and dissipative operators.

Contributions devoted to the first theme contain in particular results on the existence and finiteness of the point spectrum of Jacobi matrices, bounds for the points of spectral concentration of one-dimensional Schrödinger operators, WKB and turning points for the second order difference equations. The second theme is represented by the articles devoted to partial non-stationary perturbation determinants, self-adjointness by domination of commutators, symmetric functional models etc.

The Organizing Committee of the conference takes this opportunity to thank all session organizers for helping in putting together the scientific programm and all participants for coming to Bedlewo and making this conference into a useful scientific event. Special thanks go to M. Moszynski for helping in organization. We would like to thank

Stefan Banach International Mathematical Center

for generous financial support. It is impossible to imagine the conference without warm hospitality from everybody working at

Mathematical Research and Conference Center in Bedlewo.

The Editors would like to thank all the referees assisting in preparation of this volume and coming with numerous suggestions helping to keep the high standard of this volume.

Finally we are indebted to The Editorial Board and in particular to Professor I. Gohberg for including these Proceedings in the series *Operator Theory: Advances and Applications* and to Birkhäuser Verlag for patience and help in preparation of the volume.

Krakow-Lund-St. Petersburg, March 2004
The Editors

Operator Theory:
Advances and Applications, Vol. 154, 1–18

Partial Non-stationary Perturbation Determinants

Vadim Adamyan and Heinz Langer

Abstract. A partial non-stationary perturbation determinant $\Delta_1(t)$ is defined as follows:

$$\Delta_1(t) := \det\left(e^{itA} P_1 e^{-itH}\big|_{\mathcal{H}_1}\right), \qquad t > 0;$$

here A is a self-adjoint operator in some Hilbert space $\mathcal{H}_1$, H is a self-adjoint operator in a larger Hilbert space $\mathcal{H} \supset \mathcal{H}_1$, P_1 is the orthogonal projection in $\mathcal{H}$ onto $\mathcal{H}_1$ and $P_1(H - A)|_{\mathcal{H}_1}$ is a trace class operator in $\mathcal{H}_1$. If the operator $P_1 H(I - P_1)$ is finite-dimensional, $\Delta_1(t)$ is expressed by the resolvent kernel of a system of Fredholm integral equations on $(0, t)$ of second kind. Moreover, in a particular situation the asymtotic behavior of $\Delta_1(t)$ for $t \to \infty$ is studied.

Mathematics Subject Classification (2000). 47D06, 47A55, 47B10, 47A99; 45B05, 47A40.

Keywords. Evolution equation, groups of unitary operators, perturbation determinant, block operator matrix, spectral shift function.

1. Introduction

By a *non-stationary perturbation determinant* we mean a function of the form

$$\Delta(t) := \det\left(e^{\mathrm{i}tH_0} e^{-\mathrm{i}tH}\right), \qquad t > 0, \tag{1.1}$$

where H_0 and H are self-adjoint operators in some Hilbert space $\mathcal{H}$ such that the difference $H - H_0$ is a trace class operator. A *partial non-stationary perturbation determinant* is defined by the relation

$$\Delta_1(t) := \det\left(e^{\mathrm{i}tA} P_1 e^{-\mathrm{i}tH}\big|_{\mathcal{H}_1}\right), \qquad t > 0, \tag{1.2}$$

where A is a self-adjoint operator in some Hilbert space $\mathcal{H}_1$, H is a self-adjoint operator in a larger Hilbert space $\mathcal{H} \supset \mathcal{H}_1$, P_1 is the orthogonal projection in $\mathcal{H}$

V. Adamyan was supported by grant CRDF UM1 2090 of the U.S. Civilian Research and Development Foundation. He also thanks the Vienna University of Technology for its hospitality.

onto $\mathcal{H}_1$, and $P_1(H-A)\big|_{\mathcal{H}_1}$ is a trace class operator in $\mathcal{H}_1$. We are interested in more explicit expressions for $\Delta(t)$ and $\Delta_1(t)$ and, in particular, in the asymptotic behavior of $\Delta_1(t)$ for $t \longrightarrow \infty$.

These studies are motivated by the paper [4] which, in turn, was inspired by results of P.W. Anderson, cf. [5]. In [4] it was shown that the function $I(\omega)$, which describes the intensity of the spectral lines in the x-ray photoemission spectrum of metals plotted against the radiation frequency ω, within a certain degree of approximation can be calculated by the formula

$$I(\omega) = \lim_{\varepsilon \downarrow 0} \Re \int_0^\infty e^{i\omega t} e^{-\varepsilon t} \det\left(I + E^0_{\varepsilon_F}\left(e^{itH_0}e^{-itH} - I\right)\right) dt. \tag{1.3}$$

Here H_0 and H are the effective self-adjoint one-particle energy operators for the electron in the metal before and after the photoemission, E^0_λ is the spectral function of H_0, and ε_F is the Fermi energy level.

On the right-hand side of (1.3) there appears a partial non-stationary perturbation determinant of the form

$$\Delta_1(t) = \det\left(I + E^0_\lambda\left(e^{itH_0}e^{-itH} - I\right)\right) \tag{1.4}$$

with $\mathcal{H}_1 = \operatorname{ran} E^0_\lambda$ and $A = P_1 H_0 P_1$. In this case the operators H and H_0 in (1.4) have the following additional property: if we represent them by block operator matrices with respect to the decomposition $\mathcal{H} = \mathcal{H}_1 \oplus \mathcal{H}_1^\perp$, $\mathcal{H}_1 := P_1\mathcal{H} = E^0_\lambda\mathcal{H}$, of the space $\mathcal{H}$, then H and H_0 have the same diagonal blocks, say A and D and the spectra of these diagonal blocks are weakly separated. The latter means that there exists some real α such that

$$\max\sigma(A) \le \alpha \le \min\sigma(D) \tag{1.5}$$

and that α is not an eigenvalue of A and of D. Therefore the results from [1], [3] can be applied.

This note is organized as follows. In Section 2 we consider non-stationary perturbation determinants (1.2), actually in the slightly more general setting with the unitary groups replaced by families of solutions of two evolution equations. For the particular case of self-adjoint operators H_0 and H, $H = H_0 + V$ with a trace class operator V, it turns out that the well-known formula for the finite-dimensional case generalizes to

$$\Delta(t) \equiv \det\left(e^{itH_0}e^{-itH}\right) = e^{-i\,(\operatorname{tr} V)\,t}, \qquad t > 0.$$

In Section 3 we consider partial non-stationary perturbation determinants under the assumption that the diagonal blocks of H have weakly separated spectrum. This allows to apply the angular operator representations from [2], [3] of the spectral subspaces of H, corresponding to $(-\infty,\alpha)$ and $(\alpha,+\infty)$. As the main result of this note we show in Theorem 3.4 that if the spectrum of the perturbed operator is absolutely continuous in at least one of the intervals $(-\infty,\alpha), (\alpha,+\infty)$ then

$$\Delta_1(t) = e^{-b-iat}\left(1 + o(1)\right)$$

with non-negative numbers a and b. Moreover, under an additional assumption these numbers are zero if and only if the off-diagonal blocks of H are zero.

Finally, in Section 4 for the case of a finite-dimensional off-diagonal perturbation we express the partial non-stationary perturbation determinant by the Fredholm kernel of a system of second kind Fredholm integral equations.

2. Non-stationary perturbation determinants

Let $\mathcal{H}$ be a Hilbert space. In the sequel we often deal with integrals $\int_s^t S(\tau)\,d\tau$ for operator-valued functions $S(t)$ which are continuous in the nuclear norm $\|\cdot\|_1$ of the ideal $\mathcal{S}_1$ of trace class operators in $\mathcal{H}$. These integrals are to be understood as the limits of the corresponding Riemann sums with respect to this norm. Evidently, they define operators of the trace class $\mathcal{S}_1$.

The following theorem is formulated in a more general form than needed below. We use the notion of an *evolution system* as defined in [10, Definition 5.3], which is a two-parameter family of bounded linear operators $W(t,s)$, $0 \le s \le t < \infty$, in $\mathcal{H}$ such that the following two conditions are satisfied for $0 \le s \le r \le t < \infty$:

(i) $W(s,s) = I$, $W(t,s) = W(t,r)W(r,s)$,

(ii) the mapping $(t,s) \longrightarrow W(t,s)$ is strongly continuous.

Sufficient conditions for the fact that the operator functions $H_0(t)$ and $H(t)$ generate evolution systems can be found in [10, Chapter 5].

Theorem 2.1. *For $t \in [0,\infty)$, let $H_0(t)$ be self-adjoint and $H(t)$ be densely defined operators, such that*

$$H(t) - H_0(t) =: V(t) \in \mathcal{S}_1,$$

and that the operator function $V(t)$ is continuous on $[0,\infty)$ with respect to the norm of $\mathcal{S}_1$. Suppose further that there exist evolution systems

$$W_0(s,t),\ W(s,t), \quad 0 \le s \le t < \infty,$$

and a dense subset $\mathcal{D} \subset \mathcal{H}$, such that for $0 \le s < t$, $x \in \mathcal{D}$, it holds

$$\mathrm{i}\,\frac{\partial W_0(t,s)x}{\partial t} = H_0(t)W_0(t,s)x, \qquad \mathrm{i}\,\frac{\partial W(t,s)x}{\partial t} = H(t)W(t,s)x,$$

with the derivatives on the left-hand sides to be understood with respect to the norm of $\mathcal{H}$. Then

$$W_0(t,s)^*W(t,s) = I - \mathrm{i}\int_s^t W_0(\tau,s)^*V(\tau)W(\tau,s)\,d\tau, \tag{2.1}$$

where the integral on the right-hand side exists with respect to the norm of $\mathcal{S}_1$ and belongs to $\mathcal{S}_1$, and

$$\Delta(t,s) := \det\,(W_0(t,s)^*W(t,s)) = e^{-\mathrm{i}\int_s^t \mathrm{tr}\,[V(\tau)]\,d\tau} \tag{2.2}$$

Proof. For $x, y \in \mathcal{D}$ and $0 \le s < t < \infty$ we consider the following relation where $'$ denotes the derivative with respect to t:

$$\begin{aligned}(W(t,s)x, W_0(t,s)y)' &= (W'(t,s)x, W_0(t,s)y) + (W(t,s)x, W_0'(t,s)y)\\ &= (-\mathrm{i}H(t)W(t,s)x, W_0(t,s)y) + (W(t,s)x, -\mathrm{i}H_0(t)W_0(t,s)y)\\ &= -\mathrm{i}\,((H(t)-H_0(t))W(t,s)x, W_0(t,s)y)\\ &= -\mathrm{i}\,(V(t)W(t,s)x, W_0(t,s)y)\,.\end{aligned}$$

Since the expression on the right-hand side is a continuous function of t for $s \le t$ and the set $\mathcal{D}$ is dense in $\mathcal{H}$ the relation (2.1) follows. From the special choice $H(t) = H_0(t)$ we find that

$$W_0(t,s)^*W_0(t,s) = I, \qquad 0 \le s \le t < \infty.$$

Further,

$$\begin{aligned}W_0(t+\delta,s)^*W(t+\delta,s) &= W_0(t,s)^*W_0(t+\delta,t)^*W(t+\delta,t)W(t,s)\\ &= [W_0(t,s)^*W_0(t+\delta,t)^*W(t+\delta,t)W_0(t,s)]\,[W_0(t,s)^*W(t,s)]\end{aligned}$$

and hence

$$\det[W_0(t+\delta,s)^*W(t+\delta,s)] = \det[W_0(t+\delta,t)^*W(t+\delta,t)]\,\det[W_0(t,s)^*W(t,s)]$$

or

$$\Delta(t+\delta,s) = \Delta(t+\delta,t)\Delta(t,s). \tag{2.3}$$

It follows that

$$\frac{\Delta(t+\delta,s)-\Delta(t,s)}{\delta} = \Delta(t,s)\frac{\Delta(t+\delta,t)-1}{\delta}. \tag{2.4}$$

We denote $S(\delta) := \int_t^{t+\delta} W_0(\tau,t)^*V(\tau)W(\tau,t)\,d\tau$. Then $\|S(\delta)\|_1 < 1$ if $|\delta|$ is sufficiently small. The relation (2.1) implies for these values of δ

$$\begin{aligned}\Delta(t+\delta,t) &= \det\,(W_0(t+\delta,t)^*W(t+\delta,t))\\ &= \det\,(I-\mathrm{i}\,S(\delta))\\ &= \exp\,(\mathrm{tr}\,[\ln(I-\mathrm{i}\,S(\delta))])\\ &= \exp\,(-\mathrm{i}\,\mathrm{tr}\,[S(\delta)] + \mathrm{O}(\|S(\delta)\|_1^2))\,,\end{aligned}$$

and we find for $\delta \longrightarrow 0$:

$$\begin{aligned}\frac{\Delta(t+\delta,t)-1}{\delta} &= \frac{\exp\,(-\mathrm{i}\,\mathrm{tr}[S(\delta)] + \mathrm{O}(\|S(\delta)\|_1^2)) - 1}{\delta}\\ &= \frac{(-\mathrm{i}\,\mathrm{tr}[S(\delta)] + \mathrm{O}(\|S(\delta)\|_1^2))}{\delta}\\ &= -\mathrm{i}\,\mathrm{tr}\left[\frac{1}{\delta}\int_t^{t+\delta} W_0(t;s)^*V(s)W(t;s)\,ds\right] + \frac{\mathrm{O}(\|S(\delta)\|_1^2)}{\delta}\\ &\longrightarrow -\mathrm{i}\,\mathrm{tr}\,V(t),\end{aligned}$$

and, finally,

$$\frac{d\,\Delta(t,s)}{dt} = -\mathrm{i}\,\mathrm{tr}\,[V(t))]\,\Delta(t,s).$$

Observing that $\Delta(s,s) = 1$ the relation (2.2) follows.

□

If, in Theorem 2.1, $H_0(t) \equiv H_0$ and $H(t) \equiv H$ are constant self-adjoint operators, then with the unitary groups generated by H_0 and H we have, e.g.,

$$W_0(t,s) = e^{-\mathrm{i}(t-s)H_0},\ 0 \le s \le t,$$

and the relation (2.2) implies the following corollary.

Corollary 2.2. *If H_0 and H are self-adjoint operators and $H - H_0 \in \mathcal{S}_1$, then*

$$\det\left(e^{\mathrm{i}tH_0}e^{-\mathrm{i}tH}\right) = e^{-\mathrm{i}t\ \mathrm{tr}(H-H_0)}. \tag{2.5}$$

A discrete analog is as follows. Let H_0 and H be as in Corollary 2.2. For $z \neq \overline{z}$ we consider the Cayley transforms

$$U_0(z) := (H_0 - z)(H_0 - \overline{z})^{-1}, \quad U(z) := (H - z)(H - \overline{z})^{-1}, \tag{2.6}$$

which are unitary operators. Since $V = H - H_0$ is a trace class operator the unitary operator $U_0(z)^{-1}U(z)$ has the property

$$\begin{aligned} U_0(z)^{-1}U(z) - I &= U_0(z)^{-1}\left[U(z) - U_0(z)\right] \\ &= 2i\,\Im z\; U_0(z)^{-1}(H-\overline{z})^{-1}V(H_0-\overline{z})^{-1} \in \mathcal{S}_1. \end{aligned}$$

Hence for $n \in \mathbb{N}$ we have

$$U_0(z)^{-n}U(z)^n - I = \sum_{k=0}^{n-1} U_0(z)^{-k}\left(U_0(z)^{-1}U(z) - I\right)U(z)^k \in \mathcal{S}_1.$$

By $D_{H/H_0}(z)$ we denote the perturbation determinant of the ordered pair H, H_0, cf. [7]:

$$D_{H/H_0}(z) = \det\left(I + (H_0 - z)^{-1}V\right) = \det\left((H_0 - z)^{-1}(H - z)\right).$$

Theorem 2.3. *If H_0 and H are self-adjoint operators, $H - H_0 \in \mathcal{S}_1$ and $U_0(z)$, $U(z)$ are the Caley transforms given by (2.6) then*

$$\det\left[U_0(z)^{-n}U(z)^n\right] = e^{2n\mathrm{i}\arg D_{H/H_0}(z)}. \tag{2.7}$$

Proof. The properties of determinants (cf. [7]) imply

$$\begin{aligned} \widetilde{d}_n : &= \det\left[U_0(z)^{-n}U(z)^n\right] = \det\left[U_0(z)^{-(n-1)}U(z)^{n-1}U(z)U_0(z)^{-1}\right] \\ &= \widetilde{d}_{n-1}\det\left[U(z)U_0(z)^{-1}\right] = \widetilde{d}_{n-1}\det\left[U_0(z)^{-1}U(z)\right] = \cdots \\ &= \left(\det\left[U_0(z)^{-1}U(z)\right]\right)^n. \end{aligned}$$

Further,

$$\begin{aligned}\det\left[U_0(z)^{-1}U(z)\right] &= \det\left[(H_0-\overline{z})(H_0-z)^{-1}(H-z)(H-\overline{z})^{-1}\right]\\ &= \det\left[(H-\overline{z})^{-1}(H_0-\overline{z})(H_0-z)^{-1}(H-z)\right]\\ &= D_{H_0/H}(\overline{z})D_{H/H_0}(z).\end{aligned}$$

Taking into account that for a pair of self-adjoint operators H_0, H with $H-H_0 \in \mathcal{S}_1$ we have $D_{H_0/H}(\overline{z}) = \left(\overline{D_{H/H_0}(z)}\right)^{-1}$ the claim follows. □

3. Asymptotic behavior of $\Delta_1(t)$

Let again A be a self-adjoint operator in the Hilbert space $\mathcal{H}_1$, and consider in the Hilbert space $\mathcal{H} = \mathcal{H}_1 \oplus \mathcal{H}_2$ the operator

$$H = \begin{pmatrix} A+V & B \\ B^* & D \end{pmatrix} \tag{3.1}$$

which is supposed to be self-adjoint. We assume that $V, B \in \mathcal{S}_1$ and, additionally, that the spectra of $A+V$ and D are weakly separated, that is, for some $\alpha \in \mathbb{R}$,

$$\max\sigma(A+V) \le \alpha \le \min\sigma(D) \tag{3.2}$$

and α is neither an eigenvalue of $A+V$ nor of D. By $\mathcal{L}_-$ and $\mathcal{L}_+$ we denote the spectral subspaces of H corresponding to the intervals $(-\infty, \alpha)$ and $[\alpha, \infty)$, respectively.

According to the results of [1], [3] there exists a contraction X, $\|X\| \le 1$, from $\mathcal{H}_1$ into $\mathcal{H}_2$ such that $\mathcal{L}_-$ and $\mathcal{L}_+$ admit graph representations with an angular operator X:

$$\mathcal{L}_- = \left\{\begin{pmatrix} x \\ Xx \end{pmatrix} \middle| \, x \in \mathcal{H}_1\right\}, \ \mathcal{L}_+ = \left\{\begin{pmatrix} -X^*y \\ y \end{pmatrix} \middle| \, y \in \mathcal{H}_2\right\}. \tag{3.3}$$

Moreover, X is a trace class operator: $X \in \mathcal{S}_1$, $\operatorname{ran} X \subset \operatorname{dom} D$, $\operatorname{ran} X^* \subset \operatorname{dom} A$, and X is the unique contractive solution of the Riccati equation

$$XBX + X(A+V) - DX - B^* = 0. \tag{3.4}$$

The operator $H\big|_{\mathcal{L}_-}$ is isomorphic to the self-adjoint operator $A+V+BX$ in the Hilbert space $(\mathcal{H}_1, ((I+X^*X)\,\cdot\,, \,\cdot\,))$, and the operator $H\big|_{\mathcal{L}_+}$ is isomorphic to the self-adjoint operator $D - B^*X^*$ in the Hilbert space $(\mathcal{H}_2, ((I+XX^*)\,\cdot\,, \,\cdot\,))$. The operator T in $\mathcal{H}$, given by the matrix

$$T = \begin{pmatrix} I & -X^* \\ X & I \end{pmatrix}, \tag{3.5}$$

is invertible,

$$T^{-1} = \begin{pmatrix} (I+X^*X)^{-1} & (I+X^*X)^{-1}X^* \\ -(I+XX^*)^{-1}X & (I+XX^*)^{-1} \end{pmatrix}, \tag{3.6}$$

and it diagonalizes H:

$$H = T\begin{pmatrix} A+V+BX & 0 \\ 0 & D-B^*X^* \end{pmatrix} T^{-1}. \tag{3.7}$$

It follows that also the operator $e^{-\mathrm{i}tH}$ admits the representation

$$e^{-\mathrm{i}tH} = T\begin{pmatrix} e^{-\mathrm{i}t(A+V+BX)} & 0 \\ 0 & e^{-\mathrm{i}t(D-B^*X^*)} \end{pmatrix} T^{-1}. \tag{3.8}$$

If P_1 denotes again the projection onto $\mathcal{H}_1$ in $\mathcal{H}$, then

$$P_1 e^{-\mathrm{i}Ht}\big|_{\mathcal{H}_1} = e^{-\mathrm{i}t(A+V+BX)}(I+X^*X)^{-1} + X^* e^{-\mathrm{i}t(D-B^*X^*)}(I+XX^*)^{-1}X. \tag{3.9}$$

Lemma 3.1. *Under the assumptions at the beginning of this section we have*

$$\Delta_1(t) \equiv \det e^{-\mathrm{i}tA} P_1 e^{-\mathrm{i}tH}\big|_{\mathcal{H}_1} = \frac{e^{-\mathrm{i}t\,\mathrm{tr}(V+BX)}}{\det(I+X^*X)}\widetilde{\Delta}(t), \tag{3.10}$$

where

$$\widetilde{\Delta}(t) := \det\left(I + e^{\mathrm{i}t(A+V+BX)} X^* e^{-\mathrm{i}t(D-B^*X^*)} X\right).$$

Proof. Using the identity

$$(I+XX^*)^{-1}X = X(I+X^*X)^{-1},$$

it follows from (3.9) that

$$\begin{aligned} & e^{\mathrm{i}tA} P_1 e^{-\mathrm{i}tH}\big|_{\mathcal{H}_1} \\ &= e^{\mathrm{i}tA} e^{-\mathrm{i}t(A+V+BX)}\left(I + e^{\mathrm{i}t(A+V+BX)} X^* e^{-\mathrm{i}t(D-B^*X^*)} X\right)(I+X^*X)^{-1} \end{aligned}$$

and hence

$$\det\left(e^{\mathrm{i}tA} P_1 e^{-\mathrm{i}tH}\big|_{\mathcal{H}_1}\right) = \det\left(e^{\mathrm{i}tA} e^{-\mathrm{i}t(A+V+BX)}\right)\frac{\widetilde{\Delta}(t)}{\det(I+X^*X)}$$

By Theorem 2.1

$$\det\left(e^{\mathrm{i}tA} e^{-\mathrm{i}t(A+V+BX)}\right) = e^{-\mathrm{i}t\,\mathrm{tr}\,(V+BX)}, \tag{3.11}$$

and (3.10) is proved. □

Lemma 3.2. *Let $(U(t))_{-\infty<t<+\infty}$ be a group of unitary operators in the Hilbert space $\mathcal{H}_1$, let Y be a trace class operator from $\mathcal{H}_1$ into the Hilbert space $\mathcal{H}_2$, and let $W(t)$, $t>0$, be a function whose values are bounded operators in $\mathcal{H}_2$ and which is bounded in the operator norm: $\|W(t)\| \le c$, $t>0$. If the infinitesimal generator of the unitary group $U(t)$ has absolutely continuous spectrum then*

$$\lim_{t\to\infty} \det\,(I + U(t)Y^*W(t)Y) = 1.$$

Proof. First we observe that

$$\det(I + U(t)Y^*W(t)Y) = \det(I + YU(t)Y^*W(t)).$$

The trace class operator Y can be represented as

$$Y = \sum_\nu \tau_\nu(\,\cdot\,,\psi_\nu)\varphi_\nu, \tag{3.12}$$

where (ψ_ν) and (φ_ν) are orthonormal systems of elements in $\mathcal{H}_1$ and $\mathcal{H}_2$, respectively, and $\tau_\nu > 0$, $\sum\limits_\nu \tau_\nu < \infty$. It follows that

$$YU(t)Y^*W(t) = \sum_{\mu,\nu} \tau_\mu\tau_\nu(U(t)\psi_\nu,\psi_\mu)(\,\cdot\,,W(t)^*\varphi_\nu)\varphi_\mu$$

and

$$\begin{aligned} \|YU(t)Y^*W(t)\|_1 &\le \sum_{\mu,\nu} \tau_\mu\tau_\nu\,|(U(t)\psi_\nu,\psi_\mu)|\,\|(\,\cdot\,,W(t)^*\varphi_\nu)\varphi_\mu\|_1 \\ &\le c\sum_{\mu,\nu} \tau_\mu\tau_\nu\,|(U(t)\psi_\nu,\psi_\mu)|\,. \end{aligned} \tag{3.13}$$

Since the spectrum of the infinitesimal generator of the unitary group $(U(t))$ is absolutely continuous we have

$$(U(t)\psi_\nu,\psi_\mu) = \int_{-\infty}^{\infty} e^{i\lambda t} f_{\mu\nu}(\lambda)d\lambda, \qquad -\infty < t < +\infty,$$

where $f_{\mu\nu} \in \mathbb{L}_1(-\infty,\infty)$. The Riemann-Lebesgue lemma yields

$$(U(t)\psi_\nu,\psi_\mu) \longrightarrow 0, \quad t \to \infty.$$

Observing that

$$\sum_{\mu,\nu} \tau_\mu\tau_\nu < \infty$$

we get from (3.13)

$$\lim_{t\to\infty} \|YU(t)Y^*W(t)\|_1 = 0,$$

and hence

$$\lim_{t\to\infty} \det(I + YU(t)Y^*W(t)) = 1, \quad t \to \infty.$$ □

For the proof of the main theorem of this section we need one more result which may be of independent interest. Here $\xi_{H/H_0}(\lambda)$ denotes the spectral shift function of the pair of self-adjoint operators H, H_0, cf. [6].

Theorem 3.3. *For the pair of self-adjoint operators*

$$H_0 := \begin{pmatrix} A & 0 \\ 0 & D \end{pmatrix}, \qquad H := \begin{pmatrix} A & B \\ B^* & D \end{pmatrix} \tag{3.14}$$

with $B \in \mathcal{S}_1$ the relation

$$\xi_{H/H_0}(\lambda) = 0 \quad \text{a.e. on } \mathbb{R}$$

is equivalent to $B = 0$.

Proof. If $B = 0$ then, evidently, $\xi_{H/H_0}(\lambda) = 0$ a.e.

To prove the converse, recall that by [9] with $Q := H - H_0$ we have

$$\begin{aligned} \ln D_{H/H_0}(z) &= \ln \det \left(I + (H_0 - z)^{-1}Q\right) \\ &= \operatorname{tr} \ln \left(I + (H_0 - z)^{-1}Q\right) \\ &= \int_{-\infty}^{\infty} \frac{\xi_{H/H_0}(\lambda)}{\lambda - z} d\lambda, \qquad \Im z \neq 0, \end{aligned}$$

where ln denotes the continuous branch of the logarithm with $\ln 1 = 0$. Therefore $\xi_{H/H_0}(\lambda) = 0$ a.e. implies

$$\ln D_{H/H_0}(z) = \operatorname{tr} \ln \left(I + (H_0 - z)^{-1}Q\right) = 0, \qquad \Im z \neq 0. \tag{3.15}$$

For the self-adjoint operator H_0 and $Q \in \mathcal{S}_1$ we have

$$\left\|(H_0 - iy)^{-1}Q\right\|_1 \leq \frac{1}{y} \left\|Q\right\|_1, \qquad y > 0, \tag{3.16}$$

and

$$\lim_{y \to +\infty} -y^2 \operatorname{tr} \left((H_0 - iy)^{-1}Q(H_0 - iy)^{-1}Q\right) = \operatorname{tr} Q^2, \tag{3.17}$$

and the special form of the operators H_0, H in (3.14) implies for non-real z

$$\operatorname{tr} \left((H_0 - z)^{-1}Q\right) = \operatorname{tr} \begin{pmatrix} 0 & (A - z)^{-1}B \\ (D - z)^{-1}B^* & 0 \end{pmatrix} = 0, \tag{3.18}$$

From (3.15), (3.16), (3.18) we obtain

$$\begin{aligned} 0 = \ln D_{H/H_0}(iy) &= \operatorname{tr} \left((H_0 - iy)^{-1}Q\right) \\ &\quad - \frac{1}{2}\operatorname{tr} \left((H_0 - iy)^{-1}Q(H_0 - iy)^{-1}Q\right) + O\Big(\frac{1}{y^3}\Big) \\ &= -\frac{1}{2}\operatorname{tr} \left((H_0 - iy)^{-1}Q(H_0 - iy)^{-1}Q\right) + O\Big(\frac{1}{y^3}\Big), \qquad y \uparrow \infty, \end{aligned}$$

and, using (3.17), $\operatorname{tr} Q^2 = 0$, which because of the self-adjointness of Q implies $Q = 0$. □

Note that in general for a pair of self-adjoint operators H, H_0 with $H - H_0 \in \mathcal{S}_1$ the relation $\xi_{H/H_0}(\lambda) = 0$ a.e. does not imply that $H = H_0$. For example, let H_0 be a bounded self-adjoint operator and let $I + T$ with $T \in \mathcal{S}_1$ be a unitary operator:

$$I + T^* = (I + T)^{-1},$$

such that

$$H := (I + T^*)H_0(I + T) \neq H_0.$$

Then for z with $\Im z \neq 0$

$$\begin{aligned} D_{H/H_0}(z) &= \det\left(I + (H_0 - z)^{-1}(H - H_0)\right) \\ &= \det\left((I+T)^{-1}(H_0 - z)(I+T)(H_0 - z)^{-1}\right) = 1, \end{aligned}$$

and hence $\xi_{H/H_0}(\lambda) = 0$ a.e.

Theorem 3.4. *Suppose that, in addition to the assumptions in the first paragraph of this section, at least one of the operators $H\big|_{\mathcal{L}_-}$, $H\big|_{\mathcal{L}_+}$ has absolutely continuous spectrum. Then*

$$\Delta_1(t) = e^{-b-iat}(1 + \mathrm{o}(1)), \quad t \to \infty, \tag{3.19}$$

where

$$a = \operatorname{tr}(V + BX) \le 0, \quad b = \ln\det(I + X^*X) \ge 0.$$

The relation $b = 0$ is equivalent to $B = 0$; if $V \le 0$ then $a \le 0$, and $a = 0$ is equivalent to $V = B = 0$.

Proof. We consider the case that $H\big|_{\mathcal{L}_-}$ has absolutely continuous spectrum. Then the same is true for the isomorphic self-adjoint operator $A+V+BX$ in the Hilbert space $(\mathcal{H}_1, ((I + X^*X)\,\cdot\,,\,\cdot\,))$. Because of (3.10) the relation (3.19) will follow if we show that

$$\widetilde{\Delta}(t) = \det\left(I + e^{it(A+V+BX)}X^* e^{-it(D-B^*X^*)}X\right) = 1 + \mathrm{o}(1), \quad t \to \infty. \tag{3.20}$$

The self-adjointness of the operator $A+V+BX$ in $(\mathcal{H}_1, ((I + X^*X)\,\cdot\,,\,\cdot\,))$ means that

$$(I + X^*X)(A + V + BX) = (A + V + BX)^*(I + X^*X),$$

or that the operator

$$\widetilde{A} := (I + X^*X)^{1/2}(A + V + BX)(I + X^*X)^{-1/2}$$

is self-adjoint in $\mathcal{H}_1$. Then

$$e^{it\widetilde{A}} = (I + X^*X)^{1/2} e^{it(A+V+BX)}(I + X^*X)^{-1/2}, \qquad -\infty < t < \infty,$$

is a group of unitary operators in $\mathcal{H}_1$ with generator $\widetilde{A}$ which has absolutely continuous spectrum. Properties of determinants allow to represent $\widetilde{\Delta}(t)$ in the form

$$\widetilde{\Delta}(t) = \det\left(I + e^{it\widetilde{A}}(I + X^*X)^{1/2}X^* e^{-it(D-B^*X^*)}X(I + X^*X)^{-1/2}\right),$$

and (3.20) follows from Lemma 3.2 if we observe that $X \in \mathcal{S}_1$ and that the group $e^{-it(D-B^*X^*)}$, $-\infty < t < \infty$, is bounded in $\mathcal{H}_1$ since also the operator $D - B^*X^*$ is similar to a self-adjoint operator in $\mathcal{H}_1$.

If $B = 0$ then $X = 0$, $\det(I + X^*X) = 1$, and $b = \ln\det(I + X^*X) = 0$. Conversely, $b = 0$ implies $X = 0$, and the Riccati equation (3.4) yields $B = 0$.

In order to prove the last claim of the theorem we introduce the operator

$$\widetilde{H}_0 := \begin{pmatrix} A+V & 0 \\ 0 & D \end{pmatrix}.$$

Then, because of [1, Theorem 3.3],

$$a = \operatorname{tr}(V + BX) = \int_{-\infty}^{\alpha} \xi_{H/\widetilde{H}_0}(\lambda)\, d\lambda + \operatorname{tr} V. \tag{3.21}$$

In [1, Theorem 2.1] it was shown that $\xi_{H/\widetilde{H}_0}(\lambda) \leq 0$ a.e. on $(-\infty, \alpha)$ and $\xi_{H/\widetilde{H}_0}(\lambda) \geq 0$ a.e. on (α, ∞). In particular, the integral in (3.21) is ≤ 0, and since $V \leq 0$ by assumption, also the expression on the right-hand side of (3.21) is ≤ 0 and hence $a \leq 0$.

Suppose now that $a = 0$. Then (3.21), $V \leq 0$ and $\xi_{H/\widetilde{H}_0}(\lambda) \leq 0$ a.e. on $(-\infty, \alpha)$ imply that

$$\int_{-\infty}^{\alpha} \xi_{H/\widetilde{H}_0}(\lambda)\, d\lambda = \operatorname{tr} V = 0$$

and hence

$$V = 0, \qquad \xi_{H/\widetilde{H}_0}(\lambda) = 0 \quad \text{a. e. on } (-\infty, \alpha).$$

Since $V = 0$ we have $H_0 = \widetilde{H}_0$, and also

$$0 = \operatorname{tr}(H - H_0) = \int_{-\infty}^{\infty} \xi_{H/H_0}(\lambda) d\lambda = \int_{\alpha}^{\infty} \xi_{H/H_0}(\lambda) d\lambda,$$

which, because of $\xi_{H/H_0}(\lambda) \geq 0$ a.e. on (α, ∞), yields $\xi_{H/H_0}(\lambda) = 0$ a.e. Finally, Theorem 3.3 implies $B = 0$. □

In the following for the self-adjoint operator

$$H = \begin{pmatrix} A & B \\ B^* & D \end{pmatrix} \tag{3.22}$$

in $\mathcal{H} = \mathcal{H}_1 \oplus \mathcal{H}_2$ the invertibility of the operator

$$W(t) := P_1 e^{-\mathrm{i}tH}\Big|_{\mathcal{H}_1}$$

plays some role.

Lemma 3.5. *Suppose that the spectra of the self-adjoint operators A and D in (3.22) are separated:*

$$\max \sigma(A) \leq \alpha \leq \min \sigma(D)$$

with some $\alpha \in \mathbb{R}$ which is neither an eigenvalue of A nor of D, and suppose that B is compact. Then for all $t \in \mathbb{R}$ the operator $P_1 e^{-\mathrm{i}tH}\big|_{\mathcal{H}_1}$ is invertible.

Proof. We consider the representation (3.3) of the spectral subspaces $\mathcal{L}_-$ and $\mathcal{L}_+$ of H corresponding to the intervals $(-\infty,\alpha)$ and $[\alpha,\infty)$, respectively. Since α is not an eigenvalue of A and of D we have $\|Xx\| < \|x\|$ for all $x \in \mathcal{H}_1$, $x \neq 0$, and the compactness of X implies that $\|X\| < 1$. According to (3.9), the operator $W(t)$ can be written as

$$W(t) = e^{-it(A+BX)}\widetilde{W}(t)(I+X^*X)^{-1},$$

where

$$\widetilde{W}(t) := I + e^{it(A+BX)}X^*e^{-it(D-B^*X^*)}X.$$

The operator $A+BX$ is similar to a self-adjoint operator, therefore the group $\left(e^{-it(A+BX)}\right)_{t\in\mathbb{R}}$ is similar to a group of unitary operators and hence invertible. Now the invertibility of $W(t)$ follows if we show the invertibility of $\widetilde{W}(t)$.

If $\widetilde{W}(t)$ would not be invertible, because of the compactness of X there would exist an element $x \in \mathcal{H}_1$, $x \neq 0$, such that

$$x = -e^{it(A+BX)}X^*e^{-it(D-B^*X^*)}Xx.$$

Applying the operator $(I+X^*X)^{1/2}$, using the identities

$$(I+X^*X)^{1/2}X^* = X^*(I+XX^*)^{1/2}, \qquad (I+XX^*)^{1/2}X = X(I+X^*X)^{1/2}$$

and taking into account that the operators $(I+X^*X)^{1/2}e^{-it(A+BX)}(I+X^*X)^{-1/2}$ and $(I+XX^*)^{1/2}e^{it(D-B^*X^*)}(I+XX^*)^{-1/2}$ are unitary, we find

$$\begin{aligned}
&\|(I+X^*X)^{1/2}x\| \\
&= \|(I+X^*X)^{1/2}e^{it(A+BX)}(I+X^*X)^{-1/2}(I+X^*X)^{1/2}X^*e^{-it(D-B^*X^*)}Xx\| \\
&= \|X^*(I+XX^*)^{1/2}e^{-it(D-B^*X^*)}Xx\| \\
&< \|(I+XX^*)^{1/2}e^{-it(D-B^*X^*)}(I+XX^*)^{-1/2}(I+XX^*)^{1/2}Xx\| \\
&= \|X(I+X^*X)^{1/2}x\| < \|(I+X^*X)^{1/2}x\|,
\end{aligned}$$

a contradiction. □

4. Partial perturbation determinants and Fredholm resolvents

In this section we consider the partial perturbation determinant

$$\Delta_1(t) = \det\left(e^{itA}P_1e^{-itH}|_{\mathcal{H}_1}\right).$$

for the operator

$$H = \begin{pmatrix} A & B \\ B^* & D \end{pmatrix}$$

in the Hilbert space $\mathcal{H} = \mathcal{H}_1 \oplus \mathcal{H}_2$, where A and D are self-adjoint operators in $\mathcal{H}_1$ and $\mathcal{H}_2$, respectively, and the operator B is finite-dimensional. We represent B as

$$B = \sum_{\nu=1}^{n} \tau_\nu(\,\cdot\,, \psi_\nu)\varphi_\nu,$$

with orthonormal systems $(\varphi_\nu)_1^n$ and $(\psi_\nu)_1^n$ in $\mathcal{H}_1$ and $\mathcal{H}_2$, respectively, and positive numbers τ_ν, $\nu = 1, 2, \ldots, n$.

First we recall the following formulas which hold for any self-adjoint operator T:

$$e^{-\mathrm{i}tT-\varepsilon t} = \frac{1}{2\pi\mathrm{i}} \int_{-\infty}^{\infty} (T-\lambda-\mathrm{i}\varepsilon)^{-1} e^{-\mathrm{i}\lambda t}\, d\lambda, \quad t>0,\ \varepsilon>0, \tag{4.1}$$

$$(T-\lambda+\mathrm{i}\varepsilon)^{-1} = \mathrm{i}\int_{0}^{\infty} e^{-\mathrm{i}t(T-\lambda-\mathrm{i}\varepsilon)} dt, \quad \varepsilon>0; \tag{4.2}$$

the integrals are defined in the strong operator topology. It follows that

$$\begin{aligned} e^{-\varepsilon t} P_1 e^{-\mathrm{i}tH}\Big|_{\mathcal{H}_1} &= \frac{1}{2\pi\mathrm{i}} P_1 \int_{-\infty}^{\infty} (H-\lambda+\mathrm{i}\varepsilon)^{-1} e^{-\mathrm{i}\lambda t} d\lambda\Big|_{\mathcal{H}_1} \\ &= \frac{1}{2\pi\mathrm{i}} \int_{-\infty}^{\infty} W(\lambda+\mathrm{i}\varepsilon)^{-1} e^{-\mathrm{i}\lambda t} d\lambda, \quad t>0,\ \varepsilon>0, \end{aligned} \tag{4.3}$$

where

$$W(z) := A - z - B(D-z)^{-1}B^*, \qquad \Im z \neq 0.$$

On the resolvent sets $\rho(A)$ and $\rho(D)$ we introduce the holomorphic $n \times n$-matrix functions

$$G_1(z) = \left(\left((A-z)^{-1}\varphi_\nu, \varphi_\mu\right)\right)_{\mu,\nu=1}^n, \quad G_2(z) = \left(\tau_\mu \tau_\nu \left((D-z)^{-1}\psi_\nu, \psi_\mu\right)\right)_{\mu,\nu=1}^n$$

A straightforward calculation leads to the representation

$$W(z)^{-1} = (A-z)^{-1} + \sum_{\mu,\nu=1}^{n} g_{\mu\nu}(z)\left(\,\cdot\,, (A-\overline{z})^{-1}\varphi_\mu\right)(A-z)^{-1}\varphi_\nu, \tag{4.4}$$

where the functions $g_{\mu\nu}(z)$ are the entries of the $n \times n$-matrix function

$$G(z) := G_2(z)\left(I - G_1(z)\, G_2(z)\right)^{-1} = \left(g_{\mu\nu}(z)\right)_{\mu,\nu=1}^n.$$

Recall that a *Nevanlinna function* is a function which is analytic in the upper half-plane and has a non-negative imaginary part there.

Lemma 4.1. *The functions $G_1(z)$, $G_2(z)$ and $G(z)$ are Nevanlinna functions. Moreover, $G(z)$ admits the representation*

$$G(z) = \int_{-\infty}^{\infty} \frac{1}{\lambda - z}\, d\Sigma(\lambda), \qquad \Im z \neq 0, \tag{4.5}$$

with a non-decreasing bounded $n \times n$-matrix function $\Sigma(\lambda)$ on $\mathbb{R}$. Consequently, with the positive definite $n \times n$-matrix function

$$Q(t) := \int_{-\infty}^{\infty} e^{-i\lambda t}\, d\Sigma(\lambda), \qquad t \in \mathbb{R}, \tag{4.6}$$

the representation

$$G(z) = \mathrm{i} \int_{0}^{\infty} e^{izt} Q(t)\, dt, \qquad \Im z > 0, \tag{4.7}$$

holds.

Proof. That $G_1(z)$ and $G_2(z)$ are Nevanlinna functions is clear, for $G(z)$ this follows from the relation

$$-G(z)^{-1} = G_1(z) + (-G_2(z)^{-1})$$

and the fact that both summands on the right-hand side are Nevanlinna functions. Since

$$G_1(\mathrm{i}y),\ G_2(\mathrm{i}y) \longrightarrow 0, \qquad y \uparrow \infty,$$

and

$$\sup_{y>0} y\|G_2(\mathrm{i}y)\| < \infty,$$

the function $G(z)$ admits the representation (4.5), cf. [8, Theorem 1.4.1]. □

Inserting (4.4) with $z = \lambda + \mathrm{i}\varepsilon$ into the right-hand side of (4.3), using (4.1), (4.2), (4.7) and the convolution theorem, and setting $Q(t) =: (q_{\mu\nu}(t))_{\mu,\nu=1}^{n}$ we find

$$\begin{aligned} P_1 e^{-\mathrm{i}tH}\big|_{\mathcal{H}_1} &= e^{-\mathrm{i}tA} - \sum_{\mu,\nu=1}^{n} \int_0^t \int_0^\eta q_{\mu\nu}(\eta - s)\Big(\cdot\,, e^{\mathrm{i}(t-\eta)A}\varphi_\nu\Big)\, e^{-\mathrm{i}sA}\varphi_\mu\, ds\, d\eta \\ &= e^{-\mathrm{i}tA} - \sum_{\nu,\mu=1}^{n} \int_0^t \Big(\cdot\,, e^{\mathrm{i}(t-\eta)A}\varphi_\nu\Big) \int_0^\eta q_{\mu\nu}(\eta - s) e^{-\mathrm{i}sA}\varphi_\mu\, ds\, d\eta \\ &= e^{-\mathrm{i}tA} - R(t)e^{-\mathrm{i}tA} \end{aligned} \tag{4.8}$$

with

$$R(t) := \sum_{\nu=1}^{n} \int_0^t \big(\cdot\,, e^{-\mathrm{i}\eta A}\varphi_\nu\big) \sum_{\mu=1}^{n} \int_0^\eta q_{\mu\nu}(\eta - s) e^{-\mathrm{i}s}\varphi_\mu\, ds\, d\eta, \tag{4.9}$$

and hence

$$e^{\mathrm{i}tA}P_1e^{-\mathrm{i}tH}|_{\mathcal{H}_1} = I - e^{\mathrm{i}tA}R(t)e^{-\mathrm{i}tA}, \tag{4.10}$$

$$\Delta_1(t) := \det\left(e^{\mathrm{i}tA}P_1e^{-\mathrm{i}tH}|_{\mathcal{H}_1}\right) = \det\left(I - R(t)\right). \tag{4.11}$$

Observe that $R(t)$ and $R'(t) = \sum_{\mu,\nu=1}^{n} (\,\cdot\,, \varphi_\nu(t))\,\psi_\nu(t)$ are trace class operators.

For $\nu = 1, 2, \ldots, n$ we set

$$\varphi_\nu(\eta) := e^{-\mathrm{i}\eta A}\varphi_\nu, \quad k_{\mu\nu}(\eta) := \int_0^\eta q_{\mu\nu}(\eta-\xi)e^{-\mathrm{i}\xi A}\varphi_\mu\,d\xi, \quad \psi_\nu(\eta) := \sum_{\mu=1}^{n} k_{\mu\nu}(\eta).$$

Then $\varphi_\nu(\eta), \psi_\nu(\eta)$, $\mu, \nu = 1, 2, \ldots, n$, are norm-continuous $\mathcal{H}$-valued functions and $R(t)$ becomes

$$R(t) = \sum_{\nu=1}^{n}\int_0^t (\,\cdot\,, \varphi_\nu(\eta))\sum_{\mu=1}^{n} k_{\mu\nu}(\eta)\,d\eta = \sum_{\nu=1}^{n}\int_0^t (\,\cdot\,, \varphi_\nu(\eta))\,\psi_\nu(\eta)\,d\eta. \tag{4.12}$$

To find a more explicit expression for $\Delta_1(t)$ we use the following lemma, which is a slight extension of [7, (IV.1.14)].

Lemma 4.2. *Let $R(t)$ be a function which is defined in a real neighborhood of t_0 and with values in $\mathcal{S}_1$, which is differentiable in t_0 with respect to the nuclear norm and such that $(I - R(t_0))^{-1}$ exists. Then the function $\Theta(t) := \det(I - R(t))$ is differentiable in t_0 and the following relation holds:*

$$\frac{\Theta'(t_0)}{\Theta(t_0)} = \frac{d}{dt}\ln\Theta(t)\Big|_{t=t_0} = -\mathrm{tr}\,\left((I - R(t_0))^{-1}R'(t_0)\right). \tag{4.13}$$

Proof. We have for $t \to t_0$

$$\begin{aligned}\Theta(t) &= \det\left(I - R(t_0) - R'(t_0)(t-t_0) + \mathrm{o}(t-t_0)\right)\\ &= \det(I - R(t_0))\det\left(I - (I-R(t_0))^{-1}R'(t_0)(t-t_0) + \mathrm{o}(t-t_0)\right)\\ &= \Theta(t_0)\left(I - \mathrm{tr}\,\left((I-R(t_0))^{-1}R'(t_0))(t-t_0) + \mathrm{o}(t-t_0)\right)\right),\end{aligned}$$

where the symbol $\mathrm{o}(t-t_0)$ is to be understood with respect to the nuclear norm. It follows that

$$\frac{\Theta(t)-\Theta(t_0)}{(t-t_0)\Theta(t_0)} = -\mathrm{tr}\,\left((I-R(t_0))^{-1}R'(t_0) + \frac{\mathrm{o}(t-t_0)}{t-t_0}\right), \quad t \to t_0,$$

which implies (4.13). □

In order to calculate the trace on the right-hand side of (4.13) for the operator $R(t)$ from (4.9), with

$$k^t_{\mu\nu}(\xi,\eta) := (\psi_\nu(\eta), \varphi_\mu(\xi)), \quad \mu,\nu = 1, 2, \ldots, n,\ 0 \le \xi, \eta \le t,$$

we denote by $\mathbf{K}^t(\xi,\eta)$ the $n\times n$-matrix kernel

$$\mathbf{K}^t(\xi,\eta) := \left(k^t_{\mu\nu}(\xi,\eta)\right)^n_{\mu,\nu=1},$$

and by $\mathbf{\Gamma}^t(\xi,\eta) := \left(\gamma^t_{\mu,\nu}(\xi,\eta)\right)^n_{\mu,\nu=1}$ the corresponding Fredholm resolvent kernel. We recall that this means that

$$\mathbf{K}^t(\xi,\eta) + \int_0^t \mathbf{K}^t(\xi,\tau)\mathbf{\Gamma}^t(\tau,\eta)d\tau = \mathbf{\Gamma}^t(\xi,\eta), \quad 0 \le \xi,\eta \le t. \tag{4.14}$$

Lemma 4.3. *Let the operator $R(t)$ be given by (4.9) or (4.12). If the inverse $(I - R(t))^{-1}$ exists then the relation*

$$\operatorname{tr}\left((I - R(t))^{-1}R'(t)\right) = \operatorname{tr}\mathbf{\Gamma}^t(t,t),$$

holds, where on the right-hand side tr *denotes the matrix trace.*

Proof. Consider for $f \in \mathcal{H}$ the equation

$$x - R(t)x = f,$$

that is

$$x - \sum_{\nu=1}^n \int_0^t (x, \varphi_\nu(\eta))\, \psi_\nu(\eta)d\eta = f. \tag{4.15}$$

For an element $y \in \mathcal{H}$ we denote

$$\mathsf{y}_\nu(\eta) := (y, \varphi_\nu(\eta)), \quad 0 \le \eta \le t, \ \nu = 1, 2, \dots, n,$$

and $\mathsf{y}(\eta) := (\mathsf{y}_1(\eta), \mathsf{y}_2(\eta), \cdots, \mathsf{y}_n(\eta))^{\mathrm{t}}$. Taking the $L^2(0,t)$-inner product of equation (4.15) with $\varphi_\rho(\xi)$ for $\rho = 1, 2, \dots, n$ we obtain

$$\mathbf{x}(\xi) - \int_0^t \mathbf{K}^t(\xi,\eta)\mathbf{x}(\eta)d\eta = \mathbf{f}(\xi), \quad 0 \le \xi \le t.$$

Hence

$$\mathbf{x}(\eta) = \mathbf{f}(\eta) + \int_0^t \mathbf{\Gamma}^t(\eta,\omega)\mathbf{f}(\omega)d\omega, \quad 0 \le \eta \le t,$$

and we get from (4.15)

$$x = (I - R(t))^{-1}f = f + \int_0^t \left\langle \mathbf{f}(\eta) + \int_0^t \mathbf{\Gamma}^t(\eta,\omega)\mathbf{f}(\omega)d\omega,\, \boldsymbol{\psi}(\eta) \right\rangle_{\mathbb{C}^n} d\eta.$$

We have to apply this formula to an element of the form

$$f = R'(t)g = \sum_{\nu=1}^n (g, \varphi_\nu(t))\psi_\nu k(t) = \sum_{\nu=1}^n g_\nu(t)\psi_\nu(t).$$

It follows that

$$\begin{aligned}(I-R(t))^{-1}R'(t)g &= \sum_{\nu=1}^{n} g_\nu(t)\Bigg(\psi_\nu(t)+\int_0^t\Bigg(\sum_{\rho=1}^{n}(\psi_\nu(t),\varphi_\rho(\eta))\psi_\rho(\eta)\\ &\quad+\sum_{\rho=1}^{n}\int_0^t\sum_{\mu=1}^{n}\gamma_{\rho\mu}^t(\eta,\omega)(\psi_\nu(t),\varphi_\mu(\omega))d\omega\psi_\rho(\eta)\Bigg)d\eta\Bigg)\\ &= \sum_{\nu=1}^{n} g_\nu(t)\Bigg(\psi_\nu(t)+\sum_{\rho=1}^{n}\int_0^t\Bigg(k_{\rho\nu}^t(\eta,t)\\ &\quad+\int_0^t\sum_{\mu=1}^{n}\gamma_{\rho\mu}^t(\eta,\omega)k_{\mu\nu}^t(\omega,t)d\omega\Bigg)\psi_\rho(\eta)d\eta\Bigg)\\ &= \sum_{\nu=1}^{n}(g,\varphi_\nu(t))\left(\psi_\nu(t)+\sum_{\rho=1}^{n}\int_0^t\gamma_{\rho\nu}^t(\eta,t)\psi_\rho(\eta)d\eta\right)\end{aligned}$$

and hence

$$\begin{aligned}\operatorname{tr}\left((I-R(t))^{-1}R'(t)\right) &= \sum_{\nu=1}^{n}\left(\psi_\nu(t)+\sum_{\rho=1}^{n}\int_0^t\gamma_{\rho\nu}^t(\eta,t)\psi_\rho(\eta)d\eta,\varphi_\nu(t)\right)\\ &= \sum_{\nu=1}^{n}\left(k_{\nu\nu}^t(t,t)+\sum_{\rho=1}^{n}\int_0^t\gamma_{\rho\nu}^t(\eta,t)k_{\nu\rho}^t(t,\eta)d\eta\right)\\ &= \sum_{\nu=1}^{n}\gamma_{\nu\nu}^t(t,t)\\ &= \operatorname{tr}\left(\Gamma^t(t,t)\right).\end{aligned}$$

□

Combing Lemma 4.2 and Lemma 4.3 we obtain the following result.

Theorem 4.4. *Let the self-adjoint operator H in the Hilbert space $\mathcal{H}=\mathcal{H}_1\oplus\mathcal{H}_2$ be given by the matrix*

$$H=\begin{pmatrix} A & B\\ B^* & D\end{pmatrix}$$

with self-adjoint operators A and D in the Hilbert spaces $\mathcal{H}_1$ and $\mathcal{H}_2$, respectively, and a finite-dimensional operator B:

$$B=\sum_{\nu=1}^{n}\tau_\nu(\,\cdot\,,\psi_\nu)\varphi_\nu,$$

where $(\varphi_\nu)_1^n$ and $(\psi_\nu)_1^n$ are orthonormal systems in $\mathcal{H}_1$ and $\mathcal{H}_2$, respectively, and τ_ν, $\nu=1,2,\ldots,n$, are positive numbers. If, for $0\le s\le t$, the operator $I-R(s)$ is invertible and $\mathbf{\Gamma}^t(\xi,\eta)$, $0\le\xi,\eta\le t$, denotes the Fredholm resolvent kernel as *defined by* (4.14) *then for the partial perturbation determinant* $\Delta_1(t)$ *it holds*

$$\Delta_1(t) \equiv \det\left(e^{itA} P_1 e^{-itH}|_{\mathcal{H}_1}\right) = \exp\left(-\int_0^t \operatorname{tr}\left(\mathbf{\Gamma}^s(s,s)\right) ds\right).$$

Remark. A sufficient condition for the invertiblity of the operators $(I - R(t))^{-1}$ is the separation of the spectra of the diagonal operators A and D, see Lemma 3.5.

References

[1] V. Adamyan and H. Langer: Spectral properties of a class of rational operator-valued functions. *J. Operator Theory* **33** (1995), 259–277.

[2] V. Adamyan and H. Langer: The spectral shift function for certain block operator matrices. *Math. Nachr.* **21** (2000), 5–24.

[3] V. Adamyan, H. Langer, R. Mennicken, and R. Saurer: Spectral components of self-adjoint block operator matrices with unbounded entries. *Math. Nachr.* **178** (1996), 43–80.

[4] V. Adamyan, J. Ortner, A. Salistra, and I. Tkachenko: X-Ray absorption problem in metals within one-electron approximation. *Phys. Rev. B* **52** (1995), 13827–13837.

[5] P.W. Anderson: Infrared catastrophe in Fermi gases with local scattering potentials. *Phys. Rev. Lett.* **18** (1967), 1049–1051.

[6] M.S. Birman and D.R. Yafajev: The spectral shift function. The papers of M.G. Krein and their further development. *Algebra i Analiz* **4** (1992), 1–44 (Russian).

[7] I.C. Gohberg and M.G.Krein: Introduction to the Theory of Linear Non-self-adjoint Operators. *Translations of Mathematical Monographs* **18**, AMS, Providence, R.I., 1969.

[8] I.S. Kac and M.G.Krein: R-functions – analytic functions mapping the upper half-plane into itself. American Mathematical Society Translations, Series 2, **103**, 1974.

[9] M.G. Krein: On the trace formula in perturbation theory. *Mat. Sbornik N.S.* **33** (75) (1953), 597–626 (Russian).

[10] A. Pazy: Semigroups of linear operators and applications to partial differential equations. *Applied Mathematical Sciences*, **44**, Springer-Verlag, New York, 1983.

Vadim Adamyan
Department of Theoretical Physics
Odessa National University
65026 Odessa, Ukraine
e-mail: `vadamyan@paco.net`

Heinz Langer
Institute for Analysis and Scientific Computing
Vienna University of Technology
Wiedner Hauptstr. 8–10
A-1040 Vienna, Austria
e-mail: `hlanger@email.tuwien.ac.at`

Operator Theory:
Advances and Applications, Vol. 154, 19–33

Noncommuting Domination

Dariusz Cichoń, Jan Stochel and Franciszek Hugon Szafraniec

Abstract. We extend and refine the approach of [1] concerning essential self-adjointness (normality) of a densely defined operator subject to some domination condition involving the first or the second commutator.

Mathematics Subject Classification (2000). Primary 47B25; Secondary 47B47.

Keywords. Symmetric operator, self-adjoint operator, domination, commutator.

Introduction

Essential self-adjointness of symmetric operators is an important issue both from the theoretical point of view as well as for applications. There are different methods of achieving this. One of them is to use an auxiliary operator, already essentially self-adjoint, which interacts with the operator in question. The way the essentially self-adjoint operator interacts with the candidate operator is twofold. The first is a kind of domination, the other splits in two different approaches depending on if the operators commute in a sense (from the pioneering work of [2] to recent [5] with some intermediate references like [3]) or not (here [2] again, the treatise [4] as well as [1] with further references therein). The latter consists in replacing the zero commutators by those which are controlled somehow. For instance, in [1] the commutators are relatively bounded. In the present paper we refine the technique of [1] working under more subtle growth conditions for the first or the second commutator. Our approach (avoiding the form language of [1]), besides providing substantially more general results, turns out to be better suited for making the arguments more precise and simpler. In particular, so simple a case of bounded perturbation of a self-adjoint operator fits in our results but fails to satisfy the requirements of [1] (cf. Example 17).

This work was supported by the KBN grant 2 P03A 037 024.

1. Preparatory facts

Throughout the paper $\mathcal{H}$ stands for a complex Hilbert space, $\boldsymbol{B}(\mathcal{H})$ for the C^*-algebra of all bounded linear operators *on* $\mathcal{H}$ and I for the identity operator on $\mathcal{H}$. Given a linear operator A *in* $\mathcal{H}$, we denote by $\mathcal{D}(A)$, $\mathcal{N}(A)$, A^*, $\bar{A}$ and $\sigma(A)$ the domain, the kernel, the adjoint, the closure and the spectrum of A, respectively. Set $\mathcal{D}^\infty(A) = \bigcap_{n=1}^\infty \mathcal{D}(A^n)$. The graph norm of A is denoted by $\|\cdot\|_A$, i.e., $\|f\|_A^2 = \|f\|^2 + \|Af\|^2$ for $f \in \mathcal{D}(A)$. Recall that a linear subspace $\mathcal{E}$ of $\mathcal{D}(A)$ is said to be a *core* of A if the graph of A is contained in the closure of the graph of $A|_\mathcal{E}$. We say that a symmetric operator A in $\mathcal{H}$ is *essentially self-adjoint on* $\mathcal{E}$ if $\mathcal{E}$ is a dense linear subspace of $\mathcal{D}(A)$ and $(A|_\mathcal{E})^* = \overline{A|_\mathcal{E}}$. By maximality of self-adjoint operators, a symmetric operator A is essentially self-adjoint on $\mathcal{E}$ if and only if $\bar{A}$ is self-adjoint and $\mathcal{E}$ is a core of A. If A and B are linear operator in $\mathcal{H}$, then $[A, B]$ stands for the commutator of A and B, i.e., the operator $[A, B] \stackrel{\text{df}}{=} AB - BA$ with the domain $\mathcal{D}(AB) \cap \mathcal{D}(BA)$. For a given integer $k \geqslant 0$, we define the k-th commutator $(\operatorname{ad} A)^k(B)$ via

$$(\operatorname{ad} A)^0(B) = B \text{ and } (\operatorname{ad} A)^{k+1}(B) = [A, (\operatorname{ad} A)^k(B)] \text{ for } k \geqslant 0.$$

An induction argument based on $\binom{k}{j-1} + \binom{k}{j} = \binom{k+1}{j}$ shows that[1]

$$(\operatorname{ad} A)^k(B) \supseteq \sum_{j=0}^{k} \binom{k}{j} A^j B(-A)^{k-j}, \quad k \geqslant 0. \tag{1}$$

We say that B *dominates* A on $\mathcal{E}$ if $\mathcal{E}$ is a linear subspace of $\mathcal{D}(A) \cap \mathcal{D}(B)$ and there exits $c \geqslant 0$ such that $\|Af\| \leqslant c(\|f\| + \|Bf\|)$ for all $f \in \mathcal{E}$ (cf. [5] for more details).

Suppose that H is a closed densely defined linear operator in $\mathcal{H}$. Throughout the whole paper $R(z)$ stands for the resolvent of H, i.e., $R(z) = (z - H)^{-1} \in \boldsymbol{B}(\mathcal{H})$ for $z \in \mathbb{C} \setminus \sigma(H)$. We show by induction on m that

$$R(z)\mathcal{D}(H^m) \subseteq \mathcal{D}(H^{m+1}) \text{ for all integers } m \geqslant 0. \tag{2}$$

Indeed, if (2) holds for a fixed $m \geqslant 0$ (the case $m = 0$ is trivial), then

$$\begin{aligned} R(z)\mathcal{D}(H^{m+1}) &= R(z)\mathcal{D}((z-H)^{m+1}) = R(z)(R(z)^{m+1}\mathcal{H}) \\ &= R(z)^{m+2}\mathcal{H} = \mathcal{D}((z-H)^{m+2}) = \mathcal{D}(H^{m+2}). \end{aligned}$$

Condition (2) implies $R(z)\mathcal{D}^\infty(H) \subseteq \mathcal{D}^\infty(H)$. In consequence, if A is a linear operator in $\mathcal{H}$ such that $\mathcal{D}^\infty(H) \subseteq \mathcal{D}(A)$, then $\mathcal{D}^\infty(H) \subseteq \mathcal{D}([A, R(z)^k])$ and $\mathcal{D}^\infty(H) \subseteq \mathcal{D}((\operatorname{ad} R(z))^k(A))$ for all integers $k \geqslant 0$. Likewise, if $\mathcal{D}(H^m) \subseteq \mathcal{D}(A)$, then $\mathcal{D}(H^m) \subseteq \mathcal{D}([A, R(z)^k])$ and $\mathcal{D}(H^m) \subseteq \mathcal{D}((\operatorname{ad} R(z))^k(A))$ for all integers $k, m \geqslant 0$.

Proposition 1. *Let A be a closed operator in $\mathcal{H}$ and H be a self-adjoint operator in $\mathcal{H}$. Then $\mathcal{D}^\infty(H) \subseteq \mathcal{D}(A)$ if and only if there exists an integer $m \geqslant 0$ such that*

[1] Notice that if $\mathcal{D}(A) = \mathcal{H}$, then the inclusion (1) becomes an equality.

(i) $\mathcal{D}(H^m) \subseteq \mathcal{D}(A)$.

Suppose that $m \geqslant 0$ is an integer, $\mathcal{E}$ is a core[2] *of H^m and $\mathcal{F} \subseteq \mathcal{D}(AR(z)^m)$ is a linear subspace such that $\mathcal{D}(AR(z)^m) \subseteq \overline{\mathcal{F}}$. Then* (i) *is equivalent to any of the following conditions:*

(ii) *$\mathcal{E} \subseteq \mathcal{D}(A)$ and there exists $c \geqslant 0$ such that*

$$\|Af\| \leqslant c(\|f\| + \|H^m f\|), \quad f \in \mathcal{E}, \tag{3}$$

(iii) *the operator $AR(z)^m|_{\mathcal{F}}$ is bounded and densely defined in $\mathcal{H}$ for some $z \in \mathbb{C} \setminus \sigma(H)$ (equivalently: for every $z \in \mathbb{C} \setminus \sigma(H)$).*

Furthermore, if (ii) *holds, then $\mathcal{D}(H^m) \subseteq \mathcal{D}((\operatorname{ad} R(z))^k(A))$ and*

$$\|(\operatorname{ad} R(z))^k(A)f\| \leq c\,2^k \|R(z)\|^k(\|f\| + \|H^m f\|), \quad f \in \mathcal{D}(H^m), \tag{4}$$

for all integers $k \geqslant 0$ and $z \in \mathbb{C} \setminus \sigma(H)$; what is more, $\overline{A|_{\mathcal{E}}} = \overline{A|_{\mathcal{D}(H^m)}}$.
If (iii) *holds, then $\mathcal{D}(H^m) \subseteq \mathcal{D}(A)$ and $AR(z)^m \in \boldsymbol{B}(\mathcal{H})$ for every $z \in \mathbb{C} \setminus \sigma(H)$.*

Proof. First we show that if $\mathcal{D}^\infty(H) \subseteq \mathcal{D}(A)$, then there exists an integer $m \geqslant 0$ such that (ii) holds for $\mathcal{E} = \mathcal{D}^\infty(H)$. For this, observe that $\mathcal{D}^\infty(H)$ is a Fréchet space with the topology given by the system of graph norms $\{\|\cdot\|_{H^j}\}_{j=0}^\infty$ (because each operator H^j, $j \geqslant 0$, is closed) and $(\mathcal{D}(A), \|\cdot\|_A)$ is a Banach space. By the closed graph theorem the inclusion $\mathcal{D}^\infty(H) \ni f \longmapsto f \in \mathcal{D}(A)$ is a continuous linear operator, and hence there exist $c' \geqslant 0$ and an integer $m \geqslant 0$ such that

$$\|Af\|^2 \leqslant c' \sum_{j=0}^{m} \|H^j f\|^2, \quad f \in \mathcal{D}^\infty(H).$$

Since there exists $d_m \geqslant 0$ such that $1 + x^2 + \cdots + x^{2m} \leqslant d_m(1 + x^{2m})$ for all $x \in \mathbb{R}$, we get (by the spectral theorem)

$$\|Af\|^2 \leqslant c' d_m(\|f\|^2 + \|H^m f\|^2), \quad f \in \mathcal{D}^\infty(H),$$

which proves our claim.

If (ii) holds, then $\mathcal{E}$ being a core of H^m and the closedness of A imply

$$\mathcal{D}(H^m) \subseteq \mathcal{D}(A), \quad \|Af\| \leqslant c(\|f\| + \|H^m f\|) \text{ for all } f \in \mathcal{D}(H^m) \tag{5}$$

and $\overline{A|_{\mathcal{E}}} = \overline{A|_{\mathcal{D}(H^m)}}$. Combining (5) with the previous paragraph, we see that $\mathcal{D}^\infty(H) \subseteq \mathcal{D}(A)$ if and only if $\mathcal{D}(H^m) \subseteq \mathcal{D}(A)$ for some $m \geqslant 0$.

Applying the closed graph theorem, we see that (i) implies (ii) with $\mathcal{E} = \mathcal{D}(H^m)$, which by the previous paragraph justifies also the equivalence (i)⇔(ii).

In the next step we show the equivalence (ii)⇔(iii) for a fixed $z \in \mathbb{C} \setminus \sigma(H)$.

(ii)⇒(iii) Fix $g \in \mathcal{H}$. Since (ii)⇒(5), we can substitute $f \stackrel{\text{df}}{=} R(z)^m g$ into (5) and apply the identity $H^m R(z)^m g = (HR(z))^m g = (zR(z) - I)^m g$. In consequence, we get $AR(z)^m \in \boldsymbol{B}(\mathcal{H})$.

---

[2] Recall that $\mathcal{D}^\infty(H)$ is a core of H^m for every integer $m \geqslant 0$.

(iii)⇒(ii) Since A is closed and $R(z)^m \in \boldsymbol{B}(\mathcal{H})$, the operator $AR(z)^m$ is closed as well. By (iii), $AR(z)^m \in \boldsymbol{B}(\mathcal{H})$. This means that $\mathcal{D}(H^m) \subseteq \mathcal{D}(A)$ and there exists $c' \geqslant 0$ such that

$$\|AR(z)^m f\| \leqslant c'\|f\|, \quad f \in \mathcal{H}. \tag{6}$$

Take $g \in \mathcal{D}(H^m)$. Plugging $f = (z-H)^m g$ into (6), we have $\|Ag\| \leqslant c'\|(z-H)^m g\|$. Since there exists $d'_m \geqslant 0$ such that $|z-x|^{2m} \leqslant d'_m(1+x^{2m})$ for all $x \in \mathbb{R}$, the spectral theorem implies

$$\|Ag\|^2 \leqslant (c')^2 d'_m(\|g\|^2 + \|H^m g\|^2), \quad g \in \mathcal{D}(H^m),$$

which manifestly yields (ii).

Suppose now that (ii) holds. Then, by (5), inequality (4) is valid for $k = 0$. Assuming (4) for a fixed $k \geqslant 0$, we can proceed as follows (use (2))

$$\begin{aligned}
\|(\mathrm{ad} R(z))^{k+1}(A)f\| &= \|[R(z), (\mathrm{ad}\, R(z))^k(A)]f\| \\
&\leqslant \|R(z)(\mathrm{ad}\, R(z))^k(A)f\| + \|(\mathrm{ad}\, R(z))^k(A)R(z)f\| \\
&\leqslant c\,2^k\|R(z)\|^{k+1}(\|f\| + \|H^m f\|) + c\,2^k\|R(z)\|^k(\|R(z)f\| + \|R(z)H^m f\|) \\
&\leqslant c\,2^{k+1}\|R(z)\|^{k+1}(\|f\| + \|H^m f\|), \quad f \in \mathcal{D}(H^m).
\end{aligned}$$

This completes the proof. □

The set $\Omega_{c,d}(H)$ defined below plays a pivotal role in our paper. Given a closed densely defined operator H in $\mathcal{H}$, we put

$$\Omega_{c,d}(H) = \{z \in \mathbb{C} : \mathrm{dist}(z, \sigma(H)) \geqslant c|z| \text{ and } |z| > d\}, \quad c > 0, d \geqslant 0,$$

where $\mathrm{dist}(z, \sigma(H)) = \inf\{|z-w| : w \in \sigma(H)\}$. It is clear that $\Omega_{c,d}(H) \subseteq \mathbb{C}\backslash\sigma(H)$. Suppose that H is self-adjoint. Then, due to $\|R(z)\| = \mathrm{dist}(z, \sigma(H))^{-1}$, we have

$$\Omega_{c,d}(H) = \{z \in \mathbb{C} \setminus \sigma(H) : \|zR(z)\| \leqslant \frac{1}{c} \text{ and } |z| > d\}, \quad c > 0,\, d \geqslant 0. \tag{7}$$

One can show that if H is self-adjoint, then $\Omega_{c,d}(H)$ is unbounded if and only if $c \leqslant 1$; if this is the case, then $\{\mathrm{i}t : t \in \mathbb{R}, |t| > d\} \subseteq \Omega_{c,d}(H)$. If H is self-adjoint and bounded below (resp. above), $c \in (0,1)$ and $d \geqslant 0$, then $(-\infty, a) \subseteq \Omega_{c,d}(H)$ (resp. $(a, \infty) \subseteq \Omega_{c,d}(H)$) for some $a \in \mathbb{R}$. Remark 6 contains further properties of the set $\Omega_{c,d}(H)$. The reader should be aware that $\mathrm{dist}(z_n, \sigma(H)) \to \infty$ may not imply the boundedness of $\{\|z_n R(z_n)\|\}_{n=1}^\infty$, e.g., if H is self-adjoint, $\sigma(H) = \{n^2 : n = 1, 2, 3, \ldots\}$ and $z_n = \frac{1}{2}(n^2 + (n+1)^2)$. This shows that the set $\Omega_{c,d}(H)$ is optimal taking into account uniform boundedness of the function $zR(z)$.

The following result collects some properties of resolvents and commutators.

Proposition 2. *Let A be a linear operator in $\mathcal{H}$, H be a closed densely defined linear operator in $\mathcal{H}$ and $m, k \geqslant 0$ be integers.*
If $\mathcal{D}(H^m) \subseteq \mathcal{D}(A)$, then

(i) $[A, R(z)^k]f = \sum_{j=0}^{k-1} R(z)^j[A, R(z)]R(z)^{k-1-j}f$ *for all $f \in \mathcal{D}(H^m)$ and $z \in \mathbb{C} \setminus \sigma(H)$,*

(ii) *the operator* $(\operatorname{ad} R(z))^k(A)\big|_{\mathcal{D}(H^m)}$ *is bounded for every* $z \in \mathbb{C}\setminus\sigma(H)$*, provided* $(\operatorname{ad} R(w))^k(A)\big|_{\mathcal{D}(H^m)}$ *is bounded for some* $w \in \mathbb{C}\setminus\sigma(H)$.

If H *is self-adjoint, then for all* $c \in (0,1]$ *and* $d \geqslant 0$,

(iii) $\lim_{\Omega_{c,d}(H)\ni z\to\infty}(zR(z))^m f = f$ *for all* $f \in \mathcal{H}$,

(iv) $\sup_{z\in\Omega_{c,d}(H)} \|zR(z)\| \leqslant \frac{1}{c}$,

(v) $\sup_{z\in\Omega_{c,d+\varepsilon}(H)} \|(\operatorname{ad} R(z))^k(A)|_{\mathcal{D}(H^m)}\| < \infty$, *provided* $\varepsilon > 0$, $\mathcal{D}(H^m) \subseteq \mathcal{D}(A)$ *and* $(\operatorname{ad} R(w))^k(A)|_{\mathcal{D}(H^m)}$ *is bounded for some* $w \in \mathbb{C}\setminus\sigma(H)$.

If A *is closable,* H *is self-adjoint and* $\mathcal{D}(H^m) \subseteq \mathcal{D}(A)$*, then*

(vi) *the operator* $(\operatorname{ad} R(z))^k(A)|_{\mathcal{D}(H^m)}$ *is bounded, provided* $(\operatorname{ad} R(z))^k(A)$ *is bounded on a core* $\mathcal{E}$ *of* H^m, $z \in \mathbb{C}\setminus\sigma(H)$.

If A *is symmetric,* H *is self-adjoint and* $\mathcal{D}(H^m) \subseteq \mathcal{D}(A)$*, then*

(vii) *the operators* $R(z)^m A$ *and* $AR(z)^m$ *are bounded for every* $z \in \mathbb{C}\setminus\sigma(H)$,

(viii) *the operator* $(\operatorname{ad} R(z))^k(A)$ *is closable for every* $z \in \mathbb{C}\setminus\sigma(H)$,

(ix) $\mathrm{i}^k(\operatorname{ad} R(x))^k(A)$ *is symmetric for every* $x \in \mathbb{R}\setminus\sigma(H)$.

Proof. (i) Fixing $k \geqslant 1$ and $f \in \mathcal{D}(H^m)$, we compute

$$\begin{aligned}
&\sum_{j=0}^{k-1} R(z)^j[A,R(z)]R(z)^{k-1-j}f \\
&= \sum_{j=0}^{k-1} R(z)^j AR(z)^{k-j}f - \sum_{j=0}^{k-1} R(z)^{j+1}AR(z)^{k-1-j}f \\
&= \sum_{j=0}^{k-1} R(z)^j AR(z)^{k-j}f - \sum_{j=1}^{k} R(z)^j AR(z)^{k-j}f = [A,R(z)^k]f.
\end{aligned}$$

(ii) Let $z, w \in \mathbb{C}\setminus\sigma(H)$. Set $C = (w-H)R(z) = I + (w-z)R(z) \in \boldsymbol{B}(\mathcal{H})$. Then $C^{-1} = (z-H)R(w) = I + (z-w)R(w)$. By (2) we have $\mathcal{D}(H^m) \subseteq \mathcal{D}([R(\xi),A])$ for every $\xi \in \mathbb{C}\setminus\sigma(H)$, which together with the resolvent identity gives us

$$[R(w),A]f - [R(z),A]f = (z-w)\big([R(w),A]R(z) + R(w)[R(z),A]\big)f$$

for all $f \in \mathcal{D}(H^m)$. This implies

$$[R(w),A](I+(w-z)R(z))f = (I+(z-w)R(w))[R(z),A]f, \quad f \in \mathcal{D}(H^m),$$

which leads to

$$[R(z),A]f = C[R(w),A]Cf, \quad f \in \mathcal{D}(H^m). \tag{8}$$

We show by induction that for all integers $k \geqslant 0$,

$$(\operatorname{ad} R(z))^k(A)f = C^k(\operatorname{ad} R(w))^k(A)C^k f, \quad f \in \mathcal{D}(H^m). \tag{9}$$

Indeed, suppose (9) holds for a fixed $k \geqslant 0$ (the case $k = 0$ is trivial).

Then, applying (8) to the operator $(\operatorname{ad} R(w))^k(A)$ in place of A and (9), we get

$$\begin{aligned}(\operatorname{ad} R(z))^{k+1}(A)f &= [R(z), (\operatorname{ad} R(z))^k(A)]f \\ &= [R(z), C^k(\operatorname{ad} R(w))^k(A)C^k]f \\ &= C^k[R(z), (\operatorname{ad} R(w))^k(A)]C^k f \\ &= C^{k+1}[R(w), (\operatorname{ad} R(w))^k(A)]C^{k+1} f \\ &= C^{k+1}(\operatorname{ad} R(w))^{k+1}(A)C^{k+1} f, \quad f \in \mathcal{D}(H^m).\end{aligned}$$

It is clear that (9) implies (ii).

Conditions (iii) and (iv) can be deduced from the spectral theorem (cf. (7)).

(v) By (9), we have

$$(\operatorname{ad} R(z))^k(A)f = (I + (w - z)R(z))^k(\operatorname{ad} R(w))^k(A)(I + (w - z)R(z))^k f$$

for all $f \in \mathcal{D}(H^m)$. This and (iv) implies (v) (because $\sup_{z\in\Omega_{c,d+\varepsilon}(H)} \|R(z)\| \leqslant \frac{1}{c\varepsilon}$).

(vi) If $f \in \mathcal{D}(H^m)$, then there exists a sequence $\{f_n\}_{n=0}^{\infty} \subseteq \mathcal{E}$ such that $\lim_{n\to\infty} \|f - f_n\|_{H^m} = 0$. By (4), $(\operatorname{ad} R(z))^k(A)f_n \to (\operatorname{ad} R(z))^k(A)f$ as $n \to \infty$, which implies the boundedness of $(\operatorname{ad} R(z))^k(A)|_{\mathcal{D}(H^m)}$.

(vii) By Proposition 1, $AR(z)^m = \bar{A}R(z)^m \in \boldsymbol{B}(\mathcal{H})$ for every $z \in \mathbb{C} \setminus \sigma(H)$. Taking adjoints, we get $R(z)^m A \subseteq R(z)^m A^* \subseteq (AR(\bar{z})^m)^*$ for every $z \in \mathbb{C}\setminus\sigma(H)$, which completes the proof of (vii).

(viii) By the von Neumann theorem it suffices to show that $\mathcal{D}(H^m)$ is contained in the domain of the adjoint of $(\operatorname{ad} R(z))^k(A)$. The case $k = 0$ is obvious. Suppose that our claim is true for a fixed $k \geqslant 0$. Since the operator $(\operatorname{ad} R(z))^{k+1}(A)$ is densely defined (because its domain contains $\mathcal{D}(H^m)$), we can calculate

$$\big((\operatorname{ad} R(z))^{k+1}(A)\big)^* \supseteq \big((\operatorname{ad} R(z))^k(A)\big)^* R(\bar{z}) - R(\bar{z})\big((\operatorname{ad} R(z))^k(A)\big)^*,$$

which together with (2) completes the proof of (viii).

An induction argument similar to that in (viii) enables us to prove (ix). □

2. Symmetric operators dominated by self-adjoint ones

We now formulate one of the main results of the paper, which generalizes Theorem 1 of [1] (see Section 5).

Theorem 3. *Let A be a symmetric operator in $\mathcal{H}$, H be a self-adjoint operator in $\mathcal{H}$ and Ω be an unbounded subset of $\Omega_{c,d}(H)$, where $c \in (0,1]$ and $d \geqslant 0$.*

(i) *If $\mathcal{D}(H^m) \subseteq \mathcal{D}(A)$ for an integer $m \geqslant 0$ and $\sup_{z\in\Omega} |z|\|[R(z), A]\| < \infty$, then A is essentially self-adjoint on any core of H^m.*

(ii) *If $\mathcal{D}^\infty(H) \subseteq \mathcal{D}(A)$, $\sup_{z\in\Omega} |z|\|[R(z), A]\| < \infty$ and $\mathcal{E}$ is a core[3] of H^m for every integer $m \geqslant 0$, then A is essentially self-adjoint on $\mathcal{E}$.*

---

[3] For instance, $\mathcal{E} = \mathcal{D}^\infty(H)$ or $\mathcal{E} =$ the set of all bounded vectors of H, cf. [5].

Remark 4. Since the operator $[R(z), A]$ is closable (cf. Proposition 2 (viii)), it is bounded if and only if so is its restriction to a dense linear subspace $\mathcal{F}$ of $\mathcal{D}([R(z), A])$; moreover, the following equality holds $\|[R(z), A]\| = \|[R(z), A]|_{\mathcal{F}}\|$.

Proof of Theorem 3. There is no loss of generality in assuming that the operator A is closed.

(i) By the closed graph theorem, we can assume that $m \geqslant 1$. According to our assumption $\alpha \overset{\mathrm{df}}{=} \sup_{z\in\Omega} \|[A, zR(z)]\| < \infty$. Let $\mathcal{E}$ be a core of H^m. It follows from Proposition 1 that $\overline{A|_{\mathcal{E}}} = \overline{A|_{\mathcal{D}(H^m)}}$. Hence $\mathcal{N}((A|_{\mathcal{E}})^* + \bar{z}) = \mathcal{N}((A|_{\mathcal{D}(H^m)})^* + \bar{z})$ which is equivalent to

$$((A+z)\mathcal{E})^{\perp} = ((A+z)\mathcal{D}(H^m))^{\perp}, \quad z \in \mathbb{C}. \tag{10}$$

We show that

$$((A+\mathrm{i}\,u)\mathcal{E})^{\perp} = \{0\} \text{ for every } u > \alpha\, m\, c^{1-m}. \tag{11}$$

To prove (11), take $f \in ((A+\mathrm{i}\,u)\mathcal{E})^{\perp}$ and $z \in \Omega$, and put[4] $f_z = \bar{z}^m R(\bar{z})^m f \in \mathcal{D}(H^m)$. Since by (2) $R(z)^m f_z \in \mathcal{D}(H^{2m})$, (10) gives us

$$\langle f, (A+\mathrm{i}\,u)R(z)^m f_z\rangle = 0,$$

which can be rewritten as $\langle f, (R(z)^m(A+\mathrm{i}\,u) + [A, R(z)^m])f_z\rangle = 0$. Hence, we have

$$\mathrm{i}\,u\langle f, R(z)^m f_z\rangle = \langle f, (R(z)^m A + [A, R(z)^m])f_z\rangle. \tag{12}$$

Since $\bar{z}^m\,\mathrm{i}\,u\langle f, R(z)^m f_z\rangle = \mathrm{i}\,u\|f_z\|^2$ and $\bar{z}^m\langle f, R(z)^m A f_z\rangle = \langle f_z, A f_z\rangle$, (12) yields

$$\mathrm{i}\,u\|f_z\|^2 = \langle f_z, A f_z\rangle + \bar{z}^m\langle f, [A, R(z)^m]f_z\rangle. \tag{13}$$

Comparing imaginary parts of both sides of (13), using $\langle f_z, Af_z\rangle \in \mathbb{R}$ and employing conditions (i) and (iv) of Proposition 2, we get

$$\begin{aligned} u\|f_z\|^2 &= \mathfrak{Im}\,(\bar{z}^m\langle f, [A, R(z)^m]f_z\rangle) \leqslant |\langle f, [A, (zR(z))^m]f_z\rangle| \\ &\leqslant \sum_{j=0}^{m-1} |\langle f, (zR(z))^j[A, zR(z)](zR(z))^{m-1-j}f_z\rangle| \\ &\leqslant m\|f\|\,\|f_z\|\,\|zR(z)\|^{m-1}\|[A, zR(z)]|_{\mathcal{D}(H^m)}\| \\ &\leq \alpha\, m\, c^{1-m}\|f\|\,\|f_z\|, \quad z \in \Omega. \end{aligned} \tag{14}$$

Since the set Ω is unbounded, there exists a sequence $\{z_n\}_{n=1}^{\infty} \subseteq \Omega$ such that $\lim_{n\to\infty}|z_n| = \infty$. As $\bar{z}_n \in \Omega_{c,d}(H)$ for all $n \geqslant 1$, we infer from part (iii) of Proposition 2 that $\lim_{n\to\infty} f_{z_n} = f$. This, when combined with (14), gives us $u\|f\|^2 \leqslant \alpha\, m\, c^{1-m}\|f\|^2$. Since $u > \alpha\, m\, c^{1-m}$, it must be $f = 0$, which proves (11).

Applying (11) to $-A$, we see that $((A-\mathrm{i}\,u)\mathcal{E})^{\perp} = \{0\}$ for every $u > \alpha\, m\, c^{1-m}$, which means that both deficiency indices of $A|_{\mathcal{E}}$ are equal to zero. Hence A is essentially self-adjoint on $\mathcal{E}$.

[4] Notice that $\bar{z} \in \Omega_{c,d}(H) \subseteq \mathbb{C} \setminus \sigma(H)$.

(ii) By Proposition 1, $\mathcal{D}(H^m) \subseteq \mathcal{D}(A)$ for some integer $m \geqslant 0$. Applying (i) completes the proof. □

Theorem 3 (as well as Theorems 8 and 11) can be combined with commuting domination results of [5]. What follows is a sample application of Corollary 19 of [5] and Theorem 3.

Corollary 5. *Let A be a symmetric operator in $\mathcal{H}$, H be a self-adjoint operator in $\mathcal{H}$ and p be a real polynomial in one variable of degree $n \geqslant 1$. Assume that $\mathcal{D}(H^m) \subseteq \mathcal{D}(A^n)$ for some integer $m \geqslant 0$ and $\sup_{z\in\Omega} |z| \| [R(z), p(A)] \| < \infty$, where Ω is an unbounded subset of $\Omega_{c,d}(H)$ for some $c \in (0,1]$ and $d \in [0,\infty)$. Then for every real polynomial q in one variable with $\deg q \leqslant n$, the operator $q(A)$ is essentially self-adjoint, $\mathcal{D}(A^n)$ is a core of $q(A)$ and $\overline{q(A)} = q(\bar{A})$.*

Remark 6. The set Ω appearing in Theorem 3 (as well as in Theorems 8 and 11) can be specified in many ways not excluding the choice of unbounded sequences in $\Omega_{c,d}(H)$. For example one can always take Ω of the form $\Omega_\pm = \{\pm i t : t > d\} \subseteq \Omega_{c,d}(H)$. In the case when H is semibounded, Ω can be chosen as an unbounded interval on the real line, which is disjoint from the spectrum of H. Let us discuss two examples of semibounded self-adjoint operators. Consider first H with $\sigma(H) = \{0, 1, 2, \ldots\}$. Then $\Omega_{c,d}(H) \cap (a, \infty)$ is bounded for all $c \in (0,1]$, $d \geqslant 0$ and $a \in \mathbb{R}$. Indeed, otherwise there exists $\{x_n\}_{n=1}^\infty \subseteq \Omega_{c,d}(H) \cap (a,\infty)$ such that $0 < x_n \nearrow \infty$. Then $c\,x_n \leqslant \operatorname{dist}(x_n, \sigma(H)) \leqslant \frac{1}{2}$ for all $n \geqslant 1$, which is impossible. This shows that, for this particular H, there is no way of choosing a subset Ω of (a, ∞) satisfying the assumptions of Theorem 3 (though for each $c \in (0,1)$ there exists $a \in \mathbb{R}$ such that $(-\infty, a) \subseteq \Omega_{c,d}(H)$). However, the other example will show the opposite. Let now $\{\alpha^1, \alpha^2, \alpha^3, \ldots\}$ be the spectrum of H, where $\alpha > 1$. Set $x_n = \frac{1}{2}(\alpha^n + \alpha^{n+1})$ for $n \geqslant 1$. Then $\Omega \overset{\mathrm{df}}{=} \{x_1, x_2, x_3, \ldots\} \subseteq \Omega_{\frac{\alpha-1}{\alpha+1},1}(H) \cap (0,\infty)$. Finally, the self-adjoint operator H with $\sigma(H) = \{\ldots, -\alpha^3, -\alpha^2, -\alpha^1, \alpha^1, \alpha^2, \alpha^3, \ldots\}$ is not semibounded, whereas the set $\Omega_{\frac{\alpha-1}{\alpha+1},1}(H) \cap \mathbb{R}$ is unbounded.

Remark 7. A careful reader can ensure himself that Theorem 3 is valid if the self-adjoint operator H is replaced by a normal one. However in this case, contrary to the self-adjoint one, it may happen that the set $\Omega_{c,d}(H)$ is bounded or even empty regardless of the choice of c, d.

3. Semibounded symmetric operators dominated by self-adjoint ones

In this section we formulate a criterion for essential self-adjointness of semibounded operators, which generalizes Theorem 4 of [1] (see Section 5).

Theorem 8. *Let A be a semibounded symmetric operator in $\mathcal{H}$, H be a self-adjoint operator in $\mathcal{H}$ and Ω be an unbounded subset of $\Omega_{c,d}(H) \cap \mathbb{R}$, where $c \in (0,1]$ and $d \geqslant 0$.*

(i) *If $\mathcal{D}(H^m) \subseteq \mathcal{D}(A)$ for an integer $m \geqslant 0$ and $\sup_{x\in\Omega} x^2\|(\mathrm{ad}\, R(x))^2(A)\| < \infty$, then A is essentially self-adjoint on any core of H^m.*

(ii) *If $\mathcal{D}^\infty(H) \subseteq \mathcal{D}(A)$, $\sup_{x\in\Omega} x^2\|(\mathrm{ad}\, R(x))^2(A)\| < \infty$ and $\mathcal{E}$ is a core of H^m for every integer $m \geqslant 0$, then A is essentially self-adjoint on $\mathcal{E}$.*

Remark 9. By Proposition 2 (viii), the operator $(\mathrm{ad}\, R(x))^2(A)$ is bounded if and only if so is its restriction to a dense linear subspace $\mathcal{F}$ of $\mathcal{D}((\mathrm{ad}\, R(x))^2(A))$; if this is the case, then $\|(\mathrm{ad}\, R(x))^2(A)\| = \|(\mathrm{ad}\, R(x))^2(A)\big|_{\mathcal{F}}\|$.

Proof of Theorem 8. Assuming A is closed involves no loss of generality.

(i) The case $m = 0$ forces $A \in \boldsymbol{B}(\mathcal{H})$. Suppose $m \geqslant 1$. Without loss of generality we can assume that $A \leqslant 0$. Let $\mathcal{E}$ be a core of H^m. As in the proof of Theorem 3, one can show that (10) holds. By our assumption $\beta \overset{\mathrm{df}}{=} \sup_{x\in\Omega} \|C_x\| < \infty$, where $C_x \overset{\mathrm{df}}{=} \big((\mathrm{ad}(xR(x)))^2(A)\big)^- \in \boldsymbol{B}(\mathcal{H})$ (cf. Remark 9). We show that

$$\big((u - A)\mathcal{E}\big)^\perp = \{0\} \text{ for every } u > \frac{1}{2}\beta m^2 c^{2(1-m)}. \tag{15}$$

Take $f \in \big((u - A)\mathcal{E}\big)^\perp$ and $x \in \Omega$, and put $f_x = x^m R(x)^m f \in \mathcal{D}(H^m)$. Then by (2) we have $R(x)^m f_x \in \mathcal{D}(H^{2m})$. This and (10) give us

$$0 = \langle f, (u - A)R(x)^m f_x\rangle = u\langle f, R(x)^m f_x\rangle - \langle f, R(x)^m A f_x\rangle - \langle f, [A, R(x)^m] f_x\rangle.$$

Since $x^m u\langle f, R(x)^m f_x\rangle = u\|f_x\|^2$ and $x^m\langle f, R(x)^m A f_x\rangle = \langle f_x, A f_x\rangle \leqslant 0$, we get

$$u\|f_x\|^2 = \langle f_x, A f_x\rangle + \langle f, [A, (xR(x))^m] f_x\rangle.$$

Comparing real parts of both sides of the above equality, we obtain

$$\begin{aligned} u\|f_x\|^2 &= \langle f_x, A f_x\rangle + \mathfrak{Re}\langle f, [A, (xR(x))^m] f_x\rangle \\ &\le x^m\, \mathfrak{Re}\langle f, [A, (xR(x))^m] R(x)^m f\rangle. \end{aligned} \tag{16}$$

Set $D_x = \frac{1}{2}\sum_{j=0}^{m-1}\sum_{k=0}^{m-1} (xR(x))^{j+k} C_x (xR(x))^{2(m-1)-(j+k)} \in \boldsymbol{B}(\mathcal{H})$. Applying formula (i) of Proposition 2 to $[A, R(x)^m]$ and $[[A, R(x)], R(x)^m]$, we conclude

$$\begin{aligned} & x^m\, \mathfrak{Re}\langle g, [A, (xR(x))^m] R(x)^m g\rangle \\ &\quad = \frac{1}{2} x^{2m}\big(\langle g, [A, R(x)^m] R(x)^m g\rangle - \langle g, R(x)^m [A, R(x)^m] g\rangle\big) \\ &\quad = \frac{1}{2} x^{2m} \sum_{j=0}^{m-1} \langle g, R(x)^j [[A, R(x)], R(x)^m] R(x)^{m-1-j} g\rangle \\ &\quad = \frac{1}{2}\sum_{j=0}^{m-1}\sum_{k=0}^{m-1} \langle g, (xR(x))^{j+k} (\mathrm{ad}(xR(x)))^2(A) (xR(x))^{2(m-1)-(j+k)} g\rangle \\ &\quad = \frac{1}{2}\sum_{j=0}^{m-1}\sum_{k=0}^{m-1} \langle g, (xR(x))^{j+k} C_x (xR(x))^{2(m-1)-(j+k)} g\rangle \\ &\quad = \langle g, D_x g\rangle, \quad g \in \mathcal{D}(H^m). \end{aligned} \tag{17}$$

However by part (vii) of Proposition 2, $[A,(xR(x))^m]R(x)^m \in \boldsymbol{B}(\mathcal{H})$, which together with (17) leads to

$$x^m\,\mathfrak{Re}\langle h,[A,(xR(x))^m]R(x)^m h\rangle = \langle h, D_x h\rangle, \quad h\in\mathcal{H}. \tag{18}$$

Combining (16), (18) and part (iv) of Proposition 2, we have

$$\begin{aligned} u\|f_x\|^2 &\leqslant x^m\,\mathfrak{Re}\langle f,[A,(xR(x))^m]R(x)^m f\rangle = \langle f, D_x f\rangle \leqslant \|f\|^2\|D_x\| \\ &\leqslant \frac{1}{2}m^2\|f\|^2\|xR(x)\|^{2(m-1)}\|C_x\| \le \frac{1}{2}\beta m^2 c^{2(1-m)}\|f\|^2, \quad x\in\Omega. \end{aligned} \tag{19}$$

Since, by part (iii) of Proposition 2, $f_x \to f$ as $x\in\Omega$ and $|x|\to\infty$, we infer from (19) that $u\|f\|^2 \leqslant \frac{1}{2}\beta m^2 c^{2(1-m)}\|f\|^2$, which implies (15). As the operator $A|_{\mathcal{E}}$ is semibounded, its deficiency indices are both equal to 0, which means that A is essentially self-adjoint on $\mathcal{E}$.

(ii) is a direct consequence of (i) and Proposition 1. □

The next result is an application of Theorem 25 of [5] and Theorem 8.

Corollary 10. *Let $A_1,\dots,A_\kappa$ be formally normal operators in $\mathcal{H}$ and let H be a self-adjoint operator in $\mathcal{H}$ such that the domain of H^m is contained in the domain of $A \stackrel{\mathrm{df}}{=} A_1^*A_1+\cdots+A_\kappa^*A_\kappa$ for some integer $m\geqslant 0$. Assume that $\mathcal{E}$ is a core of H^m and*

(i) *$\mathcal{E}\subseteq\mathcal{D}(A_iA_j)\cap\mathcal{D}(A_jA_i)$ and $A_iA_jf = A_jA_if$, $f\in\mathcal{E}$, for $1\le i<j\le\kappa$,*
(ii) *$\langle A_i^*A_if, A_jg\rangle = \langle A_j^*f, A_i^*A_ig\rangle$, $f,g\in\mathcal{E}$, for $i,j=1,\dots,\kappa$,*
(iii) *$\sup_{x\in\Omega} x^2\|(\mathrm{ad}\,R(x))^2(A)\|<\infty$ for an unbounded subset Ω of $\Omega_{c,d}(H)\cap\mathbb{R}$, where $c\in(0,1]$ and $d\geqslant 0$.*

Then $\bar{A}_1,\dots,\bar{A}_\kappa$ are spectrally commuting normal operators and $\mathcal{E}$ is a joint core of any subsystem of $\{\bar{A}_1,\dots,\bar{A}_\kappa\}$.

4. More on symmetric operators dominated by self-adjoint ones

Given a densely defined operator A in $\mathcal{H}$, we set $\mathfrak{Im}\,A = \frac{1}{2\mathrm{i}}(A-A^*)$. If the operator $\mathfrak{Im}\,A$ is densely defined, then it is symmetric and consequently closable.

Theorem 11. *Let A be a symmetric operator in $\mathcal{H}$, H be a self-adjoint operator in $\mathcal{H}$ and Ω be an unbounded subset of $\Omega_{c,d}(H)\cap\mathbb{R}$, where $c\in(0,1]$ and $d\geqslant 0$. If $\mathcal{D}(H^m)\subseteq\mathcal{D}(A)$ for an integer $m\geqslant 0$ and $\sup_{x\in\Omega}\left\|x^{m+1}\,\mathfrak{Im}\left([R(x),A]R(x)^m\right)\right\|<\infty$, then A is essentially self-adjoint on any core of H^m.*

Remark 12. By (2) and part (vii) of Proposition 2, $[R(x),A]R(x)^m\in\boldsymbol{B}(\mathcal{H})$ and consequently $\mathfrak{Im}([R(x),A]R(x)^m)\in\boldsymbol{B}(\mathcal{H})$. Since

$$-R(x)^m[R(x),A] \subseteq \left([R(x),A]R(x)^m\right)^*,$$

we get $\frac{1}{2\mathrm{i}}\left([R(x),A]R(x)^m + R(x)^m[R(x),A]\right)\subseteq\mathfrak{Im}([R(x),A]R(x)^m)$.

Proof of Theorem 11. Let $\mathcal{E}$ be a core of H^m. As in the proof of Theorem 3 we can assume that A is closed, $m \geqslant 1$ and (10) holds. According to Remark 12, for every $x \in \Omega$, $C_x \stackrel{\text{df}}{=} x^{m+1}\,\mathfrak{Im}([R(x), A]R(x)^m) \in \boldsymbol{B}(\mathcal{H})$ and

$$x^{m+1}\big([R(x), A]R(x)^m + R(x)^m[R(x), A]\big) \subseteq 2\,\mathrm{i}\,C_x.$$

By our assumption $\gamma \stackrel{\text{df}}{=} \sup_{x\in\Omega} \|C_x\| < \infty$. We show that

$$\big((A + \mathrm{i}\,u)\mathcal{E}\big)^{\perp} = \{0\} \text{ for every } u > \gamma\, m\, c^{1-m}. \tag{20}$$

Take $f \in \big((A + \mathrm{i}\,u)\mathcal{E}\big)^{\perp}$ and $x \in \Omega$, and put $f_x = x^m R(x)^m f \in \mathcal{D}(H^m)$. Then by (2) we have $R(x)^m f_x \in \mathcal{D}(H^{2m})$. This and (10) give us

$$\begin{aligned} 0 &= \langle f, (A + \mathrm{i}\,u)R(x)^m f_x\rangle \\ &= -\,\mathrm{i}\,u\langle f, R(x)^m f_x\rangle + \langle f, R(x)^m A f_x\rangle + \langle f, [A, R(x)^m] f_x\rangle, \end{aligned}$$

which leads to

$$\mathrm{i}\,u\|f_x\|^2 = \langle f_x, A f_x\rangle + \langle f, [A, (xR(x))^m] f_x\rangle.$$

Comparing imaginary parts of both sides of the above equality, we obtain

$$u\|f_x\|^2 = x^{2m}\,\mathfrak{Im}\langle f, [A, R(x)^m]R(x)^m f\rangle. \tag{21}$$

Set $D_x = \sum_{j=0}^{m-1} (xR(x))^j C_x (xR(x))^{m-1-j} \in \boldsymbol{B}(\mathcal{H})$. Applying formula (i) of Proposition 2, we get

$$\begin{aligned} &x^{2m}\,\mathfrak{Im}\langle g, [A, R(x)^m]R(x)^m g\rangle \\ &\quad= \frac{1}{2\mathrm{i}} x^{2m}\big(\langle g, [A, R(x)^m]R(x)^m g\rangle + \langle g, R(x)^m[A, R(x)^m] g\rangle\big) \\ &\quad= \frac{1}{2\mathrm{i}} x^{2m} \sum_{j=0}^{m-1} \langle g, R(x)^j([A, R(x)]R(x)^m + R(x)^m[A, R(x)])R(x)^{m-1-j} g\rangle \\ &\quad= \langle g, D_x g\rangle, \quad g \in \mathcal{D}(H^m). \end{aligned} \tag{22}$$

However by part (vii) of Proposition 2, $[A, R(x)^m]R(x)^m \in \boldsymbol{B}(\mathcal{H})$, which together with (22) leads to

$$x^{2m}\,\mathfrak{Im}\langle h, [A, R(x)^m]R(x)^m h\rangle = \langle h, D_x h\rangle, \quad h \in \mathcal{H}. \tag{23}$$

Combining (21), (23) and part (iv) of Proposition 2, we have

$$u\|f_x\|^2 = \langle f, D_x f\rangle \leqslant m\|f\|^2\|xR(x)\|^{m-1}\|C_x\| \leq \gamma\, m\, c^{1-m}\|f\|^2, \quad x \in \Omega. \tag{24}$$

Since, by part (iii) of Proposition 2, $f_x \to f$ as $x \in \Omega$ and $|x| \to \infty$, we infer from (24) that $u\|f\|^2 \leqslant \gamma\, m\, c^{1-m}\|f\|^2$, which implies (20). Applying (20) to $-A$, we conclude that A is essentially self-adjoint on $\mathcal{E}$. □

Note that Theorem 11 implies the version of Theorem 3 in which the set Ω is additionally assumed to be contained in $\mathbb{R}$.

5. Comments and examples

The first two remarks emphasize the role played by domination in our paper.

Remark 13. The inclusion $\mathcal{D}(H^m) \subseteq \mathcal{D}(A)$ is one of the main assumptions of Theorems 3, 8 and 11. In virtue of Proposition 1, $\mathcal{D}(H^m) \subseteq \mathcal{D}(A)$ if and only if A is dominated by H^m on a core of H^m (provided A is closed). The choice of this core does not determine in any way that appearing in the respective conclusions.

Remark 14. Conclusions of Theorems 3, 8 and 11 can be supplied with a joint core assertion. Namely any core $\mathcal{E}$ of H^m is always a joint core of (A, H^m), which means that the joint graph[5] $\mathcal{G}(A, H^m)$ of the pair (A, H^m) is contained in the closure of $\mathcal{G}(A|_{\mathcal{E}}, H^m|_{\mathcal{E}})$. This can be deduced directly from domination (use (5)).

Remark 15. Typical assumption in Theorems 3 and 8 is of the form

$$\sup_{z\in\Omega} |z|^k \|(\mathrm{ad}\, R(z))^k(A)\| < \infty$$

with $k = 1$ or 2. Consequently

$$\lim_{\Omega\ni z\to\infty} \|(\mathrm{ad}\, R(z))^k(A)\| = 0$$

and this carries further information about the behavior of the commutator $(\mathrm{ad}\, R(z))^k(A)$ at infinity (compare with part (v) of Proposition 2).

We now concentrate on relating Theorems 1 and 4 of [1] to our Theorems 3 and 8.

Remark 16. We begin with proving that Theorem 1 of [1] follows from Theorem 3. For this, let us assume that A and H satisfy the assumptions of Theorem 1 in [1], i.e., A is a closed symmetric operator in $\mathcal{H}$ and H is a self-adjoint operator in $\mathcal{H}$ such that for some $w \in \mathbb{C}\setminus\sigma(H)$, $\|AR(w)^m|_{\mathcal{D}^\infty(H)}\| < \infty$ for some integer $m \geqslant 0$ (the latter is equivalent to $\mathcal{D}(H^m) \subseteq \mathcal{D}(A)$, cf. Proposition 1) and the sesquilinear form φ associated with the expression $(\mathrm{ad}\, H)(A)R(w)$ is bounded on $\mathcal{D}^\infty(H)$. Set $C_z = I+(w-z)R(z)$. Then the equalities $R(w)C_z = R(z)$ and $HR(\xi) = \xi R(\xi) - I$, $\xi \in \mathbb{C}\setminus\sigma(H)$, imply

$$\begin{aligned}\varphi(C_z f, R(\bar z)g) &= \langle AR(w)C_z f, HR(\bar z)g\rangle - \langle AHR(w)C_z f, R(\bar z)g\rangle\\ &= \langle AR(z)f, HR(\bar z)g\rangle - \langle AHR(z)f, R(\bar z)g\rangle \qquad (25)\\ &= \langle [R(z), A]f, g\rangle, \quad f, g \in \mathcal{D}^\infty(H).\end{aligned}$$

By Remark 4 and part (iv) of Proposition 2, we have

$$\sup_{z\in\Omega_{c,d}(H)} |z|\|[R(z),A]\| \le \|\varphi\| \sup_{z\in\Omega_{c,d}(H)} \|zR(z)\|\|I+(w-z)R(z)\| < \infty,$$

for all $c \in (0,1]$ and $d > 0$. By Theorem 3, A is essentially self-adjoint on any core of H^m.

[5] Cf. [5].

Theorem 4 in [1] can be deduced from Theorem 8 by proving that if A is a closed semibounded symmetric operator in $\mathcal{H}$ and H is a semibounded self-adjoint operator in $\mathcal{H}$ such that for some $w \in \mathbb{C} \setminus \sigma(H)$, $\|AR(w)^m|_{\mathcal{D}^\infty(H)}\| < \infty$ for some integer $m \geqslant 0$ and the sesquilinear form ψ associated with the expression $R(w)(\operatorname{ad} H)^2(A)R(w)$ is bounded on $\mathcal{D}^\infty(H)$, then

$$\sup_{z \in \Omega_{c,d}(H)} |z|^2 \|(\operatorname{ad} R(z))^2(A)\| < \infty, \quad c \in (0,1),\, d > 0.$$

Reasoning similar to that in (25) enables us to show that

$$\psi(C_z R(z) f, C_z^* R(\bar{z}) g) = \langle (\operatorname{ad} R(z))^2(A) f, g \rangle, \quad f, g \in \mathcal{D}^\infty(H),$$

which in view of Remark 9 and part (iv) of Proposition 2 gives us

$$\sup_{z \in \Omega_{c,d}(H)} |z|^2 \|(\operatorname{ad} R(z))^2(A)\| \leqslant \|\psi\| \sup_{z \in \Omega_{c,d}(H)} \|zR(z)\|^2 \|I + (w-z)R(z)\|^2 < \infty.$$

Hence, by Theorem 8, we get the essential self-adjointness of A on any core of H^m.

We conclude this paper with an example of a pair (A, H) which is covered by our theorems but fails to satisfy assumptions of theorems in [1].

Example 17. Fix an integer $m \geqslant 1$. Let H be an unbounded self-adjoint operator in $\mathcal{H}$ such that $H \geqslant I$. Write E for the spectral measure of H. First we show that there exists a vector $e \in \mathcal{H}$ of norm 1 such that

$$e \notin \bigcup_{t \in (0,\infty)} \mathcal{D}(H^t). \tag{26}$$

Indeed, since the operator $\log H \stackrel{\text{df}}{=} \int_1^\infty \log x E(\mathrm{d}x)$ is unbounded (because the closed support of E is unbounded), we can find a unit vector e which is not in $\mathcal{D}(\log H)$. This means that $\int_1^\infty (\log x)^2 \langle E(\mathrm{d}x)e, e\rangle = \infty$. Noticing that for every $t \in (0,\infty)$ there exits $c(t) \in (0,\infty)$ such that[6] $\log x \leqslant c(t) x^t$ for all $x \in [1,\infty)$, we see that $e \notin \bigcup_{t\in(0,\infty)} \mathcal{D}(H^t)$.

Define $e \otimes e \in \boldsymbol{B}(\mathcal{H})$ by

$$(e \otimes e)(f) = \langle f, e\rangle e, \quad f \in \mathcal{H}.$$

It is clear that the operator $A_e \stackrel{\text{df}}{=} H^m + e \otimes e$ is positive and self-adjoint, and $\mathcal{D}(H^m) = \mathcal{D}(A_e)$. Moreover, $[R(z), A_e] \subseteq [R(z), e \otimes e]$ and, in consequence,

$$(\operatorname{ad} R(z))^k (A_e) \subseteq (\operatorname{ad} R(z))^k (e \otimes e)$$

for all $z \in \mathbb{C} \setminus \sigma(H)$. Hence, by part (iv) of Proposition 2, $\sup_{z\in\Omega} \|z[R(z), A_e]\| < \infty$, $\sup_{x\in\Omega} \|x^2 (\operatorname{ad} R(x))^2(A_e)\| < \infty$ and $\sup_{x\in\Omega} \|x^{m+1} \mathfrak{Im}\,([R(x), A_e]R(x)^m)\| < \infty$ for $\Omega = \Omega_{c,d}(H)$, which means that the pair (A_e, H) satisfies all the assumptions of our Theorems 3, 8 and 11.

It turns out that the pair (A_e, H) does not satisfy the assumption (b) of Theorem 1 even in a weakened version proposed in the remark following Theorem 1

---

[6] E.g., expanding the series defining e^{cx^t} one can calculate $c(t) = (n(t)!)^{1/n(t)}$, where $n(t)$ is the least integer greater than or equal to $1/t$.

in [1]. We first show that for all $\alpha, \beta \in [0, \infty)$ and $z \in \mathbb{C} \backslash \sigma(H)$, the sesquilinear form φ associated with the expression $|R(z)|^\alpha [H, A_e] |R(z)|^\beta$ is unbounded on $\mathcal{D}^\infty(H)$ if and only if either $\alpha < 1$ or $\beta < 1$. For $f, g \in \mathcal{D}^\infty(H)$, we compute[7]

$$\begin{aligned}\varphi(f,g) &= \langle A_e|R(z)|^\beta f, H|R(z)|^\alpha g\rangle - \langle A_e H|R(z)|^\beta f, |R(z)|^\alpha g\rangle \\ &= \langle (e\otimes e)|R(z)|^\beta f, H|R(z)|^\alpha g\rangle - \langle (e\otimes e)H|R(z)|^\beta f, |R(z)|^\alpha g\rangle \\ &= \langle |R(z)|^\beta f, e\rangle\langle e, H|R(z)|^\alpha g\rangle - \langle H|R(z)|^\beta f, e\rangle\langle e, |R(z)|^\alpha g\rangle. \qquad (27)\end{aligned}$$

Notice that if $t \in [0, \infty)$, then $H|R(z)|^t = \int_1^\infty \frac{x}{|z-x|^t} E(\mathrm{d}x)$ and consequently the operator $H|R(z)|^t$ is bounded if and only if $t \geqslant 1$. This implies that if $\alpha, \beta \geqslant 1$, then by (27) the form φ is bounded. Consider now the case $\alpha \in [0, 1)$ and suppose contrary to our claim that φ is bounded. Since $|R(z)|^\beta$ is injective, we may choose $f \in \mathcal{D}^\infty(H)$ such that $\langle |R(z)|^\beta f, e\rangle \neq 0$. It follows from (27) that the linear functional $g \longmapsto \langle H|R(z)|^\alpha g, e\rangle$ is bounded on $\mathcal{D}^\infty(H)$. As $\mathcal{D}^\infty(H)$ is a core[8] of $H|R(z)|^\alpha$ we see that $e \in \mathcal{D}(H|R(z)|^\alpha) = \mathcal{D}(H^{1-\alpha})$, which contradicts (26). Interchanging the roles of α and β (as well as f and g), we settle the case $\beta \in [0, 1)$. Similar argument shows that the sesquilinear forms associated with $[H, A_e]R(z)^k$ and $R(z)^k[H, A_e]$ are both unbounded on $\mathcal{D}^\infty(H)$ for every integer $k \geqslant 0$. In particular, the pair (A_e, H) does not satisfy the assumption (c) of Theorem 3 of [1].

Our next aim is to show that the pair (A_e, H) does not satisfy the assumption (b) of Theorems 3 and 4 of [1] (as well as Remarks following them). More precisely, we prove that for all $z \in \mathbb{C} \setminus \sigma(H)$ and $\alpha, \beta \in [0, \infty)$, the sesquilinear form ψ associated with the expression $|R(z)|^\alpha (\mathrm{ad}\, H)^2 (A_e) |R(z)|^\beta$ is unbounded on $\mathcal{D}^\infty(H)$ if and only if either $\alpha < 2$ or $\beta < 2$. For $f, g \in \mathcal{D}^\infty(H)$, we can calculate

$$\begin{aligned}\psi(f,g) &= \langle A_e|R(z)|^\beta f, H^2|R(z)|^\alpha g\rangle - 2\langle A_e H|R(z)|^\beta f, H|R(z)|^\alpha g\rangle \\ &\qquad + \langle A_e H^2|R(z)|^\beta f, |R(z)|^\alpha g\rangle \\ &= \langle (e\otimes e)|R(z)|^\beta f, H^2|R(z)|^\alpha g\rangle - 2\langle (e\otimes e)H|R(z)|^\beta f, H|R(z)|^\alpha g\rangle \\ &\qquad + \langle (e\otimes e)H^2|R(z)|^\beta f, |R(z)|^\alpha g\rangle \\ &= \langle |R(z)|^\beta f, e\rangle\langle e, H^2|R(z)|^\alpha g\rangle - 2\langle H|R(z)|^\beta f, e\rangle\langle e, H|R(z)|^\alpha g\rangle \qquad (28) \\ &\qquad + \langle H^2|R(z)|^\beta f, e\rangle\langle e, |R(z)|^\alpha g\rangle.\end{aligned}$$

Since $H|R(z)|^t, H^2|R(z)|^t \in \boldsymbol{B}(\mathcal{H})$ for $t \in [2, \infty)$, we infer from (28) that the form ψ is bounded for all $\alpha, \beta \in [2, \infty)$. Consider now the case $\alpha \in [0, 2)$ and suppose contrary to our claim that ψ is bounded. Fix $f \in \mathcal{D}^\infty(H)$ such that $a \stackrel{\mathrm{df}}{=} \langle e, |R(z)|^\beta f\rangle \neq 0$. Putting $b = -2\langle e, H|R(z)|^\beta f\rangle$, we deduce from (28) that the linear functional $g \longmapsto \langle (aH^2 + bH)|R(z)|^\alpha g, e\rangle$ is bounded on $\mathcal{D}^\infty(H)$. Since $\mathcal{D}^\infty(H)$ is a core of $(aH^2 + bH)|R(z)|^\alpha$ (cf. footnote 8), we obtain

$$e \in \mathcal{D}((aH^2 + bH)|R(z)|^\alpha) = \mathcal{D}(H^{2-\alpha}),$$

[7] Notice that the spectral theorem yields $|R(z)|^t \mathcal{D}^\infty(H) \subseteq \mathcal{D}^\infty(H)$ for $t \in [0, \infty)$.

[8] Because for $g \in \mathcal{D}(H|R(z)|^\alpha)$, $g_n \stackrel{\mathrm{df}}{=} E([1,n])g \in \mathcal{D}^\infty(H)$ and $\lim_{n\to\infty} \|g - g_n\|_{H|R(z)|^\alpha} = 0$.

which again contradicts (26). The proof of the remaining case $\beta \in [0,2)$ goes through as for $\alpha \in [0,2)$. Likewise, one can verify that the sesquilinear form associated with the expression $R(z)(\mathrm{ad}\, H)^2(A_e)R(z)$ is unbounded on $\mathcal{D}^\infty(H)$ for every $z \in \mathbb{C} \setminus \sigma(H)$.

Summarizing, we have argued that our Theorems 3 and 8 essentially generalize Theorems 1 and 4 in [1] and, consequently, extend their applicability; the subsequent paper will be devoted to that.

References

[1] W. Driessler, S.J. Summers, On commutators and self-adjointness, *Lett. Math. Phys.* **7** (1983), 319–326.

[2] E. Nelson, Analytic vectors, *Ann. Math.* **70** (1959), 572–615.

[3] A.E. Nussbaum, A commutativity theorem for unbounded operators in Hilbert space, *Trans. Amer. Math. Soc.* **140** (1969), 485–491

[4] M. Reed, B. Simon, *Methods of Modern Mathematical Physics*; II, *Fourier analysis, self-adjointness*, Academic Press, New York, 1975.

[5] J. Stochel, F.H. Szafraniec, Domination of unbounded operators and commutativity, *J. Math. Soc. Japan* **55** (2003), 405–437.

Dariusz Cichoń
Instytut Matematyki
Uniwersytet Jagielloński
ul. Reymonta 4
PL-30059 Kraków
e-mail: cichon@im.uj.edu.pl

Jan Stochel
Instytut Matematyki
Uniwersytet Jagielloński
ul. Reymonta 4
PL-30059 Kraków
e-mail: stochel@im.uj.edu.pl

Franciszek Hugon Szafraniec
Instytut Matematyki
Uniwersytet Jagielloński
ul. Reymonta 4
PL-30059 Kraków
e-mail: fhszafra@im.uj.edu.pl

Operator Theory:
Advances and Applications, Vol. 154, 35–50

The Problem of the Finiteness of the Point Spectrum for Self-adjoint Operators. Perturbations of Wiener-Hopf Operators and Applications to Jacobi Matrices

Petru Cojuhari

Abstract. The purpose of this paper is to study the point spectrum for the case of self-adjoint operators. Based on the direct methods of the perturbation theory, sufficient conditions for the finiteness of the point spectrum of some self-adjoint operators are established. The problem is treated for the general case of abstract operators and then, as applications, some concrete classes of operators are investigated. In particular, operators associated to Jacobi type matrices are also considered.

Mathematics Subject Classification (2000). Primary 47A55; Secondary 47A75.

Keywords. Spectral theory, compact perturbation, Wiener-Hopf operators, Jacobi matrices.

1. Introduction

The paper is devoted to the problem of the finiteness of the point spectrum for the case of self-adjoint operators. Within the framework of abstract operators the problem can be formulated as follows. Let A and B denote symmetric operators on a Hilbert space $\mathcal{H}$, and let Λ be an interval of the real axis which is free of eigenvalues of A. Let us consider the operator B as an additive perturbation of the operator A. Our purpose is to find conditions under which the set of perturbed eigenvalues (counted according to their multiplicities) from Λ is at most finite. This problem as well as any other one connected with the study of the structure of the spectrum of an operator represents one of the most important problems of spectral analysis and its applications. Such a problem frequently appears in various principle situations from such domains as theoretical and mathematical physics (especially quantum mechanics), theory of differential (and pseudo-differential) operators, eigenfunction expansions theory and, in the general, spectral operator theory itself, etc. A good deal of background material on the development and perspectives of the problem can be found in [25], [15] (see also the references quoted there). It should be pointed out the works [21], [22] (see also [23],[26]) dedicated to scattering theory, from which the necessity to study the structure of the point spectrum in some basic questions becomes clear. The main results of this paper are obtained from the same point of view based on the perturbation theory of operators. First, we state an abstract result on the finiteness of the point spectrum for some perturbations of symmetric operators. This abstract result is given by Theorem 2.1 (see also Corollary 2.2) from Section 2. Then, combining the theory of Wiener-Hopf type operators developed as in [18] with the general results from Section 2, in Section 3 we study the point spectrum of perturbed Wiener-Hopf abstract operators. It is worth to emphasize the role of the abstract Hardy type inequalities from [13] used in the proof of the main results of Section 3. The corresponding results concerned perturbations of Wiener-Hopf discrete operators are presented in Section 3. Finally, in Section 4 the spectra of operators generated by band Jacobi matrices are investigated. The main results of this section are derived from those of Section 3 as direct applications.

2. The abstract results

In this section we cite an abstract result on the finiteness of the point spectrum (i.e., the set of eigenvalues, including those contained in the continuous spectrum) for self-adjoint operators. The detailed prove of it together with some applications can be found in our earlier work [3] (see also [8] for applications to Dirac operators). On the basis of this abstract approach there are proved the main results considered in the next sections.

In the sequel, $\mathcal{H}$ will denote a complex Hilbert space. We denote by $\mathbb{B}(\mathcal{H})$ the space of all bounded operators on $\mathcal{H}$ and by $\mathbb{B}_\infty(\mathcal{H})$ the subspace of $\mathbb{B}(\mathcal{H})$consisting of all compact operators in $\mathcal{H}$. The domain and the range of an operator A are denoted by $\mathrm{Dom}(A)$ and $\mathrm{Ran}(A)$, respectively. The resolvent set of A is denoted by $\rho(A)$ and the spectrum by $\sigma(A)$. $\sigma_p(A)$ stands for the point spectrum of A. The resolvent operator $(A-zI)^{-1}, z\in\rho(A)$, will be denoted briefly by $R(z;A)$.

Theorem 2.1. *Let A and B be symmetric operators in a space $\mathcal{H}$ and let the operator A has no eigenvalues on a closed interval Λ of the real axis. Suppose that there exists a operator-valued function $T(\lambda)$ defined on the interval Λ having the properties that*

(i) $T(\lambda)\in\mathbb{B}_\infty(\mathcal{H})$ $(\lambda\in\Lambda)$,

(ii) *$T(\lambda)$ is continuous on Λ in the uniform norm topology, and*

(iii) *for each $\lambda\in\Lambda$ and for each $u\in\mathrm{Dom}(B)$ such that $Bu\in\mathrm{Ran}(A-\lambda I)$ there holds the following inequality*

$$\| (A-\lambda I)^{-1}Bu \| \leq \| T(\lambda)u \| . \tag{2.1}$$

Then the point spectrum of the perturbed operator $A+B$ on the interval Λ consists only of finite number of eigenvalues of finite multiplicity.

As we already mentioned this theorem (in a slightly different form) is contained in our work [3] (see also [8]).

In particular, an immediate consequence of Theorem 2.1 is the following result often useful for concrete applications.

Corollary 2.2. *Let A and B be symmetric operators in $\mathcal{H}$, and let Λ denote a closed interval of the real axis such that the set $\sigma_p(A) \cap \Lambda$ is empty. Suppose that the operator B can be represented as $B = ST$, where T is a compact operator from $\mathcal{H}$ to another Hilbert space $\mathcal{H}_1$ and S is a closed operator from $\mathcal{H}_1$ into $\mathcal{H}$, respectively. In addition, if the following estimate holds*

$$\| (A - \lambda I)^{-1} Su \| \leq c \| u \| \ (c = const; \lambda \in \Lambda),$$

whenever $u \in \mathrm{Dom}(S)$ and $Su \in \mathrm{Ran}(A - \lambda I)$, then $\sigma_p(A + B) \cap \Lambda$ is a finite set and each possible eigenvalues of the perturbed operator $A + B$ from Λ has a finite multiplicity.

We will apply Theorem 2.1 just as Corollary 2.2 to the study of the problem of the finiteness of the perturbed eigenvalues for the concrete classes of operators considered in the forthcoming sections. We mainly study the eigenvalues contained in the continuous spectrum of a given operator. However, Theorem 2.1 and Corollary 2.2 can be also applied to the study of the discrete part of the spectrum. For instance, in view of this remark, Theorem 2.1 (more exactly, Corollary 2.2) implies the following perturbation theorem.

Theorem 2.3. *Let A be a self-adjoint operator in $\mathcal{H}$, $\Lambda = (a, b) \subset \rho(A)$, $a \notin \sigma_p(A)$ and B be a symmetric operator with $\mathrm{Dom}(B) \supset \mathrm{Dom}(A)$. If the operator $(A - aI)^{-1}B$ is densely defined and compact in $\mathcal{H}$, then the perturbed operator $H = A + B$ with $\mathrm{Dom}(H) = \mathrm{Dom}(A)$ is self-adjoint in $\mathcal{H}$, the spectrum of H on Λ is only discrete and a is not an accumulation point for the set $\sigma(H) \cap \Lambda$.*

In order to show how this theorem can be obtained from Corollary 2.2 denote by T the complete extension (which evidently is unique) of the operator $(A - aI)^{-1}B$. Then T^* is an extension of $B(A - aI)^{-1}$, and it is easy to verify that

$$BR(\lambda; A) = T^* + (\lambda - a)T^* R(\lambda; A) (\lambda \in \rho(A)).$$

Consequently, since T^* is compact, also $BR(\lambda; A)$ is compact for each $\lambda \in \rho(A)$, and so a result from [16] can now be applied to conclude that the operator H is self-adjoint and its spectrum on Λ is only discrete.

Next, let $S = A - aI$. Then the operator B can be expressed as $B = ST$ and, since

$$\| R(\lambda; A)Su \| = \| R(\lambda; A)(A - aI)u \| \leq c \| u \| \ (c = const)$$

holds for each $u \in \mathrm{Dom}(A)$ and for each λ nearly of a from the right, it follows that all of hypotheses of Corollary 2.2 are fulfilled.

The above theorem is given in [2]. Among other applications, in [2] it has been studied the spectrum for an equation of radiation energy transfer. In this context we note the works [4], [7] (see also [5] and [10]) for further results and other applications. We also note the discussion undertaken in [4] (see also [7]) on the connection with related results from the works [1], [20].

3. Perturbations of Wiener-Hopf abstract operators

In this section we study the point spectrum of the operators obtained by perturbations of Wiener-Hopf abstract operators. The concept of the Wiener-Hopf abstract operators has been given by I. C. Gohberg and I. A. Feldman [18], see also S. Prössdorf [24]. According to the theory developed in [18], Wiener-Hopf type operators can be regarded in a certain sense as functions of one-sided invertible operators. As our case is that of Hilbert's space, the considered operators will be presented as functions of an isometric operator. Moreover, these functions (so-called symbols of the corresponding operators) will be assumed to be continuous on the unit circle of the complex plane.

In order to specify the notion connected with the foregoing discussion, let $\mathcal{H}$ be an arbitrary Hilbert space and consider on it an isometric operator denoted by V. In what follows, it will be always assumed that the operator V^* has no eigenvalues on the unit circle $\mathbb{T} = \{z \in \mathbb{C}/|z| = 1\}$. Then, it is clear that the point spectrum of V is empty and, hence, there exist the closed and unbounded operators $(V - zI)^{-1}$ and $(V^* - zI)^{-1}$ for all $z \in \mathbb{T}$. We denote, as in [18], by $\Re(V)$ the closed hull of all operators $V^n (n = 0, \pm 1, \cdots)$, where $V^n = (V^*)^{-n} (n = -1, -2, \ldots)$ and the closure is taken with respect to the operator norm of $\mathbb{B}(\mathcal{H})$. In line with approaches from [18], it follows that to each $A \in \Re(V)$ corresponds a complex-valued function $A(z)$ continuous on the unit circle $\mathbb{T}$. This function $A(z), z \in \mathbb{T}$, is called the symbol of the operator A. We allow ourselves to write (formally for the moment at least) $A = A(V)$. The set of symbols which correspond to all operators from $\Re(V)$ coincide with the set $C(\mathbb{T})$ of all complex-valued functions continuous on $\mathbb{T}$ (see [18], p. 34). If $A(\cdot) \in C(\mathbb{T})$ is a real-valued function, then the corresponding operator $A \in \Re(V)$ is self-adjoint and the spectrum of A is the set of all values attained by $A(z), z \in \mathbb{T}$. In spite of the fact that the most results presented below can be adapted effortlessly for the general case, we will be concerned exclusively with the case of operators with rational symbols. However, this will be entirely enough for our applications from the next sections.

Thus, in what follows, we consider an operator $A \in \Re(V)$ of the following form

$$A = \sum_{j=-n}^{n} a_j V^j, \tag{3.1}$$

where $a_j (j = 0, \pm 1, \cdots, \pm n)$ are fixed complex numbers. As it was already mentioned the symbol of A is the polynomial $A(z) = \sum_{k=-n}^{n} a_k z^k$, and if $A(z)$ represents a real-valued function on $\mathbb{T}$ or, equivalently, $a_j = \bar{a}_{-j} (j = 0, 1, \ldots, n)$, then A is a self-adjoint operator on $\mathcal{H}$. Henceforth this property will be always assumed. Note that the spectrum of A is a closed interval, namely $\sigma(A) = [a, b]$, where $a = \min A(z)$ and $b = \max A(z)$ on $\mathbb{T}$.

Before formulating the main result of this section it will be convenient to recall some notions and definitions.

Let C be a symmetric operator in $\mathcal{H}$. The operator C is called semi-bounded from below, if there exists a number $\gamma, \gamma \in \mathbb{R}$, such that $(Cu,u) \geq \gamma \parallel u \parallel^2$ for each $u \in \mathrm{Dom}(C)$. For the later, we denote by $\gamma(C)$ the greatest lower bound of C, i.e.,

$$\gamma(C) = \inf\{(Cu,u)/u \in \mathrm{Dom}(C), \parallel u \parallel = 1\}.$$

The operator C is said to be nonnegative ($C \geq 0$) if $\gamma(C) \geq 0$. If $\gamma(C) \geq 0$, but $(Cu,u) > 0$ for each $u \in \mathrm{Dom}(C), u \neq 0$, the operator C is said to be positive. Finally, C is said to be positive definite ($C >> 0$) if $\gamma(C) > 0$. In a similar manner can be understood the notion of an operator semi-bounded from above and all the corresponding notions related with it.

Next, let us consider an operator J on the space $\mathcal{H}$ for which the following properties are assumed.

(i) J is a bounded and positive operator ;
(ii) $V^*(\mathrm{Ran}(J)) \subset \mathrm{Ran}(J)$ and there is a definite operator C (so that either $C \gg 0$ or $C \ll 0$) such that $C = J^{-1} - VJ^{-1}V^*$ on the set $\mathrm{Ran}(J)$.

Remark 3.1. *The existence of an operator J for which the properties* (i) *and* (ii) *are satisfied assures applicability of obtained results in* [13] (*see also* [14]) *on abstract Hardy type inequalities. It turns out that , the following inequality holds*

$$c \parallel Ju \parallel \leq \parallel (V - zI)u \parallel \tag{3.2}$$

for all $u \in H$ and all $z \in \mathbb{C}, \mid z \mid \geq 1$, where $c = \frac{1}{2} \mid z \mid \gamma(C)$. In addition, we note that from the inequality (3.2) *it can be extracted some useful information. For instance, it follows that the operator V has no eigenvalue on the unit circle $\mathbb{T}$. Indeed, if $Vu = zu$ for some $z \in \mathbb{T}$, then* (3.2) *implies $Ju = 0$ and, since $J > 0$, we get $u = 0$. Moreover, the unbounded operator $(V - zI)^{-1}$, where $z \in \mathbb{T}$,becomes to be a bounded one by multiplying it with J from left. More precisely, for every $z \in \mathbb{T}$ the operator $J(V - zI)^{-1}$, which is densely defined, has a bound extension on H. This fact will be systematically used in the proof of the main results given below* (*see Theorems* 3.2 *and* 3.3).

Now, let us consider an operator $A \in \Re(V)$ of the form (3.1), and denote by $A(z)$, $z \in \mathbb{T}$, its symbol, and let, as before, $a = \min A(z)$ and $b = \max A(z)$ on $\mathbb{T}$. It was also noted that $\sigma(A) = [a,b]$. It is well known, but this fact will be also clear from the proof of Theorem 3.2 given below, that the operator A has no eigenvalues.

Next, let $n(\lambda)$ be the number of all zeros of the polynomial $A(z) - \lambda(a \leq \lambda \leq b)$, which belong to the unit circle $\mathbb{T}$ (and counted according to their multiplicities), and let $m(\lambda)$ be the maximal multiplicity of them. It is clear that $n(\lambda)$ is a piecewise constant function on the interval $[a,b]$, and the set $\mathcal{N}$ of its discontinuous points is finite.

Let B be another self-adjoint operator on $\mathcal{H}$ and let us consider it as a perturbation of the operator A. Let Λ be an interval contained in the continuous spectrum of A. Our purpose is to find conditions under which the set of perturbed eigenvalues (counted according to their multiplicities) from Λ is only finite. We start by considering the situation in which $m(\lambda) = 1$ on Λ.

Theorem 3.2. *Let A be an operator of the form (3.1), let $\lambda_0 \in (a,b) \setminus \mathcal{N}$ and $m(\lambda_0) = 1$. In addition, let J be an operator on $\mathcal{H}$ for which the above properties (i) and (ii) are satisfied. If the operator $B \in \mathbb{B}(\mathcal{H})$ is self-adjoint and the operator $J^{-1}B$ is compact in $\mathcal{H}$, then λ_0 is not an accumulation point for the set of eigenvalues of the perturbed operator $A + B$. Each possible eigenvalue of $A + B$ in a neighborhood of λ_0 has a finite multiplicity.*

Proof. It is clear that, since $m(\lambda_o) = 1$, also $m(\lambda) = 1$ and $n(\lambda) = n(\lambda_0)$ for λ belonging to a closed neighborhood Λ of λ_0. Then the symbol $A(z) - \lambda$ can be represented as follows

$$A(z) - \lambda = \sum_{j=1}^{n_0} (z^{-1} - \alpha_j(\lambda)) A_\lambda(z) (\lambda \in \Lambda), \tag{3.3}$$

where $n_0 = n(\lambda_0)$, $\alpha_j(\lambda)(|\alpha_j(\lambda)| = 1; j = 1, \dots, n_0)$ are continuous functions on Λ, $\alpha_j(\lambda) \neq \alpha_k(\lambda)(\lambda \in \Lambda; j \neq k, j, k = 1, \dots, n_0)$, $A_\lambda(z)$ is a polynomial in z and depending continuously on λ, and such that $A_\lambda(z) \neq 0$ for $|\ z \ | = 1$ and $\lambda \in \Lambda$. Clearly, for each fixed $\lambda \in \Lambda$, $A_\lambda(z)$ can be regarded as a symbol of an operator, let it be denoted by $A_\lambda(V)$, from $\Re(V)$. From the above factorization (3.3) of $A(z) - \lambda$ it follows that

$$A - \lambda I = \prod_{j=1}^{n_0} (V^* - \alpha_j(\lambda) I) A_\lambda(V) (\lambda \in \Lambda). \tag{3.4}$$

According to the restrictions V, the operators $V^* - \alpha_j(\lambda)I$ $(j = 1, \dots, n_0)$ are one-to-one, and their corresponding inverses $R_j(\lambda) = (V^* - \alpha_j(\lambda)I)^{-1}$ $(j = 1, \dots, n_0)$ are closed and unbounded operators in the space $\mathcal{H}$. Henceforth, for convenience, we let $R(\lambda) = \prod_{j=1}^{n_0} R_j(\lambda)$ $(\lambda \in \Lambda)$. Also, it is easy to see that the operator $A_\lambda = A_\lambda(V)$ is invertible on the left, and let $A_\lambda^{(-1)} (\in \mathbb{B}(\mathcal{H}))$ denote its left inverse. Since in addition the operators $V^* - \alpha_j(\lambda)I$ $(j = 1, \dots, n_0)$ are commutative it follows from (3.4) that

$$(A - \lambda I)^{-1} u = A_\lambda^{(-1)} R(\lambda) u (\lambda \in \Lambda) \tag{3.5}$$

for every $u \in \operatorname{Ran}(A - \lambda I)$.

Next, it will be shown that all of hypotheses of Theorem 2.1 are satisfied. To this end, note that since the values $\alpha_j(\lambda)$ $(j = 1, \dots, n_o)$ are pairwise distinct for every $\lambda \in \Lambda$, there exist some functions $a_j(\lambda)$ $(j = 1, \dots, n_0)$, which evidently can be supposed to be continuous, such that

$$\prod_{j=1}^{n_0} (z - \alpha_j(\lambda))^{-1} = \sum_{j=1}^{n_0} a_j(\lambda)(z - \alpha_j(\lambda))^{-1} (\lambda \in \Lambda). \tag{3.6}$$

It turns out that the relation (3.6) implies that

$$\mathrm{Dom}(R(\lambda)) = \bigcap_{j=1}^{n_0} \mathrm{Dom}(R_j(\lambda))(\lambda \in \Lambda), \tag{3.7}$$

and, respectively,

$$R(\lambda)u = \sum_{j=1}^{n_0} a_j(\lambda) R_j(\lambda) u(\lambda \in \Lambda) \tag{3.8}$$

for every $u \in \mathrm{Dom}(R(\lambda))$.

In order to show that , we rewrite the relation (3.6) as follows

$$\sum_{j=1}^{n_0} a_j(\lambda) \prod_{k \neq j} (z - \alpha_k(\lambda)) = 1(\lambda \in \Lambda)$$

from which, we readily get

$$\sum_{j=1}^{n_0} a_j(\lambda) \prod_{k \neq j} (V^* - \alpha_k(\lambda) I) = I(\lambda \in \Lambda). \tag{3.9}$$

Now,let $u \in \bigcap_{j=1}^{n_0} \mathrm{Dom}(R_j(\lambda))$, i.e., for each $j = 1, \ldots, n_0$ there exists $u_j \in \mathcal{H}$ such that $u = (V^* - \alpha_j(\lambda) I) u_j$. Then, by (3.9),the element u can be expressed in the form

$$u = \prod_{j=1}^{n_0} (V^* - \alpha_j(\lambda) I) v,$$

where $v = \sum_{j=1}^{n_0} a_j(\lambda) u_j$. Hence $u \in \mathrm{Dom}(R(\lambda))$, and so $\bigcap_{j=1}^{n_0} \mathrm{Dom}(R_j(\lambda)) \subset \mathrm{Dom}(R(\lambda))$. The opposite inclusion is evident, and thus the relation (3.7) is established. Now, it is clear that the relation (3.8) is a simple consequence of (3.9).

As has been mentioned above (see Remark 3.1), the operator J is related to V in such way that for arbitrary $z \in \mathbb{T}$ the densely defined operator $J(V - zI)^{-1}$ has a bound extension on $\mathcal{H}$. Moreover, the norm of each such extension is less then $2\gamma^{-1}$, where $\gamma = \gamma(C)$ (cf. the property (ii) for J). It then follows by duality that the operator $(V^* - zI)^{-1} J$, for each fixed $z \in \mathbb{T}$, is bounded on its domain. Namely, there holds the following estimate

$$\| (V^* - zI)^{-1} J u \| \leq 2\gamma^{-1} \| u \| \ (z \in \mathbb{T}), \tag{3.10}$$

for all $u \in \mathcal{H}$ such that $Ju \in \mathrm{Ran}(V^* - zI)$.

Next, let u be an arbitrary element in $\mathcal{H}$ such that $Ju \in \mathrm{Ran}(A - \lambda I)$. Then, as is clear from (3.4), also $Ju \in \mathrm{Dom}(R(\lambda))$, and further from (3.8), by virtue of (3.10), one obtains

$$\| R(\lambda) J u \| \leq 2\gamma^{-1} \Big(\sum_{j=1}^{n_0} | a_j(\lambda) |\Big) \| u \| \ (\lambda \in \Lambda). \tag{3.11}$$

But $a_j(\lambda)$ $(j = 1, \ldots, n_0)$, as continuous functions on a closed interval Λ, are uniformly bounded on Λ. The uniformly boundedness of operators $A_\lambda^{(-1)}$ as λ

ranges over Λ is also clear. Consequently, by (3.5) and (3.11), we find that there exists a finite positive constant c independent of λ and u such that

$$\| (A-\lambda I)^{-1}Ju \| \leq c \| u \| \ (\lambda \in \Lambda).$$

Thus, taking $S = J$ and $T = J^{-1}B$, all hypotheses of Theorem 2.1 (or Corollary 2.2) are satisfied, and therefore, the proof of our theorem is complete. □

Let us now consider the case $m(\lambda_0) > 1$, where $\lambda_0 \in (a,b)\backslash\mathcal{N}$ as above. Let α_k $(k = 1,\dots,p)$ be all distinct roots of the polynomial $A(z)-\lambda_0$ contained on the unit circle $\mathbb{T}$, and let m_k $(k = 1,\dots,p)$ designate their corresponding multiplicities. For simplicity, we let $p = 2$. The arguments for the general case are similar. In this case a representation as was given by (3.3) also occurred. Obviously, it can meaningfully asserted that for any $\lambda, \lambda \neq \lambda_0$, from a certain closed neighborhood Λ of λ_0 the roots $\alpha_j(\lambda)$ $(j = 1,\dots,n_0)$ in that representation are simple. Moreover, $\alpha_j(\lambda)$ $(j = 1,\dots,n_0)$ can be enumerated in such way that the first m_1 of them to converge to $\alpha_1$as $\lambda \to \lambda_0$ and, respectively, the others $m_2(= n_0 - m_1)$ to α_2.

Next, it will be shown that the abstract results from the previous section can also be applied. To this end, it will be sufficient for the perturbation operator B to indicate a factorization of the form $B = ST$, in which $S \in \mathbb{B}(\mathcal{H})$, $T \in \mathbb{B}_\infty(\mathcal{H})$ and, in addition, the operator S to be chosen in such way that the following estimate

$$\| R(\lambda)Su \| \leq c \| u \| \ (\lambda \in \Lambda) \tag{3.12}$$

holds for each $u \in \mathcal{H}$ such that $Su \in \mathrm{Dom}(R(\lambda))$. Here c is a constant independent of λ and u, and it is used the notation $R(\lambda) = \prod_{j=1}^{n_0} R_j(\lambda)$ and $R_j(\lambda) = (V^* - \alpha_j(\lambda)I)^{-1}$ $(j = 1,\dots,n_0)$ as before. It turns out that an operator S suitable for our purposes can be useful $S = J^{m_o}$, where $m_o = m(\lambda_0)$, and the operator$J$ is chosen that the following estimates hold

$$c_\tau \| J^\tau u \| \leq \| J^{\tau-1}(V-zI)u \| \ (\tau = 1,\dots,m_0) \tag{3.13}$$

for all $u \in \mathcal{H}$ and $z \in \mathbb{T}$, where c_τ are some positive constants. In order to see this, we choose polynomials $p_j(z,\lambda)(j = 1,2)$ in z and depending continuously on λ for $\lambda \in \Lambda$ such that (cf. the decomposition (3.6))

$$\prod_{j=1}^{n_0}(z-\alpha_j(\lambda))^{-1} = p_1(z,\lambda)\prod_{j=1}^{m_1}(z-\alpha_j(\lambda))^{-1} + p_2(z,\lambda)\prod_{j=m_1+1}^{n_0}(z-\alpha_j(\lambda))^{-1} (\lambda \in \Lambda).$$

An argument like that leading to (3.8) now implies that

$$R(\lambda)u = p_1(V^*,\lambda)\prod_{j=1}^{m_1} R_j(\lambda)u + p_2(V^*,\lambda)\prod_{j=m_1}^{n_0} R_j(\lambda)u, (\lambda \in \Lambda), \tag{3.14}$$

whenever $u \in \mathrm{Dom}(R(\lambda))$.

Further, by means of (3.13), it is easily verified that

$$\| (V-z_1I)\cdot\dots\cdot(V-z_mI)u \| \geq c_1 \cdot \dots \cdot c_m \| J^m u \| \tag{3.15}$$

for $u \in \mathcal{H}$, $z_1,\dots,z_m \in \mathbb{T}$, $m \leq m_0$ and $c_1,\dots,c_m$ determined as in (3.13).

Using this fact, it can be obtained by some simple manipulations estimates for the operator-valued functions

$$F_1(\lambda) = \prod_{j=1}^{m_1} R_j(\lambda) \quad \text{and} \quad F_2(\lambda) = \prod_{j=m_1+1}^{n_0} R_j(\lambda)S \quad \text{for} \quad \lambda \in \Lambda.$$

Namely, for every $u \in \mathcal{H}$ such that $Su \in \text{Dom}(R(\lambda))$, one can be obtained

$$\| F_j(\lambda)u \| \leq K_j \| u \| \ (j = 1, 2; \lambda \in \Lambda)$$

with constants$K_j (j = 1, 2)$ independent of λ and u. Taking this into account and the fact that the operator-functions $p_j(V^*, \lambda)$ $(j = 1, 2)$ are uniformly bound on Λ, it follows from (3.14) the desired estimate (3.12).

Now, assuming that the operator J is one-to-one and the operator $T = J^{-m_0}B$ is compact in the space $\mathcal{H}$, we get a factorization $B = ST$, where all hypotheses of Corollary 2.2 are fulfilled.

The above discussion it is summarized by Theorem 3.4 from below. Before formulating this theorem we make the following remark.

Remark 3.3. *If an operator J has the following properties*

(i) *J is a bounded and positive operator in $\mathcal{H}$,*

(ii)$_\tau$ $V^*(\text{Ran}(J^{2\tau-1})) \subset \text{Ran}(J^{2\tau-1})(\tau = 1, \ldots, m_0)$,

and

(iii)$_\tau$ *for each $\tau = 1, \ldots, m_0$ there is a definite operator $S_{\tau,\tau-1}$ (let $S_{\tau,\tau-1} \gg 0$, for instance) such that*

$$S_{\tau,\tau-1} = J^{-1} - J^{\tau-1}VJ^{-2\tau+1}V^*J^{\tau-1}$$

on the set $\text{Ran}(J^\tau)$,

then the estimates (3.13) *hold for all $u \in \mathcal{H}$ and for all $z \in \mathbb{C}$ such that $| z | \geq 1$, where $c_\tau = 2^{-1} | z | \gamma(S_{\tau,\tau-1})$* [13].

Theorem 3.4. *Let A be an operator of the form* (3.1), *let $\lambda_0 \in (a, b) \setminus \mathcal{N}$ and $m_0 = m(\lambda_0) > 1$. In addition, let J be an operator on $\mathcal{H}$ satisfying either* (3.13) *or* (i), (ii)$_\tau$, (iii)$_\tau$ *from Remark* 3.3. *If the operator $B \in \mathbb{B}(\mathcal{H})$ is self-adjoint and the operator $J^{-m_0}B$ is compact in $\mathcal{H}$, then λ_0 is not an accumulation point for the set of eigenvalues of the perturbed operator $A + B$. Each possible eigenvalues of $A + B$ in a neighborhood of λ_0 has a finite multiplicity.*

Remark 3.5. *An argument similar to that used in proving Theorem 3.4 can be applied to prove that if $\lambda_0 \in \mathcal{N}$ and $n(\lambda) = n(\lambda_0)$ for $\lambda \in [\lambda_0 - \varepsilon, \lambda_0]$ ($\lambda \in [\lambda_0, \lambda_0 + \varepsilon]$),where $\varepsilon > 0$, and the operator A and B satisfy the conditions of Theorem 3.4, then the point spectrum of the perturbed operator $A + B$ on the interval $[\lambda_0 - \varepsilon, \lambda_0]$ ($[\lambda_0, \lambda_0 + \varepsilon]$) consists only of finite set of eigenvalues of finite multiplicities.*

As a consequence of Theorems 3.2 and 3.4 there holds the following result.

Corollary 3.6. *Let A be a self-adjoint operator of the form* (3.1) *and let J be defined as in Theorem* 3.1. *If the operator $B \in \mathbb{B}(\mathcal{H})$ is self-adjoint and the operator $J^{-1}B$ is compact in $\mathcal{H}$, then the set of all eigenvalues of the perturbed operator $A+B$ has at most a finite number of accumulation points. Each eigenvalue of $A+B$ can be only of a finite multiplicity.*

The assertion follows at once from the fact that the function $m(\lambda)$ has at most a finite number of discontinuous points.

The following result can be also regarded as an immediate consequence of Theorems 3.2 and 3.4 which however is important for the relevant results about Jacobi matrices (cf. Section 5 below).

Corollary 3.7. *Let $A = V + V^*$, let B be a bounded self-adjoint operator on $\mathcal{H}$, and let J denote an operator defined as in Theorem* 3.1.

(i) *If operator $J^{-1}B$ is compact in $\mathcal{H}$, then the set of all eigenvalues of the perturbed operator $A+B$ has only the end points of the essential spectrum of $A+B$, i.e., ± 2, as accumulation points.*

(ii) *If however the operator $J^{-2}B$ is compact in $\mathcal{H}$, then the point spectrum of $A+B$ from the essential spectrum $[-2,2]$ is only finite.*

In both cases each eigenvalue of the perturbed operator $A+B$ has a finite multiplicity.

Proof. The point is that the polynomial $A(z) = z + z^{-1}$, $z \in \mathbb{T}$, is the symbol of the unperturbed operator A. Therefore $\sigma(A) = [-2,2]$, $n(\lambda) = 2$ $(-2 \le \lambda \le 2)$, $m(\lambda) = 1$ $(-2 < \lambda < 2)$, $\mathcal{N} = \{-2, 2\}$, $m(\pm 2) = 2$. There remains then to use Theorem 3.2 for proving (i) and Theorem 3.4 or, more convenient, the assertions made in remark 3.5 for (ii). □

4. Perturbations of Wiener-Hopf discrete operators

In this section the results obtained in the previous section are adapted to study the discrete case of perturbed Wiener-Hopf operators.

1. In the sequel, it is considered instead of $\mathcal{H}$ the Hilbert space $l_2(\mathbb{N})$ of all square summable sequence $\xi = (\xi_n)$, $\xi_n \in \mathbb{C}$ $(n = 1, 2, \dots)$, and V denote the elementary shift in $l_2(\mathbb{N})$, i.e., $(V\xi)_n = \xi_{n-1}$ $(n = 1, 2, \dots; \xi_0 = 0)$. V is an isometric operator on the space $l_2(\mathbb{N})$, i.e., $V^*V = I$, and we recall that the operator V^* (and V as well) has no eigenvalue on unit circle $\mathbb{T}$. The operators of $\Re(V)$ set up the class of discrete Wiener-Hopf operators (see [18], for instance). We remark that an operator A from the class $\Re(V)$ is generated in the space $l_2(\mathbb{N})$ by a matrix $[a_{j-k}]_1^\infty$, where $a_j (j = 0, \pm 1, \cdots)$ stand for the Fourier coefficients of the symbol $A(z)$ of A. In what follows, as in previous section, it is assumed only the case in which a finite number of Fourier coefficients a_j are different of zero. In other terms this means that the symbol $A(z)$ of A represents a polynomial$A(z) = \sum_{j=-n}^{n} a_j z^j$, $z \in \mathbb{T}$.

Thus, as in Sections 3, we consider an operator $A \in \Re(V)$ of the form (3.1), where a_j $(j = 0, \pm 1, \cdots, \pm n)$ are complex number such that $a_j = \bar{a}_{-j}$ $(j = 0, 1, \ldots, n)$.

Let B denote a bounded self-adjoint operator on $l_2(\mathbb{N})$ defined by $B = [b_{jk}]_1^\infty$ and let us consider it as a perturbation of the operator A. We will preserve the notations introduced in Section 3.

Theorem 4.1. *Let A and B be defined as above, let $\lambda_0 \in (a,b) \backslash \mathcal{N}$ and $m_0 = m(\lambda_0)$. If $B_0 = [j^{m_0} b_{jk}]_1^\infty \in \mathbb{B}_\infty(l_2(\mathbb{N}))$, then λ_0 is not an accumulation point for the set of eigenvalues of the perturbed operator $A+B$. Each possible eigenvalue of $A+B$ in a neighborhood of λ_0 has a finite multiplicity.*

Proof. Consider the operator J defined on $l_2(\mathbb{N})$ by

$$(J\xi)_n = n^{-1}\xi_n, (n = 1, 2, \ldots; \xi = (\xi_n) \in l_2(\mathbb{N})).$$

In view of the Hardy inequality ([19], Theorem 326; also, for more general results, see [13]), one gets

$$\| J^\tau (V - zI)^{-\tau} \xi \| \leq 2^\tau \| \xi \|, (\xi \in \operatorname{Ran}(V - zI)^\tau; z \in \mathbb{T}, \tau > 0).$$

Therefore, the operators $J^\tau (V - zI)^{-\tau} (z \in \mathbb{T}, \tau > 0)$ have bounded extensions on $l_2(\mathbb{N})$, hence the operators $(V^* - zI)^{-\tau} J^\tau (z \in \mathbb{T}, \tau > 0)$ are also bounded on $l_2(\mathbb{N})$, and so, taking into account that $J^{-m_0} B = B_0$, for the case of elementary shift V in $l_2(\mathbb{N})$ the operator J together with $B = [b_{jk}]_1^\infty \in \mathbb{B}(l_2(\mathbb{N}))$ satisfy all the conditions of Theorem 3.4. □

The next theorem stands out as an alternate version of Corollary 3.6 for the considered discrete case of operators.

Theorem 4.2. *Let A be a self-adjoint discrete Wiener-Hopf operator of the form (3.1) and let B denote a bounded self-adjoint operator on $l_2(\mathbb{N})$ defined by $B = [b_{jk}]_1^\infty$. If $B_1 = [j b_{jk}]_1^\infty \in \mathbb{B}_\infty(l_2(\mathbb{N}))$, then the point spectrum of the perturbed operator $A+B$ has at most a finite number of accumulation points. Each eigenvalue of $A+B$ can be only of a finite multiplicity.*

It is also clear the reformulation of the results given by Corollary 3.7 concerning the concrete case in which the unperturbed operator A has the special form $A = V + V^*$. Unconditionally, the correspond results can be derived straightly from above Theorems 4.1 and 4.2.

Corollary 4.3. *Let B denote a bounded self-adjoint operator on $l_2(\mathbb{N})$ defined by $B = [b_{jk}]_1^\infty$.*

(i) *If $B_1 = [j b_{jk}]_1^\infty \in \mathbb{B}_\infty(l_2(\mathbb{N}))$, then the point spectrum of the perturbed operator $A_1 = V + V^* + B$ has only the points ± 2 as accumulation points.*
(ii) *If however $B_2 = [j^2 b_{jk}]_1^\infty \in \mathbb{B}_\infty(l_2(\mathbb{N}))$, then the set $\sigma_p(A_1) \bigcap [-2, 2]$ is only finite.*

In both cases each eigenvalue of the perturbed operator A_1 has a finite multiplicity.

Remark 4.4. *In our earlier work* [12], *we proved results on the absence of eigenvalues for perturbed Wiener-Hopf discrete operators* (*see also* [11] *for related results concerning operators generated by Jacobi matrices*). *We note that the conditions on the perturbation are in a sense close to the ones stated above* (*see Theorems* 4.1 *and* 4.2). *However, it should be noted that they depend not only on maximal multiplicity of the zeros of the symbol, but also on their number, the position of roots of the symbol in the complex plane, etc.*

2. The methods of Section 3 can be applied to obtain similar results by considering other type of isometries. The next consideration can be regarded as typical in this respect. Let us consider the case when $\mathcal{H}$ is the space $L_2(\mathbb{R}_+)$ and the operator V is the translation operator in $L_2(\mathbb{R}_+)$ defined by $(Vu)(x) = u(x-h)$ for $x > h$ $(h > 0)$ (taking $u(x-h) = 0$ for $x < h$).

It is easy to see that an operator $A \in \Re(V)$ of the form (3.1) is given in $L_2(\mathbb{R}_+)$ by

$$(Au)(x) = \sum_{j=-n}^{n} a_j u(x - jh),$$

where $a_j (j = 0, \pm 1, \cdots, \pm n)$ are complex numbers. As above, the conditions for self-adjointness of A, i.e., $a_j = \bar{a}_{-j} (j = 0, 1, \ldots, n)$ are always assumed. So, let A be a self-adjoint operator defined as above and let us consider a perturbation of A given by an integral operator

$$(Bu)(x) = \int_0^\infty b(x, y)u(y)dy,$$

where $b(x, y)$ is a measurable function with respect to both variables $x, y \in \mathbb{R}_+$. Suppose that the operator B is also self-adjoint and bounded on $L_2(\mathbb{R}_+)$.

Let us represent only a version of Theorem 3.4 in which the following situation $\lambda_0 \notin \mathcal{N}$, $m_0 = m(\lambda_0) > 1$ is considered.

Theorem 4.5. *If the integral operator with a kernel* $(1 + x)^{m_0} b(x, y)$ *is compact in the space* $L_2(\mathbb{R}_+)$, *then* λ_0 *is not an accumulation point for the set of eigenvalues of the perturbed operator* $A + B$. *Each possible eigenvalue of* $A + B$ *in a neighborhood of* λ_0 *has a finite multiplicity.*

We note that the operator J_1 defined in $L_2(\mathbb{R}_+)$ by

$$(J_1)(x) = (1 + x)^{-1} u(x)(u \in L_2(\mathbb{R}_+))$$

plays the role of J in the arguments of Section 3.

Remark 4.6. *It is worth to note that the methods developed in Section* 3 *can be also applied to study the point spectrum for perturbation Wiener-Hopf integral operators. This is the case in which the operator* V *defining the unperturbed operator is nothing but the canonical shift with respect to the system of Laguerre functions* (*see* [18], *p.* 69). *We cite our earlier work* [6] *in which results on the finiteness of the point spectrum for perturbed Wiener-Hopf integral operators are presented.*

5. Jacobi matrices

In this section we study the point spectrum of the operators generated by Jacobi matrices. The main results are obtained by applying directly the results from the previous section concerning the perturbations of Wiener-Hopf discrete operators.

1. We first focus our attention on the problem of the finiteness of the point spectrum for so-called band Jacobi matrices (or, in other terms, for generalized Jacobi matrices). We recall that a matrix $A = [a_{jk}]_1^\infty$ with $a_{jk} \in \mathbb{C}$ $(j,k = 1,2,\dots)$ is said to be a band Jacobi matrix of order $2n$ if $a_{jk} = 0$ for $\mid j-k \mid > n$, where n is a positive integer number. In the following, it is as ever considered the case of self-adjoint operator A on the space $l_2(\mathbb{N})$, that is, $a_{jk} = \bar{a}_{kj}$ $(j,k = 1,2,\dots)$.

Furthermore, we assume that

$$a_{j+r,j} \to a_r (j \to \infty; r = 0, \pm 1, \cdots, \pm n).$$

Then the operator A can be written as a sum $A = A_0 + B$, where

$$A_0 = \sum_{j=-n}^{n} a_j V^j (a_j = \bar{a}_{-j}, j = 0, 1, \dots, n) \tag{5.1}$$

and $B = [b_{jk}]_1^\infty$ with $b_{jk} = a_{jk} - a_{j-k}$ for $\mid j-k \mid \leq n$ and $b_{jk} = 0$ for $\mid j-k \mid > n$.

It is clear from (5.1) that A_0 is a Wiener-Hopf discrete operator with a rational symbol $A_0(z) = \sum_{j=-n}^{n} a_j z^j$ $(z \in \mathbb{T})$. Thus, the spectrum of the operator A_0 is the interval $\sigma(A) = [a,b]$, where $a = \min A_0(z)$ and $b = \max A_0(z)$ on $\mathbb{T}$.

In the next theorem there are used the notations similar those introduced in Section 3 before Theorem 3.2. $n(\lambda)$ designates for the number of all zeros of $A_0(z) - \lambda (a \leq \lambda \leq b)$ belonging to $\mathbb{T}$ (counted according to their multiplicities), $m(\lambda)$ stands for the maximal multiplicity of them, and $\mathcal{N}$ designates for the set of discontinuous points of the function $n(\lambda)$ on $[a,b]$.

Theorem 5.1. *Let $\lambda_0 \in (a,b) \backslash \mathcal{N}$ and $m_0 = m(\lambda_0)$. If*

$$\lim_{j\to\infty} j^{2m_0} b_{j+r,j} = 0, (j = 0, \pm 1, \cdots, \pm n),$$

then λ_0 is not an accumulation point for the set of eigenvalues of the operator A. Each possible eigenvalue of A in a neighborhood of λ_0 has a finite multiplicity.

The assertion is an immediate consequence of Theorem 4.1.

For a band Jacobi matrix A of order $2n$ there holds the following result.

Corollary 5.2. *If $\lambda_0 \notin \mathcal{N}$ and if*

$$\lim_{j\to\infty} j^{2n} b_{j+r,j} = 0, (r = 0, \pm 1, \cdots, \pm n),$$

then the point spectrum of A in a neighborhood of λ_0 is only a finite set.

Theorem 4.2 in turn implies the following result.

Theorem 5.3. *If*

$$\lim_{j\to\infty} jb_{j+r,j} = 0, (r = 0, \pm 1, \cdots, \pm n),$$

then the point spectrum of A has at most a finite set of accumulation points. Moreover, each eigenvalue of A can be only of a finite multiplicity.

2. In the case of an ordinary Jacobi matrix, i.e., when $n = 1$, the results given by Theorem 5.1 and Corollaries 5.2 and 5.3 can be refined. There will be no loss of generality in supposing that the operator A is defined in the space $l_2(\mathbb{N})$ by the following Jacobi matrix

$$A = \begin{pmatrix} 0 & a_1 & 0 & \cdots \\ a_1 & 0 & a_2 & \cdots \\ 0 & a_2 & 0 & \cdots \\ \cdot & \cdot & \cdot & \cdots \end{pmatrix},$$

where $a_j \in \mathbb{R}$ $(j = 1, 2, \ldots)$, $a_j \to 1$ $(j \to \infty)$.

In what follows we denote

$$b_j = a_j - 1, c_j = \mid b_j \mid + \mid b_{j+1} \mid (j = 1, 2, \ldots).$$

Clearly $A = A_0 + B$, where $A_0 = V + V^*$ and $B = A - A_0 \in \mathbb{B}_\infty(l_2(\mathbb{N}))$. Thus, in accord with our notations, $A_0(z) = z + z^{-1}$, $z \in \mathbb{T}$, $\sigma(A_0) = [-2, 2]$, $n(\lambda) = 2 (-2 \leq \lambda \leq 2)$, $m(\lambda) = 1$ $(-2 < \lambda < 2)$, $m(\pm 2) = 2$ and $\mathcal{N} = \{-2, 2\}$.

Theorem 5.4. *If the operator defined by*

$$C = \begin{pmatrix} c_1 & c_2 & c_3 & \cdots \\ 0 & 2c_2 & 2c_3 & \cdots \\ 0 & 0 & 3c_3 & \cdots \\ \cdot & \cdot & \cdot & \cdots \end{pmatrix},$$

is compact in the space $l_2(\mathbb{N})$, then on the continuous spectrum $[-2, 2]$ is contained at most a finite set of eigenvalues of the operator A. Moreover, each possible eigenvalue of A has a finite multiplicity.

The proof of Theorem 5.4 is given in our work [9]. Also, in this work it is proved the following result.

Theorem 5.5. *If the operator defined by*

$$C_1 = \begin{pmatrix} c_1 & c_2 & c_3 & \cdots \\ 0 & c_2 & c_3 & \cdots \\ 0 & 0 & c_3 & \cdots \\ \cdot & \cdot & \cdot & \cdots \end{pmatrix},$$

is compact in the space $l_2(\mathbb{N})$, then the accumulation points for the set of all eigenvalues of A can be only the end points of the continuous spectrum of A, i.e., $\lambda = -2$ or $\lambda = 2$.

Remark 5.6. *The methods developed in the present paper can be applied to more general operators involving matrix Wiener-Hopf operators (in this respect we note the work* [17] *where the theory of Wiener-Hopf equations for the matrix case is developed). By this approach, it is possible to study from the same point of view the point spectrum for operators generated by periodic Jacobi matrices. We note the work* [4] *for some results on the problem of the finiteness of the discrete spectrum for perturbed matrix Wiener-Hopf operators and applications to periodic Jacobi matrices.*

References

[1] M.S. Birman: On the spectrum of singular boundary-value problems *Mat. Sb. 55*, (1961), 125–174 (Russian).

[2] P.A. Cojuhari, I.A. Feldman: On the finiteness of the discrete spectrum of a self-adjoint operator *Mat. Issled., 47*, (1978), 35–40 (Russian).

[3] P.A. Cojuhari: On the discrete spectrum of a perturbed Wiener-Hopf operator *Mat. Issled., 54*, (1980), 50–55 (Russian).

[4] P.A. Cojuhari: On the spectrum of a perturbed matrix Wiener-Hopf operator *Mat. Issled., 61*, (1981), 29–39 (Russian).

[5] P.A. Cojuhari: On the spectrum of perturbed self-adjoint operators *Izv. Akad. Nauk MSSR Mat. 2*, (1982), 51–53 (Russian).

[6] P.A. Cojuhari: On the discrete spectrum of a perturbed Wiener-Hopf integral operator *Issled. Funkts. Anal. Differents. Uravn., Sht. Kishinev*, (1984), 69–82.

[7] P.A. Cojuhari: On the finiteness of the discrete spectrum of some pseudodifferential operators *Mat. Issled., 80*, (1985), 85–97 (Russian).

[8] P.A. Cojuhari: On the point spectrum of the Dirac operator *Dif. Urav., Minsk, Dep. VINITI 24.06.87., N4611-B87*, (1987), 1–17 (Russian).

[9] P.A. Cojuhari: Finiteness of the discrete spectrum of Jacobi matrices *Issled. Differents. Uravn. i Mat. Anal., 173, Sht. Kishinev*, (1988).

[10] P.A. Cojuhari: On the finiteness of the discrete spectrum of some matrix pseudo-differential operators *Izv. Vyssh. Uchebn. Zaved. Mat. 1*, (1989), 42–50 (Russian).

[11] P.A. Cojuhari: The absence of eigenvalues of operators close to operators generated by infinite Jacobi matrices *Izv. Akad. Nauk SSRM, Ser. Mat., 2*, (1990), 15–21.

[12] P.A. Cojuhari: The absence of eigenvalues of the perturbed discrete Wiener-Hopf operator *Izv. Akad. Nauk SSRM, Ser. Mat. 3*, (1990), 26–35.

[13] P.A. Cojuhari: Generalized Hardy type inequalities and some applications to spectral theory, Operator theory, operator algebras and related topics *in Proc. of OT 16 Conf. Bucharest*, (1996), 79–99.

[14] P.A. Cojuhari: Hardy type inequalities for abstract operators *Bul. Acad. de Sht. a Repub. Moldova Mat., 2(33)*, (2000), 79–84.

[15] H.L. Cycon, R.G. Froese, W. Kirsch, B. Simon: Schrödinger operators with application to quantum mechanics and global geometry *Springer-Verlag, Berlin-Heidelberg-New York*, (1987).

[16] I.C. GOHBERG, M.G. KREIN: The basic propositions on defect numbers, root numbers and indices of linear operators *Uspehi Mat. Nauk (N. S.) 12, 2(74)*, (1957), 43–118 (Russian).

[17] I.C. GOHBERG, M.G. KREIN: Systems of integral equations on a half line with kernels depending on the difference of arguments *Uspehi Mat. Nauk. 13, 2(80)*, (1958), 3–72.

[18] I.C. GOHBERG, I.A. FELDMAN: Equations in Convolutions and Projective Methods for their Solution *Nauka, Moscow*, (1971), (Russian).

[19] G.H. HARDY, J.E. LITTELWOOD, G. POLYA: Inequalities *Cambridge Univ. Press, Cambridge*, (1952).

[20] R. KONNO, S.T. KURODA: On the finiteness of perturbed eigenvalues *J. Fac. Sci. Univ. Tokyo Sect. IA Math. 13*, (1966), 55–63.

[21] S.T. KURODA: An abstract stationary approach to perturbation of continuous spectra and scattering theory *J. d'Analyse Math. 20*, (1967), 57–117.

[22] S.T. KURODA: Scattering theory for differential operators *I, J. Math. Soc. Japan, 25, no. 1*, (1973), 75–103.

[23] S.T. KURODA: An introduction to scattering theory *Aarhus Universitet, Lecture Note Series, no. 51*, (1978).

[24] S. PRÖSSDORF: Einige Klassen singulärer Gleichungen *Akademie-Verlag-Berlin*, (1974).

[25] M. REED, B. SIMON: Methods of modern mathematical physics *IV. Analysis of operators., Academic Press, New York-London*, (1978).

[26] D.R. YAFAEV: Mathematical scattering theory *Vol. 105. Transl. Math. Monographs, AMS*, (1992).

Petru Cojuhari
Department of Applied Mathematics
AGH University of Science and Technology,
Al. Mickievicza 30
30-059 Cracow
Poland
e-mail: cojuhari@uci.agh.edu.pl

Operator Theory:
Advances and Applications, Vol. 154, 51–68

Singular Perturbations as Range Perturbations in a Pontryagin Space

Vladimir Derkach, Seppo Hassi and Henk de Snoo

Abstract. When the singular finite rank perturbations of an unbounded self-adjoint operator A_0 in a Hilbert space $\mathfrak{H}_0$, formally defined by $A_{(\alpha)} = A_0 + G\alpha G^*$, are lifted to an exit Pontryagin space $\mathfrak{H}$ by means of an operator model, they become ordinary range perturbations of a self-adjoint operator H_∞ in $\mathfrak{H} \supset \mathfrak{H}_0$: $H_\tau = H_\infty - \Omega\tau^{-1}\Omega^*$. Here G is a mapping from $\mathbb{C}^d$ into some scale space $\mathfrak{H}_{-k}(A_0)$, $k \in \mathbb{N}$, of generalized elements associated with A_0, while Ω is a mapping from $\mathbb{C}^d$ into the extended space $\mathfrak{H}$, where H_τ is defined. The connection between these two perturbation formulas is studied.

Mathematics Subject Classification (2000). Primary 47A55, 47B25, 47B50; Secondary 34L40, 81Q10, 81Q15.

Keywords. Singular finite rank perturbation, extension theory, Kreĭn's formula, boundary triplet, Weyl function, generalized Nevanlinna function, operator model.

1. Introduction

Let A_0 be an unbounded self-adjoint operator in a Hilbert space $\mathfrak{H}_0$ and let $\mathfrak{H}_{-k}(A_0)$, $k \in \mathbb{N}$, be the dual space of generalized elements corresponding to the space $\mathfrak{H}_{+k}(A_0) = \operatorname{dom} |A_0|^{k/2}$ equipped with the graph norm, cf. [6]. Singular finite rank perturbations of an unbounded self-adjoint operator A_0 in a Hilbert space $\mathfrak{H}_0$ are defined formally as

$$A_{(\alpha)} = A_0 + G\alpha G^*, \tag{1.1}$$

where G is a linear mapping from $\mathcal{H} = \mathbb{C}^d$ into $\mathfrak{H}_{-k}(A_0)$ and α is a self-adjoint operator in $\mathcal{H}$. In [15], [23] an operator model for one-dimensional singular perturbations of the form (1.1) was constructed by extending the space $\mathfrak{H}_0$ with a finite-dimensional exit space $\mathfrak{H}_Q$; see also [2], [3] for the case of finite rank $\mathfrak{H}_{-2}$-perturbations. Several further references on this topic can be found in [3]. The model given in [9], [11] uses a coupling method for identifying the singular finite rank perturbations $A_{(\alpha)}$ in (1.1) with the self-adjoint extensions H_τ of a symmetric operator in $\mathfrak{H} = \mathfrak{H}_0 \oplus \mathfrak{H}_Q$ with some natural additional assumptions on G (see Theorem 3.1 below). It turns out that the extensions H_τ are in fact ordinary range perturbations of one of the extensions, namely of the self-adjoint operator H_∞ in $\mathfrak{H} \supset \mathfrak{H}_0$:

$$H_\tau = H_\infty - \Omega\tau^{-1}\Omega^*, \tag{1.2}$$

where Ω is a mapping from $\mathcal{H}$ into $\mathfrak{H}$ and τ is a self-adjoint parameter in $\mathcal{H}$. The perturbations H_τ in (1.2) induce a symmetric restriction S of H_∞ in $\mathfrak{H}$ via

$$\operatorname{dom} S = \{\, F \in \operatorname{dom} H_\infty : \Omega^* F = 0 \,\},$$

which, due to the assumption $\operatorname{ran}\Omega \subset \mathfrak{H}$, is maximally nondensely defined in $\mathfrak{H}$. Therefore, among the self-adjoint extensions of S there are linear relations which are not operators. In particular, the generalized Friedrichs extension (see [17], [18]) of S is not an operator. A classification of the perturbations H_τ by decomposing the self-adjoint parameter τ into its operator and multi-valued parts leads to intermediate symmetric extensions of S and their generalized Friedrichs extensions. These extensions of S turn out to be precisely those which are given by the so-called extremal boundary conditions and whose compressed resolvents to the original space $\mathfrak{H}_0$ are canonical, i.e., coincide with a resolvent of a self-adjoint relation in $\mathfrak{H}_0$.

The contents of this paper are now briefly described. Section 2 contains the necessary facts concerning boundary triplets and Weyl functions in a Pontryagin space. A concise introduction to finite rank singular perturbations of a self-adjoint operator in a Hilbert space is given in Section 3. Such finite rank singular perturbations are identified with self-adjoint relations in a larger Pontryagin space. They are interpreted as range perturbations in Section 4. A connection with so-called extremal boundary conditions can be found in Section 5.

2. Boundary triplets and abstract Weyl functions

Let $\mathfrak{H}$ be a Pontryagin space with negative index κ, cf. [5]. The set of all bounded everywhere defined linear operators acting on $\mathfrak{H}$ is denoted by $[\mathfrak{H}]$. If T is a linear relation in $\mathfrak{H}$, then $\operatorname{dom} T$, $\ker T$, $\operatorname{ran} T$, and $\operatorname{mul} T$ indicate the domain, kernel, range, and multi-valued part of T, respectively; moreover, $\rho(T)$ denotes the set of regular points of the linear relation T. Let S be a not necessarily densely defined closed symmetric relation in $\mathfrak{H}$ with equal defect numbers $d_+(S) = d_-(S) < \infty$ and let S^* be the adjoint linear relation of S, so that $S \subset S^*$. Recall (see [16], [7]) that a triplet $\Pi = \{\mathcal{H}, \Gamma_0, \Gamma_1\}$ of a Hilbert space $\mathcal{H}$ with $\dim \mathcal{H} = n_\pm(S)$ and two linear mappings Γ_j, $j = 0, 1$, from S^* to $\mathcal{H}$ is called a *boundary triplet* for S^*, if the mapping $\Gamma = (\Gamma_0, \Gamma_1)^\top : \widehat{f} \to (\Gamma_0\widehat{f}, \Gamma_1\widehat{f})^\top$ from S^* into $\mathcal{H}\oplus\mathcal{H}$ is surjective and the following abstract Green's identity holds for every $\widehat{f} = \{f, f'\}$, $\widehat{g} = \{g, g'\} \in S^*$:

$$(f', g) - (f, g') = (\Gamma_1\widehat{f}, \Gamma_0\widehat{g})_{\mathcal{H}} - (\Gamma_0\widehat{f}, \Gamma_1\widehat{g})_{\mathcal{H}} = i(\Gamma\widehat{g})^* J(\Gamma\widehat{f});$$

here J stands for the block operator

$$J = \begin{pmatrix} 0 & -iI_{\mathcal{H}} \\ iI_{\mathcal{H}} & 0 \end{pmatrix}.$$

The adjoint S^* of every closed symmetric relation S with equal defect numbers has a boundary triplet $\Pi = \{\mathcal{H}, \Gamma_0, \Gamma_1\}$. All other boundary triplets $\widetilde{\Pi} = \{\mathcal{H}, \widetilde{\Gamma}_0, \widetilde{\Gamma}_1\}$ are related to Π via a J-unitary transformation W: $\widetilde{\Gamma} = W\Gamma$. In particular, the *transposed boundary triplet* $\Pi^\top = \{\mathcal{H}, \Gamma_0^\top, \Gamma_1^\top\}$, is defined by $\Gamma^\top = iJ\Gamma$. When S is densely defined, S^* can be identified with its domain $\operatorname{dom} S^*$ and the boundary mappings can be interpreted as mappings from $\operatorname{dom} S^*$ onto $\mathcal{H}$.

Let $\Pi = \{\mathcal{H}, \Gamma_0, \Gamma_1\}$ be a boundary triplet for S^*. The mapping $\Gamma^\top : \widehat{f} \to \{\Gamma_1 \widehat{f}, -\Gamma_0 \widehat{f}\}$ from S^* onto $\mathcal{H} \oplus \mathcal{H}$ establishes a one-to-one correspondence between the set of all self-adjoint extensions of S and the set of all self-adjoint linear relations τ in $\mathcal{H}$ via

$$A_\tau := \ker(\Gamma_0 + \tau\Gamma_1) = \{\widehat{f} \in S^* : \{\Gamma_1 \widehat{f}, -\Gamma_0 \widehat{f}\} \in \tau\} = \{\widehat{f} \in S^* : \Gamma^\top \widehat{f} \in \tau\}. \tag{2.1}$$

When the parameter τ is an operator in $\mathcal{H}$ the equation (2.1) takes the form

$$\Gamma_0 \widehat{f} + \tau \Gamma_1 \widehat{f} = 0, \quad \widehat{f} \in S^*. \tag{2.2}$$

The identity $\tau = \infty$ is to be interpreted as $\tau^{-1} = 0$ or, more precisely, by using graph notation as $\tau = \{0, I_{\mathcal{H}}\}$; in this case the equation in (2.2) takes the form $\Gamma_1 \widehat{f} = 0$. More generally, there is a similar interpretation, when τ is decomposed orthogonally in terms of its operator part and multi-valued part. To each boundary triplet Π one may naturally associate two self-adjoint extensions of S by $A_0 = \ker \Gamma_0$, $A_1 \, (= A_\infty) = \ker \Gamma_1$, corresponding to the linear relations $\tau = 0$ and $\tau = \infty$ via (2.1).

Let $\mathfrak{N}_\lambda(S^*) = \ker(S^* - \lambda)$, $\lambda \in \widehat{\rho}(S)$, be the defect subspace of S and let $\widehat{\mathfrak{N}}_\lambda(S^*) := \{\{f_\lambda, \lambda f_\lambda\} : f_\lambda \in \mathfrak{N}_\lambda(S^*)\}$; here the notations $\mathfrak{N}_\lambda$ and $\widehat{\mathfrak{N}}_\lambda$ are used when the context is clear. Every boundary triplet Π gives rise to two operator functions defined for $\lambda \in \rho(A_0)$ $(\neq \emptyset)$ by the formulas

$$\gamma(\lambda) = p_1(\Gamma_0 \upharpoonright \widehat{\mathfrak{N}}_\lambda)^{-1} (\in [\mathcal{H}, \mathfrak{N}_\lambda]), \quad M(\lambda) = \Gamma_1(\Gamma_0 \upharpoonright \widehat{\mathfrak{N}}_\lambda)^{-1} \, (\in [\mathcal{H}]). \tag{2.3}$$

Here p_1 denotes the orthogonal projection onto the first component of $\mathcal{H} \oplus \mathcal{H}$. The functions γ and M in (2.3) are holomorphic on $\rho(A_0)$ and they are called the *γ-field* and the *Weyl function* of S corresponding to the boundary triplet Π, respectively; cf. [7], [13]. The function M is also the *Q-function* of the pair (S, A_0) in the sense of [21]). The γ-field $\gamma^\top$ and the abstract Weyl function $M^\top$ corresponding to the transposed boundary triplet $\Pi^\top$ are related to γ and M via

$$\gamma^\top(\lambda) = \gamma(\lambda) M(\lambda)^{-1}, \quad M(\lambda)^\top = -M(\lambda)^{-1}, \quad \lambda \in \rho(A_1) \, (\neq \emptyset).$$

When $\mathfrak{H}$ is a Hilbert space, a Weyl function M of S belongs to the class of Nevanlinna functions, that is, M is holomorphic in the upper half-plane $\mathbb{C}_+$, $\operatorname{Im} M(\lambda) \geq 0$ for all $\lambda \in \mathbb{C}_+$, and M satisfies the symmetry condition $M(\lambda)^* = M(\bar{\lambda})$ for $\lambda \in \mathbb{C}_+ \cup \mathbb{C}_-$. In the case where $\mathfrak{H}$ is a Pontryagin space of negative index κ, the Weyl function M of S belongs to the class $\mathbf{N}_k$, $k \le \kappa$, of *generalized Nevanlinna functions* which are meromorphic on $\mathbb{C}_+ \cup \mathbb{C}_-$, satisfy $M(\lambda)^* = M(\bar{\lambda})$, and for which the kernel

$$\mathsf{N}_M(\lambda,\mu) = \frac{M(\lambda) - M(\bar{\mu})}{\lambda - \bar{\mu}}, \quad \mathsf{N}_M(\lambda,\bar{\lambda}) = \frac{d}{d\lambda} M(\lambda), \quad \lambda,\mu \in \mathbb{C}_+,$$

has k negative squares [21]. If S is *simple*, that is,

$$\mathfrak{H} = \overline{\operatorname{span}}\,\{\, \mathfrak{N}_\lambda(S^*) : \lambda \in \rho(A_0)\,(\neq \emptyset)\,\},$$

then S is an operator without eigenvalues. In this case the Weyl function M belongs to the class $\mathbf{N}_\kappa$, i.e., $k = \kappa$, and the domain of holomorphy $\rho(M)$ of M coincides with the resolvent set $\rho(A_0)$.

The resolvent of the extension A_τ and its spectrum $\sigma(A_\tau)$ can be expressed in terms of τ and the Weyl function M via Kreĭn's formula. In the terminology of boundary triplets the result can be formulated as follows, see [12], [13], [7].

Proposition 2.1. *Let S be a closed symmetric relation in the Pontryagin space $\mathfrak{H}$ with equal defect numbers (d,d), $d < \infty$, let $\Pi = \{\mathcal{H}, \Gamma_0, \Gamma_1\}$ be a boundary triplet for S^* with the Weyl function M, let τ be a linear relation in $\mathcal{H}$ connected with A_τ via* (2.1). *Then the resolvent of A_τ is given by*

$$(A_\tau - \lambda)^{-1} = (A_0 - \lambda)^{-1} - \gamma(\lambda)(\tau^{-1} + M(\lambda))^{-1}\gamma(\bar{\lambda})^*, \quad \lambda \in \rho(A_\tau) \cap \rho(A_0).$$

Moreover, for every $\lambda \in \rho(A_0)$ the following equivalences hold:

(i) *$\lambda \in \rho(A_\tau)$ if and only if $\tau^{-1} + M(\lambda)$ is invertible;*
(ii) *$\lambda \in \sigma_p(A_\tau)$ if and only if $\ker(\tau^{-1} + M(\lambda))$ is nontrivial.*

Similarly, for a (generalized) Nevanlinna family $\widetilde{\tau}(\lambda)$ the function

$$(A_0 - \lambda)^{-1} - \gamma(\lambda)(\widetilde{\tau}(\lambda) + M(\lambda))^{-1}\gamma(\bar{\lambda})^*,$$

is the compressed resolvent of an exit space extension of S in a Hilbert (or a Pontryagin) space, cf. [21], [24], [12], [7].

3. A model for singular perturbations

In a number of papers singular rank one perturbations of A_0 generated by $\omega \in \mathfrak{H}_{-2n-2}$ have been studied by means of exit space extensions of a symmetric operator S connected with A_0, see [23], [14], [15], [22]. In this section the main ingredients for constructing a model for *finite rank singular perturbations* of A_0 generated by G with $\operatorname{ran} G \subset \mathfrak{H}_{-2n-j}$, $n > 0$, $j = 1, 2$, are given. This model was established in [9] and further used in [11], see also [10] for a special case. The model uses an orthogonal coupling of two symmetric operators and it is motivated by a perturbation result concerning the extending inner product space $\mathfrak{H} \supset \mathfrak{H}_0$: the resolvents associated with the perturbations of A_0 should be finite rank perturbations of the resolvent generated in $\mathfrak{H}$ by $(A_0 - \lambda)^{-1}$ (see Theorem 3.1 below). In a more general setting a similar model to give realizations for arbitrary scalar $\mathbf{N}_\kappa$-functions has been established recently in [8].

3.1. Some operators associated with matrix polynomials

Let q be a monic $d \times d$ matrix polynomial of the form

$$q(\lambda) = I_{\mathcal{H}}\lambda^n + q_{n-1}\lambda^{n-1} + \cdots + q_1\lambda + q_0, \tag{3.1}$$

and let r be a self-adjoint $d \times d$ matrix polynomial of the form

$$r(\lambda) = r_{2n-1}\lambda^{2n-1} + r_{2n-2}\lambda^{2n-2} + \cdots + r_1\lambda + r_0, \quad r_j = r_j^*, \quad j = 0, \ldots, 2n-1. \tag{3.2}$$

Observe, that the function Q given by

$$Q(\lambda) = \begin{pmatrix} 0 & q(\lambda) \\ q^\sharp(\lambda) & r(\lambda) \end{pmatrix}, \tag{3.3}$$

is a $2d \times 2d$ matrix polynomial whose leading coefficient is, in general, noninvertible. In fact, Q is a strict generalized matrix Nevanlinna function whose Nevanlinna kernel has dn negative (and dn positive) squares.

Associated with the matrix polynomial q there are $n \times n$ block matrices $\mathcal{B}_q$ and $\mathcal{C}_q$ defined by

$$\mathcal{B}_q = \begin{pmatrix} q_1 & q_2 & \cdots & q_{n-1} & I_{\mathcal{H}} \\ q_2 & \cdots & q_{n-1} & I_{\mathcal{H}} & 0 \\ \vdots & \cdot^{\cdot^{\cdot}} & \cdot^{\cdot^{\cdot}} & 0 & 0 \\ q_{n-1} & I_{\mathcal{H}} & \cdot^{\cdot^{\cdot}} & \cdot^{\cdot^{\cdot}} & \vdots \\ I_{\mathcal{H}} & 0 & 0 & \cdots & 0 \end{pmatrix}$$

and

$$\mathcal{C}_q = \begin{pmatrix} 0 & I_{\mathcal{H}} & 0 & \cdots & 0 \\ 0 & 0 & I_{\mathcal{H}} & \ddots & \vdots \\ \vdots & \vdots & \ddots & \ddots & 0 \\ 0 & 0 & \cdots & 0 & I_{\mathcal{H}} \\ -q_0 & -q_1 & \cdots & -q_{n-2} & -q_{n-1} \end{pmatrix}$$

Moreover, the following block matrices are needed

$$\mathcal{B} = \begin{pmatrix} 0 & \mathcal{B}_q \\ \mathcal{B}_{q^\sharp} & \mathcal{B}_r \end{pmatrix}, \quad \mathcal{C} = \begin{pmatrix} \mathcal{C}_{q^\sharp} & \mathcal{C}_{12} \\ 0 & \mathcal{C}_q \end{pmatrix}, \quad \mathcal{B}_r = (r_{j+k+1})_{j,k=0}^{n-1}, \quad \mathcal{C}_{12} = \mathcal{B}_{q^\sharp}^{-1}\mathcal{D}, \tag{3.4}$$

where

$$\mathcal{D} = \begin{pmatrix} r_n \\ r_{n+1} \\ \vdots \\ r_{2n-1} \end{pmatrix} (q_0, q_1, \ldots, q_{n-1}) - \begin{pmatrix} I_{\mathcal{H}} \\ 0 \\ \vdots \\ 0 \end{pmatrix} (r_0, r_1, \ldots, r_{n-1}).$$

In addition, the following vectors depending on $\lambda \in \mathbb{C}$ are used:

$$\Lambda = (I_{\mathcal{H}}, \lambda I_{\mathcal{H}}, \ldots, \lambda^{n-1} I_{\mathcal{H}}),$$

$$\Lambda_1 = \lambda^n \Lambda \widetilde{\mathcal{B}}_{(r)} \mathcal{B}_q^{-1}, \quad \widetilde{\mathcal{B}}_{(r)} = \begin{pmatrix} r_{n+1} & \dots & r_{2n-1} & 0 \\ \vdots & \iddots & 0 & 0 \\ r_{2n-1} & \iddots & \iddots & \vdots \\ 0 & 0 & \dots & 0 \end{pmatrix}.$$

The main objective here is the matrix polynomial Q defined in (3.3). It determines the structure of the exit space $\mathfrak{H}_Q = \mathcal{H}^n \oplus \mathcal{H}^n$ $(= \mathbb{C}^{2dn})$ used for constructing the model for singular perturbations. The inner product in $\mathfrak{H}_Q$ is defined by the block matrix $\mathcal{B}$ via $\langle \cdot, \cdot \rangle_{\mathfrak{H}_Q} = (\mathcal{B} \cdot, \cdot)$ in which case *the companion type operator* $\mathcal{C}$ in (3.4) becomes self-adjoint in $\mathfrak{H}_Q$. The restriction of $\mathcal{C}$ to the subspace

$$\operatorname{dom} S_Q = \left\{ F = \begin{pmatrix} f \\ \widetilde{f} \end{pmatrix} \in \mathfrak{H}_Q : f_1 = \widetilde{f}_1 = 0 \right\} \tag{3.5}$$

defines a closed simple symmetric operator S_Q in $\mathfrak{H}_Q$ with defect numbers $(2d, 2d)$. It is maximally nondensely defined and a straightforward calculation shows that its adjoint S_Q^* (a linear relation in $\mathfrak{H}_Q$) is given by

$$S_Q^* = \left\{ \widehat{F} = \left\{ F, \mathcal{C}F + \mathcal{B}^{-1} \begin{pmatrix} \varphi \otimes e_1 \\ \widetilde{\varphi} \otimes e_1 \end{pmatrix} \right\} : F \in \mathfrak{H}_Q, \varphi, \widetilde{\varphi} \in \mathcal{H} \right\}.$$

It is possible to associate a boundary triplet $\Pi_Q = \{\mathcal{H} \oplus \mathcal{H}, \Gamma_0^Q, \Gamma_1^Q\}$ with S_Q^* by defining the boundary mappings on S_Q^* via

$$\Gamma_0^Q \widehat{F} = \begin{pmatrix} f_1 \\ \widetilde{f}_1 \end{pmatrix}, \quad \Gamma_1^Q \widehat{F} = \begin{pmatrix} \varphi \\ \widetilde{\varphi} \end{pmatrix}, \quad \widehat{F} \in S_Q^*.$$

In this case the Weyl function of S_Q associated with the boundary triplet Π_Q coincides with the matrix polynomial Q, cf. [9].

3.2. A perturbation result for the resolvents

Let G be an injective linear mapping from $\mathcal{H} = \mathbb{C}^d$ into the scale space $\mathfrak{H}_{-2n-1}$ generated by the self-adjoint operator A_0 and let $\widetilde{A}_0$ be the $[\mathfrak{H}_{-2n+1}, \mathfrak{H}_{-2n-1}]$-continuation of A_0. The adjoint operator G^* maps $\mathfrak{H}_{2n+1}$ into $\mathcal{H}$. The case where G is a mapping into $\mathfrak{H}_{-2n}$ is similar to the present case; it can be found in [11]. Observe, that if $\operatorname{ran} G \cap \mathfrak{H}_{-2} = \{0\}$, then the restriction of A_0 to

$$\operatorname{dom} S_0 = \operatorname{dom} A_0 \cap \ker G^*$$

gives rise to an essentially self-adjoint operator S_0 whose closure coincides with A_0. In this case the vector $\widetilde{R}_\lambda G h = (\widetilde{A}_0 - \lambda)^{-1} G h$, $h \in \mathcal{H} \setminus \{0\}$, $\lambda \in \rho(A_0)$, does not belong to the space $\mathfrak{H}_0$, since G is injective. However, one can give a sense to the vector $\widetilde{R}_\lambda G h$ by extending the space $\mathfrak{H}_0$ suitably. For instance, if $0 \in \rho(A_0)$, then the vector

$$\gamma(\lambda) h := \widetilde{R}_\lambda G h = \widetilde{A}_0^{-1} G h + \dots + \lambda^{n-1} \widetilde{A}_0^{-n} G h + \lambda^n \widetilde{R}_\lambda \widetilde{A}_0^{-n} G h$$

can be considered as a vector from an extended inner product space $\mathfrak{H}$ satisfying the condition

$$\mathfrak{H} \supset \operatorname{span}\{\, \mathfrak{H}_0, \widetilde{A}_0^{-j} \operatorname{ran} G : \, j = 1, \ldots, n \,\}. \tag{3.6}$$

In this space the continuation $\widetilde{A}_0$ of A_0 generates an operator, say H_0, for which the operator function $\gamma(\lambda)$, $\lambda \in \rho(A_0)$, can be interpreted to form its γ-field in the sense that

$$\frac{\gamma(\lambda) - \gamma(\mu)}{\lambda - \mu} = (H_0 - \lambda)^{-1} \gamma(\mu), \quad \lambda, \mu \in \rho(A_0).$$

This identity implies that

$$\frac{d}{d\lambda} \gamma(\lambda) = (H_0 - \lambda)^{-1} \gamma(\lambda), \quad \lambda \in \rho(A_0).$$

The inner product $\langle u, \varphi \rangle_{\mathfrak{H}}$ in $\mathfrak{H}$ should coincide with the form (u, φ) generated by the inner product in $\mathfrak{H}_0$ if the vectors u, φ are in duality, say, $u \in \mathfrak{H}_{2(n-j)+1}$, $\varphi \in \widetilde{A}_0^{-j} \operatorname{ran} G$. Now, for the other vectors in (3.6) it will be supposed that the conditions

$$\left\langle \widetilde{A}_0^{-j} G h, \widetilde{A}_0^{-k} G f \right\rangle_{\mathfrak{H}} = (t_{j+k-1} h, f)_{\mathcal{H}}, \quad j, k = 1, \ldots, n; \; h, f \in \mathcal{H}, \tag{3.7}$$

are satisfied for some operators $t_j = t_j^* \in [\mathcal{H}]$, $j = 1, \ldots, 2n - 1$. The next result shows that under such mild conditions on the extending space the structure of perturbed resolvents becomes completely fixed under some basic assumptions on H_0. This fact gives rise to the model presented in [9] for singular finite rank perturbations of A_0.

Theorem 3.1. ([11, Theorem 4.8]) *Let* $0 \in \rho(A_0)$, *let* $\operatorname{ran} G \setminus \{0\} \subset \mathfrak{H}_{-2n-1} \setminus \mathfrak{H}_{-2n}$ *with* $\ker G = \{0\}$, *and let* $G_0 = \widetilde{A}_0^{-n} G$. *Moreover, assume that* $\mathfrak{H} \supset \mathfrak{H}_0$ *is (an isometric image of) an inner product space satisfying* (3.6), (3.7), *and let* H *and* H_0 *be self-adjoint linear relations in* $\mathfrak{H}$ *such that*

(i) $\rho(H_0) = \rho(A_0)$;
(ii) $\gamma(\lambda)' = (H_0 - \lambda)^{-1} \gamma(\lambda)$ *holds for (an isometric image of) the function* $\gamma(\lambda) = (\widetilde{A}_0 - \lambda)^{-1} G$, $\lambda \in \rho(A_0)$;
(iii) $(H - \lambda)^{-1} - (H_0 - \lambda)^{-1} = -\gamma(\lambda) \sigma(\lambda) \gamma(\bar{\lambda})^*$, $\lambda \in \rho(H) \cap \rho(H_0)$;

for some matrix function $\sigma(\lambda)$ *holomorphic and invertible for* $\lambda \in \rho(H_0) \cap \rho(H)$. *Then* $\sigma(\lambda)^{-1}$ *can be represented in the form*

$$\sigma^{-1}(\lambda) = \beta + t(\lambda) + \lambda^{2n} M_0(\lambda), \tag{3.8}$$

where $\beta = \beta^* \in [\mathcal{H}]$, $t(\lambda) = t_1 \lambda + \cdots + t_{2n-1} \lambda^{2n-1}$, *and* $M_0(\lambda) = G_0^* \widetilde{R}_\lambda G_0$ *is a Nevanlinna function in* $\mathcal{H}$.

In Theorem 3.1 the function σ^{-1} can be seen as a Weyl function (or a Q-function) of an underlying symmetric operator S. The formula for σ^{-1} in (3.8) shows that it is a generalized Nevanlinna function and therefore in general the operator S cannot be symmetric in some Hilbert space. The model constructed in [9] for S uses a coupling method resulting in a Pontryagin space $\mathfrak{H}$ such that

S becomes symmetric in $\mathfrak{H}$. The construction of the model space via the coupling method is briefly recalled in the next subsection.

Note that the condition $0 \in \rho(A_0)$, which was assumed for simplicity, leads to the particular form of $\sigma(\lambda)^{-1}$ in (3.8). Other invertibility conditions on A_0 lead to the more general form of $\sigma(\lambda)^{-1}$ in (3.9).

3.3. The model

Let S_0 be a closed symmetric operator in a Hilbert space $\mathfrak{H}_0$ with defect numbers (d,d) and the Weyl function M_0. Let S_Q be the symmetric operator in the Pontryagin space $\mathfrak{H}_Q = \mathbb{C}^{dn} \oplus \mathbb{C}^{dn}$ defined as the restriction of C to (3.5). The next theorem (cf. [9]) gives a symmetric linear relation S in the Pontryagin space $\mathfrak{H} = \mathfrak{H}_0 \oplus \mathfrak{H}_Q = \mathfrak{H} \oplus (\mathbb{C}^{dn} \oplus \mathbb{C}^{dn})$ as a coupling of the operators S_0 and S_Q, such that the following function is a Weyl function for S:

$$M(\lambda) = r(\lambda) + q^\sharp(\lambda) M_0(\lambda) q(\lambda). \tag{3.9}$$

Here the matrix polynomials q and r are as in (3.1) and (3.2).

Theorem 3.2. ([9, Theorem 4.2]) *Let S_0 be a closed symmetric operator in the Hilbert space $\mathfrak{H}_0$ and let $\Pi^0 = \{\mathcal{H}, \Gamma_0^0, \Gamma_1^0\}$ be a boundary triplet for S_0^* with the Weyl function M_0 and the γ-field γ_0. Let q, r, and Q be as in (3.1), (3.2), and (3.3), respectively. Then:*

(i) *The linear relation*

$$S = \left\{ \left\{ \begin{pmatrix} f_0 \\ f \\ \widetilde{f} \end{pmatrix}, \begin{pmatrix} f_0' \\ \mathcal{C}\begin{pmatrix} f \\ \widetilde{f} \end{pmatrix} + \mathcal{B}^{-1}\begin{pmatrix} \Gamma_0^0 \widehat{f}_0 \otimes e_1 \\ 0 \end{pmatrix} \end{pmatrix} \right\} : \begin{array}{c} \widehat{f}_0 = \{f_0, f_0'\} \in S_0^* \\ f, \widetilde{f} \in \mathbb{C}^{dn} \\ f_1 = \Gamma_1^0 \widehat{f}_0, \widetilde{f}_1 = 0 \end{array} \right\}$$

is closed and symmetric in $\mathfrak{H}_0 \oplus \mathfrak{H}_Q$ and has defect numbers (d,d).

(ii) *The adjoint S^* is given by*

$$S^* = \left\{ \left\{ \begin{pmatrix} f_0 \\ f \\ \widetilde{f} \end{pmatrix}, \begin{pmatrix} f_0' \\ \mathcal{C}\begin{pmatrix} f \\ \widetilde{f} \end{pmatrix} + \mathcal{B}^{-1}\begin{pmatrix} \Gamma_0^0 \widehat{f}_0 \otimes e_1 \\ \widetilde{\varphi} \otimes e_1 \end{pmatrix} \end{pmatrix} \right\} : \begin{array}{c} \widehat{f}_0 = \{f_0, f_0'\} \in S_0^* \\ f, \widetilde{f} \in \mathbb{C}^{dn}, \widetilde{\varphi} \in \mathbb{C}^d \\ f_1 = \Gamma_1^0 \widehat{f}_0 \end{array} \right\}.$$

(iii) *A boundary triplet $\Pi = \{\mathcal{H}, \Gamma_0, \Gamma_1\}$ for S^* is determined by*

$$\Gamma_0(\widehat{f}_0 \oplus \widehat{F}) = \widetilde{f}_1, \quad \Gamma_1(\widehat{f}_0 \oplus \widehat{F}) = \widetilde{\varphi}, \quad \widehat{f}_0 \oplus \widehat{F} \in S^*.$$

(iv) *The corresponding Weyl function M is of the form (3.9) and the γ-field γ is given by*

$$\gamma(\lambda)h = \gamma_0(\lambda) q(\lambda) h \oplus ((\Lambda^\top M_0(\lambda) q(\lambda) + \Lambda_1^\top) h + \Lambda^\top h), \quad h \in \mathcal{H}. \tag{3.10}$$

If the operator S_0 is densely defined in $\mathfrak{H}_0$, then S is an operator. When $r = 0$ the formulas for S and S^* in Theorem 3.2 can be simplified and the Weyl function takes the factorized form

$$M(\lambda) = q^\sharp(\lambda) M_0(\lambda) q(\lambda).$$

3.4. Self-adjoint extensions of the model operator

The self-adjoint extensions of the model operator S can be parametrized by the self-adjoint relations τ in the parameter space $\mathcal{H}$ via $H_\tau = \ker(\Gamma_0 + \tau\Gamma_1)$. From Theorem 3.2 one obtains the following explicit expressions for H_τ, cf. [11].

Proposition 3.3. *Let the assumptions be as in Theorem 3.2, and let γ and M be given by* (3.10) *and* (3.9), *respectively. Then:*

(i) *The self-adjoint extensions H_τ of S in $\mathfrak{H} = \mathfrak{H}_0 \oplus \mathfrak{H}_Q$ are in a one-to-one correspondence with the self-adjoint relations τ in $\mathcal{H}$ via*

$$H_\tau = \left\{ \left\{ \begin{pmatrix} f_0 \\ f \\ \widetilde{f} \end{pmatrix}, \begin{pmatrix} f_0' \\ C\begin{pmatrix} f \\ \widetilde{f} \end{pmatrix} + \mathcal{B}^{-1}\begin{pmatrix} \Gamma_0^0 \widehat{f}_0 \otimes e_1 \\ \widetilde{\varphi} \otimes e_1 \end{pmatrix} \end{pmatrix} \right\} : \begin{array}{c} \widehat{f}_0 = \{f_0, f_0'\} \in S_0^* \\ f, \widetilde{f} \in \mathbb{C}^{dn} \\ f_1 = \Gamma_1^0 \widehat{f}_0, \widetilde{f}_1 + \tau\widetilde{\varphi} = 0 \end{array} \right\}.$$

(ii) *For every $\lambda \in \rho(H_\tau) \cap \rho(H_0)$ the resolvent $(H_\tau - \lambda)^{-1}$ satisfies the relation*

$$(H_\tau - \lambda)^{-1} = (H_0 - \lambda)^{-1} - \gamma(\lambda)(\tau^{-1} + M(\lambda))^{-1}\gamma(\bar{\lambda})^*. \tag{3.11}$$

(iii) *For every $\lambda \in \rho(H_0)$ the following equivalences hold:*

$$\lambda \in \sigma_p(H_\tau) \Leftrightarrow 0 \in \sigma_p(\tau^{-1} + M(\lambda)),$$
$$\lambda \in \rho(H_\tau) \Leftrightarrow 0 \in \rho(\tau^{-1} + M(\lambda)).$$

Proof. (i) The condition $\widehat{f}_0 \oplus \widehat{F} \in \ker(\Gamma_0 + \tau\Gamma_1)$ means that $\{\widetilde{\varphi}, \widetilde{f}_1\} \in -\tau$, or equivalently, that $\widetilde{f}_1 + \tau\widetilde{\varphi} = 0$, see (iii) of Theorem 3.2. The representation of H_τ is now obtained from the formula for S^* in Theorem 3.2.

(ii) The form of the resolvent of H_τ is obtained by applying Proposition 2.1 to the data in Theorem 3.2.

(iii) This statement is immediate from Proposition 2.1. □

The operator S_0 in Theorem 3.2 is allowed to be nondensely defined in the original Hilbert space $\mathfrak{H}_0$. If S_0 is densely defined in $\mathfrak{H}_0$ then S is an operator in the model Pontryagin space $\mathfrak{H}_0 \oplus \mathfrak{H}_Q$. However, even in this case the model operator S is not densely defined in $\mathfrak{H}_0 \oplus \mathfrak{H}_Q$. Therefore, among the self-adjoint extensions of S there are linear relations which are not operators. In fact, the following result holds.

Proposition 3.4. *The multi-valued parts of S and H_τ are given by*

$$\operatorname{mul} S = \left\{ \begin{pmatrix} f_0' \\ \mathcal{B}^{-1}\begin{pmatrix} \Gamma_0^0 \widehat{f}_0 \otimes e_1 \\ 0 \end{pmatrix} \end{pmatrix} : \widehat{f}_0 = \{0, f_0'\} \in A_1 \right\}, \tag{3.12}$$

$$\operatorname{mul} H_\tau = \left\{ \begin{pmatrix} f_0' \\ \mathcal{B}^{-1}\begin{pmatrix} \Gamma_0^0 \widehat{f}_0 \otimes e_1 \\ \widetilde{\varphi} \otimes e_1 \end{pmatrix} \end{pmatrix} : \widehat{f}_0 = \{0, f_0'\} \in A_1, \widetilde{\varphi} \in \ker\tau \right\}, \tag{3.13}$$

and the equalities

$$\dim \operatorname{mul} S = \dim \operatorname{mul} A_1, \quad \dim \operatorname{mul} H_\tau = \dim \operatorname{mul} A_1 + \dim \ker \tau \tag{3.14}$$

hold. In particular, H_τ *is an operator in* $\mathfrak{H}$ *if and only if* $A_1 = \ker \Gamma_1^0$ *is an operator in* $\mathfrak{H}_0$ *and* $\ker \tau = \{0\}$. *Moreover,* H_0 *is the unique self-adjoint extension* H_τ *of* S *for which the equality* $\operatorname{mul} H_\tau = \operatorname{mul} S^*$ *holds, and it has the representation*

$$H_0 = S \hat{+} (\{0\} \oplus \operatorname{mul} S^*), \tag{3.15}$$

where $\hat{+}$ *stands for the componentwise sum of the graphs.*

Proof. The form of $\operatorname{mul} H_\tau$ is straightforward to check by using the formulas for the self-adjoint extensions H_τ in Proposition 3.3. By letting $\widetilde{\varphi} = 0$ in (3.13) one obtains the description (3.12) for $\operatorname{mul} S$. The equalities (3.14) follow from (3.12) and (3.13).

Moreover, by comparing $\operatorname{mul} H_\tau$ with the multi-valued part of the adjoint relation S^* (see [9]) one concludes that the condition $\operatorname{mul} H_\tau = \operatorname{mul} S^*$ is equivalent to $\dim \ker \tau = d$. This means that $\tau = 0$, i.e., the only self-adjoint extension with the maximal multi-valued part $\operatorname{mul} S^*$ is H_0.

The representation (3.15) of H_0 is now obvious. □

If the self-adjoint extension $A_1 = \ker \Gamma_1^0$ of S_0 is an operator in $\mathfrak{H}_0$, then $\operatorname{mul} S = \{0\}$ and H_τ is an operator in $\mathfrak{H}$ if and only if $\ker \tau = \{0\}$. In view of (3.15) the extension H_0 is always multi-valued, since S is nondensely defined in $\mathfrak{H}$. In fact, H_0 has a natural interpretation as a *generalized Friedrichs extension* of S, see [17], [18]. The representation (3.15) shows that, together with S, H_0 is maximally nondensely defined in $\mathfrak{H}$. In fact, H_0 has a nontrivial root subspace $\mathfrak{L}$ at ∞ and, moreover, the following results shows that the finite spectrum of H_0 coincides with the spectrum of the self-adjoint extension A_0 of S_0 in the original Hilbert space $\mathfrak{H}_0$. Hence, in particular the assumption (i) in Theorem 3.1 is satisfied.

Proposition 3.5. ([11, Proposition 3.4]) *Let the assumptions be as in Theorem* 3.2 *and let* $H_0 = \ker \Gamma_0$ *be as in Proposition* 3.3 *(with* $\tau = 0$*). Then:*

(i) $\rho(H_0) = \rho(A_0)$;

(ii) *the compression of the resolvent of* H_0 *to the subspace* $\mathfrak{H}_0$ *is given by*

$$P_{\mathfrak{H}_0}(H_0 - \lambda)^{-1} \restriction \mathfrak{H}_0 = (A_0 - \lambda)^{-1}, \quad \lambda \in \rho(H_0);$$

(iii) *the subspace* $\mathcal{L} = \{0\} \oplus \mathcal{H}^n \oplus \{0\}$ *of* $\mathfrak{H} = \mathfrak{H}_0 \oplus \mathfrak{H}_Q$ *is maximal neutral and invariant under the resolvent* $(H_0 - \lambda)^{-1}$. *It satisfies* $(H_0 - \lambda)^{-n}\mathcal{L} = \{0\}$, $\lambda \in \rho(H_0)$.

This result will be extended in Section 5 to a certain subclass of self-adjoint extensions of the model operator S in $\mathfrak{H}$ (i.e., for certain singular perturbations of A_0).

4. Singular perturbations as range perturbations

Let $H_\infty = \ker \Gamma_1$ be the self-adjoint extension of S corresponding to $\tau^{-1} = 0$ in Proposition 3.3. The self-adjoint extensions H_τ of S in Proposition 3.3 can be seen as "range perturbations" of H_∞ in the Pontryagin space $\mathfrak{H} = \mathfrak{H}_0 \oplus \mathfrak{H}_Q$, cf. [19], [20] for the Hilbert space case. For simplicity the results in this section are stated when $A_1 = \ker \Gamma_1^0$ is an operator in $\mathfrak{H}_0$, which is always the case when S_0 is densely defined in $\mathfrak{H}_0$. In this case H_∞ is also an operator by Proposition 3.4. Introduce $\Omega : \mathcal{H} \to \operatorname{mul} S^* \subset \mathfrak{H}_0 \oplus \mathfrak{H}_Q$ by

$$\Omega h = \begin{pmatrix} 0 \\ h \otimes e_n \\ 0 \end{pmatrix}, \quad h \in \mathcal{H}. \tag{4.1}$$

In the rest of this paper the following notations will be used

$$\mathbf{F} = \begin{pmatrix} f_0 \\ f \\ \widetilde{f} \end{pmatrix}, \; \mathbf{G} = \begin{pmatrix} g_0 \\ g \\ \widetilde{g} \end{pmatrix} \in \mathfrak{H}_0 \oplus \mathfrak{H}_Q.$$

Proposition 4.1. *Let the assumptions be as in Theorem* 3.2 *and assume that* $A_1 = \ker \Gamma_1^0$ *is an operator. Then* S *is a domain restriction of* H_∞ *given by*

$$\operatorname{dom} S = \{\, F \in \operatorname{dom} H_\infty : \Omega^* F = 0 \,\}, \tag{4.2}$$

and the self-adjoint extensions H_τ *and* H_∞ *of* S *in Proposition* 3.3 *are connected by*

$$H_\tau = H_\infty - \Omega \tau^{-1} \Omega^*, \tag{4.3}$$

where the difference is understood in the sense of relations.

Proof. Since A_1 is assumed to be an operator, Proposition 3.4 shows that H_∞ is an operator. The adjoint $\Omega^* : \mathfrak{H}_0 \oplus \mathfrak{H}_Q \to \mathcal{H}$ of Ω in (4.1) is given by

$$\Omega^* \mathbf{F} = \widetilde{f}_1. \tag{4.4}$$

The equality (4.2) is now clear from the formulas for H_∞ in Proposition 3.3 and for S in Theorem 3.2.

Let $\mathbf{F} = (f_0, f, \widetilde{f})^\top, \mathbf{G} = (g_0, g, \widetilde{g})^\top \in \mathfrak{H}$. By definition, $\{\mathbf{F}, \mathbf{G}\} \in \Omega \tau^{-1} \Omega^*$ if and only if $\{\Omega^* \mathbf{F}, \widetilde{\varphi}\} = \{\widetilde{f}_1, \widetilde{\varphi}\} \in \tau^{-1}$ and $\mathbf{G} = \Omega \widetilde{\varphi}$ for some $\widetilde{\varphi} \in \mathcal{H}$. Consequently, $\{\mathbf{F}, \mathbf{G}\} \in H_\infty - \Omega \tau^{-1} \Omega^*$ if and only if

$$\{\mathbf{F}, \mathbf{G}\} = \{\mathbf{F}, H_\infty \mathbf{F} + \Omega \widetilde{\varphi}\}, \quad \mathbf{F} \in \operatorname{dom} H_\infty, \quad \{\widetilde{\varphi}, \widetilde{f}_1\} \in -\tau. \tag{4.5}$$

Now using (4.1) and comparing (4.5) with the expression for H_τ in Proposition 3.3 the equality (4.3) follows. □

The above result depends on the fact that the model operator S in Theorem 3.2 is maximally nondensely defined in $\mathfrak{H} = \mathfrak{H}_0 \oplus \mathfrak{H}_Q$. In the case of defect numbers $(1,1)$ the extension H_0 is the only self-adjoint extension of S which is not an operator and the other extensions H_τ, $\tau \neq 0$, are given by (4.3), cf. [10]. In the special case of defect numbers $(1,1)$ a result similar to Proposition 4.1 has been obtained in [20, Theorem 3.2] for the model concerning perturbations in $\mathfrak{H}_{-2}$.

The perturbation formula (4.3) in Proposition 4.1 gives an explicit expression for the self-adjoint extensions H_τ of S. Moreover, the resolvent formula (3.11) can be obtained by a straightforward calculation from (4.3), cf., e.g., [17]. It is also clear from (4.3) that H_τ is an operator if and only if the inverse τ^{-1} is an operator in $\mathcal{H}$, in which case H_τ, $\ker\tau = \{0\}$, is an ordinary range perturbation of the operator H_∞ in the Pontryagin space $\mathfrak{H}$. An opposite extreme case is $\tau = 0$. Then the condition $\{\widetilde{\varphi}, \widetilde{f}_1\} \in -\tau$ in (4.5) is equivalent to $\mathbf{F} \in \ker\Omega^*$ which together with $\mathbf{F} \in \operatorname{dom} H_\infty$ implies that $\mathbf{F} \in \operatorname{dom} S$, while $\operatorname{mul}\Omega\tau^{-1}\Omega^* = \operatorname{ran}\Omega$. Hence, the perturbation (4.3) for $\tau = 0$ coincides with the form of H_0 given in (3.15), i.e., with the generalized Friedrichs extension of S in $\mathfrak{H}$. A more specific classification associated with the perturbation formula (4.3) is obtained by decomposing the self-adjoint parameter τ into its operator and multi-valued parts,

$$\tau = \tau_s \oplus \tau_\infty, \quad \tau_s = \{\,\{h,k\} \in \tau : k \perp \operatorname{mul}\tau\,\}, \quad \tau_\infty = \{0\} \oplus \operatorname{mul}\tau. \tag{4.6}$$

Here τ_s is a self-adjoint operator in $\mathcal{H}_s = \operatorname{dom}\tau$, τ_∞ is a self-adjoint relation in $\mathcal{H}_\infty = \operatorname{mul}\tau$, and $\mathcal{H} = \mathcal{H}_s \oplus \mathcal{H}_\infty$.

Proposition 4.2. *Let the assumptions be as in Theorem 3.2 and assume that* $A_1 = \ker\Gamma_1^0$ *is an operator. Let* P *be an orthogonal projection in* $\mathcal{H}$ *and define* $\Omega_P = \Omega{\restriction}\operatorname{ran} P$, *where* Ω *is given by (4.1). Then:*

(i) *The domain restriction* S_P *of* H_∞, *given by*

$$\operatorname{dom} S_P = \{\,\mathbf{F} \in \operatorname{dom} H_\infty : \Omega_P^*\mathbf{F} = 0\,\},$$

is a closed symmetric operator in $\mathfrak{H}$ *whose defect numbers are given by* (s,s), $s = \dim\operatorname{ran} P$.

(ii) *The adjoint of* S_P *is given by*

$$S_P^* = \{\,\{\mathbf{F},\mathbf{G}\} \in S^* : (I-P)\widetilde{\varphi} = 0\,\}.$$

(iii) *The self-adjoint extensions* H_τ *of* S *with the property* $H_\tau \cap H_\infty = S_P$ *are in one-to-one correspondence with the parameters* τ *for which* $\operatorname{mul}\tau = \ker P$ *and they are given by*

$$H_\tau = H_\infty - \Omega_P\tau_s^{-1}\Omega_P^*, \tag{4.7}$$

where $\tau = \tau_s \oplus \tau_\infty$ *is decomposed as in (4.6) and the difference is understood in the sense of relations.*

(iv) *The generalized Friedrichs extension of* S_P *corresponds to* $\tau_s = 0$ *in (4.7) and is given by*

$$S_P \hat{+} (\{0\} \oplus \operatorname{mul} S_P^*) = \{\,\{\mathbf{F},\mathbf{G}\} \in S^* : \Omega_P^*\mathbf{F}\,(= P\widetilde{f}_1) = 0,\ (I-P)\widetilde{\varphi} = 0\,\}.$$

Proof. Let $\mathbf{F} = (f_0, f, \widetilde{f})^\top$, $\mathbf{G} = (g_0, g, \widetilde{g})^\top \in \mathfrak{H}$. In view of (4.4), $\Omega_P^*\mathbf{F} = P\Omega^*\mathbf{F} = P\widetilde{f}_1$. Hence, $S_P = \{\,\{\mathbf{F},\mathbf{G}\} \in S^* : \Omega_P^*\mathbf{F} = P\widetilde{f}_1 = 0, \widetilde{\varphi} = 0\,\}$ from which the statements (i) and (ii) easily follow.

(iii) Clearly, $\{\mathbf{F},\mathbf{G}\} \in H_\infty \cap H_\tau$ if and only if $\{\mathbf{F},\mathbf{G}\} \in S^*$, and the conditions $\widetilde{\varphi}=0$ and $\{\widetilde{\varphi},\widetilde{f_1}\} \in -\tau$ are satisfied. Equivalently, $\mathbf{F} \in \operatorname{dom} H_\infty$ and $\widetilde{f_1} \in \operatorname{mul}\tau$. Comparing this with the condition $\Omega_P^*\mathbf{F} = P\widetilde{f_1} = 0$ for S_P in (i), one concludes that $\operatorname{mul}\tau = \ker P$. It is easy to check that for such τ, the equality $\Omega\tau^{-1}\Omega^* = \Omega_P\tau_s^{-1}\Omega_P^*$ holds (cf. the proof of Proposition 4.1). Hence, (4.7) follows from (4.3).

(iv) The discussion concerning H_0 above shows that the generalized Friedrichs extensions of S_P corresponds to $\tau_s = 0$ in (4.7), in which case $\{\widetilde{\varphi},\widetilde{f_1}\} \in -\tau$ is equivalent to $(I-P)\widetilde{\varphi}=0$ and $P\widetilde{f_1}=0$. □

Proposition 4.2 shows that S_P is maximally nondensely defined: $\dim \operatorname{mul} S_P^* = s$. Clearly, the perturbation formula (4.7) is an analog of (4.3). The characterization of operator extensions in (4.7) agrees with the one in (4.3), since $\ker\tau_s = \ker\tau$.

5. The class of self-adjoint extensions with extremal boundary conditions

According to Proposition 3.5 the compressed resolvent of H_0 from $\mathfrak{H}$ to the Hilbert space $\mathfrak{H}_0 \subset \mathfrak{H}$ coincides with the resolvent of the (unperturbed) operator A_0 in $\mathfrak{H}_0$. In this section the corresponding property will be proved for a certain subclass of self-adjoint extensions H_τ of the model operator S. A compressed resolvent of S in $\mathfrak{H}_0$ is said to be canonical if it coincides with the resolvent of some self-adjoint extension $\widetilde{A}$ of S_0 in the Hilbert space $\mathfrak{H}_0$.

Proposition 4.2 shows that the generalized Friedrichs extension of the intermediate symmetric extension $S_P \subset H_\infty$ is determined by the (abstract) boundary conditions

$$P\Gamma_0\widehat{\mathbf{F}} = (I-P)\Gamma_1\widehat{\mathbf{F}} = 0, \quad \widehat{\mathbf{F}} \in S^*, \quad P = P^* = P^2, \tag{5.1}$$

where $\{\mathcal{H},\Gamma_0,\Gamma_1\}$ is the boundary triplet associated with S^* in Theorem 3.2. In what follows boundary conditions of the form (5.1) are called *extremal boundary conditions* associated with S^*, since they have an interpretation as extreme points in the parameter space, see, e.g., [4]. When in the model of Theorem 3.2 the matrix polynomial $r=0$ and the matrix polynomial q is of the form $q = I_\mathcal{H} \otimes \widetilde{q}$, where $\widetilde{q}$ is a monic scalar polynomial, a different description for this class of extensions of S can be obtained by means of the compressed resolvents in $\mathfrak{H}_0$. In fact, for these extensions of S the following analog of Proposition 3.5 can be proved.

Theorem 5.1. *Let the assumptions be as in Theorem* 3.2. *Let* $r=0$ *in* (3.2), *let* $q = I_\mathcal{H}\otimes\widetilde{q}$, *where* $\widetilde{q}$ *is a monic scalar polynomial, and let* $\mathcal{C}$ *be given by* (3.4). *Then the compressed resolvent of* H_τ *to* $\mathfrak{H}_0$ *is canonical if and only if* H_τ *is given by the extremal boundary conditions of the form* (5.1). *In this case* $\tau = \{\{Ph,(I-P)h\} : h \in \mathcal{H}\}$ *and for the corresponding* H_τ *the following assertions hold:*

(i) $\rho(H_\tau) = \rho(A_\tau)\cap\rho(\mathcal{C})\,(=\rho(A_\tau)\backslash\sigma(q^\sharp q))$, *where* $A_\tau = \ker(\Gamma_0^0+\tau\Gamma_1^0)$ *and* $\tau \neq 0$.

(ii) $P_{\mathfrak{H}_0}(H_\tau-\lambda)^{-1}\!\restriction\!\mathfrak{H}_0 = (A_\tau-\lambda)^{-1}, \quad \lambda\in\rho(H_\tau)$.

(iii) *The subspace* $\mathcal{L}_\tau = \mathcal{L}_1 \dotplus \mathcal{L}_2$ *with*

$$\mathcal{L}_1 = \{0\} \oplus (\operatorname{ran} P)^n \oplus \{0\}, \quad \mathcal{L}_2 = \{0\} \oplus \{0\} \oplus (\ker P)^n$$

is maximal neutral and invariant under the resolvent $(H_\tau - \lambda)^{-1}$. *Moreover,*

$$(H_\tau - \lambda)^{-n}\mathcal{L}_1 = \{0\}, \quad (H_\tau - \lambda)^{-1}(0,0,\widetilde{g})^\top = (0,0,(\mathcal{C}_q - \lambda)^{-1}\widetilde{g})^\top, \tag{5.2}$$

where $(0,0,\widetilde{g})^\top \in \mathcal{L}_2$ *and* $\lambda \in \rho(H_\tau)$.

(iv) *The Weyl functions* $\widetilde{M}_\tau$ *of* (S, H_τ) *and* $\widetilde{M}_{0,\tau}$ *of* (S_0, A_τ) *are given by*

$$\widetilde{M}_\tau = \begin{pmatrix} \widetilde{q}_1^\sharp & 0 \\ 0 & \widetilde{q}_2^{-1} \end{pmatrix} \widetilde{M}_{0,\tau} \begin{pmatrix} \widetilde{q}_1 & 0 \\ 0 & \widetilde{q}_2^{-\sharp} \end{pmatrix}$$

and

$$\widetilde{M}_{0,\tau} = \begin{pmatrix} M_{11} - M_{12}M_{22}^{-1}M_{21} & -M_{12}M_{22}^{-1} \\ -M_{22}^{-1}M_{21} & -M_{22}^{-1} \end{pmatrix},$$

where $\widetilde{q}_1 = I_{\mathcal{H}_s} \otimes \widetilde{q}$, $\widetilde{q}_2 = I_{\mathcal{H}_\infty} \otimes \widetilde{q}$, *and the decomposition of the Weyl function* $M_0 = (M_{ij})_{i,j=1}^2$ *of* (S_0, A_0) *is according to* $\mathcal{H} = \mathcal{H}_s \oplus \mathcal{H}_\infty = \operatorname{ran} P \oplus \ker P$.

Proof. It follows from Proposition 3.3 that the compressed resolvent of H_τ is given by

$$P_{\mathfrak{H}_0}(H_\tau - \lambda)^{-1}\restriction \mathfrak{H}_0 = (A_0 - \lambda)^{-1} - \gamma_0(\lambda)(\widetilde{\tau}(\lambda)^{-1} + M_0(\lambda))^{-1}\gamma_0(\bar{\lambda})^*, \tag{5.3}$$

where $\widetilde{\tau}(\lambda) = q(\lambda)\tau q^\sharp(\lambda)$ due to assumption $r = 0$. The formula (5.3) coincides with a canonical resolvent of S_0 if and only if the function $\widetilde{\tau}$ does not depend on λ. Clearly, this condition is satisfied if and only if $\tau_s = 0$ in (4.6), i.e., H_τ is given by the extremal boundary conditions (5.1) for some $P = P^* = P^2$.

To prove (i)–(iv) introduce the following boundary mappings

$$\begin{cases} \widetilde{\Gamma}_0 = P\Gamma_0 - (I - P)\Gamma_1, \\ \widetilde{\Gamma}_1 = (I - P)\Gamma_0 + P\Gamma_1, \end{cases} \qquad \begin{cases} \widetilde{\Gamma}_0^0 = P\Gamma_0^0 - (I - P)\Gamma_1^0, \\ \widetilde{\Gamma}_1^0 = (I - P)\Gamma_0^0 + P\Gamma_1^0, \end{cases} \tag{5.4}$$

so that $H_\tau = \ker \widetilde{\Gamma}_0$ and $A_\tau = \ker \widetilde{\Gamma}_0^0$.

(i) Let $\mathbf{G} = (g_0, g, \widetilde{g})^\top \in \mathfrak{H}$, $\widehat{f}_0 = \{f_0, f_0'\} \in S_0^*$, and let $\lambda \in \rho(A_\tau) \cap \rho(\mathcal{C})$. Then by Proposition 3.3 the relation $\mathbf{G} \in \operatorname{ran}(H_\tau - \lambda)$ can be rewritten as a system of equalities

$$\begin{cases} f_0' - \lambda f_0 = g_0, \\ (\mathcal{C}_{q^\sharp} - \lambda)f + \widetilde{\varphi} \otimes e_n = g, \\ (\mathcal{C}_q - \lambda)\widetilde{f} + \Gamma_0^0\widehat{f}_0 \otimes e_n = \widetilde{g}, \quad f_1 = \Gamma_1^0\widehat{f}_0, \quad P\widetilde{f}_1 = 0, \quad (I - P)\widetilde{\varphi} = 0. \end{cases} \tag{5.5}$$

Now, one can solve $P\Gamma_0^0\widehat{f}_0$ from the third equality in (5.5) by means of the companion operator $\mathcal{C}_q$ (cf. [11]). Since $\lambda \in \rho(\mathcal{C}_{q^\sharp})$ the second equality in (5.5) gives $(I - P) \otimes f = (\mathcal{C}_{q^\sharp} - \lambda)^{-1}(I - P) \otimes g$. In particular, $(I - P)\Gamma_1^0\widehat{f}_0 = (I - P)f_1$ and consequently $\widetilde{\Gamma}_0^0\widehat{f}_0$ has been solved. Let $\widetilde{\gamma}_\tau$ be the γ-field for (S_0, A_τ) associated with the boundary triplet $\{\mathcal{H},\widetilde{\Gamma}_0^0,\widetilde{\Gamma}_1^0\}$ in (5.4). Then one can write f_0 and f_0' in the form

$$f_0=(A_\tau-\lambda)^{-1}g_0+\widetilde{\gamma}_\tau(\lambda)\widetilde{\Gamma}_0^0\widehat{f}_0,\quad f_0'=\lambda f_0+g_0.$$

Now $\widetilde{f}$ can be solved from the third equation in (5.5) and, since $f_1=\Gamma_1^0\widehat{f}_0$, the vectors $(f_2,\dots,f_n)$ and $\widetilde{\varphi}$ can be solved from the second equality in (5.5). This proves $\rho(A_\tau)\cap\rho(\mathcal{C})\subset\rho(H_\tau)$.

To prove the reverse inclusion it is first shown that $\sigma(\mathcal{C})\subset\sigma_p(H_\tau)$ holds for every $\tau\neq 0$. In view of

$$(\mathcal{C}_q-\lambda)\Lambda^\top h=(0,\dots,0,-q(\lambda)h)^\top,\quad \lambda\in\mathbb{C},\ h\in\mathcal{H},$$

the eigenspace of $\mathcal{C}_q$ at λ is given by

$$\ker(\mathcal{C}_q-\lambda)=\{\,\Lambda^\top h:\ h\in\ker q(\lambda)\,\}. \tag{5.6}$$

Assume that $\lambda\in\sigma(\mathcal{C}_q)$. Since $\tau\neq 0$, one has $P\neq I$ and hence in view of (5.6) and the assumption $q=I_{\mathcal{H}}\otimes\widetilde{q}$ one can find $\widetilde{f}\neq 0$ such that $P\widetilde{f}_1=0$ and $(\mathcal{C}_q-\lambda)\widetilde{f}=0$. It is easy to check that $(0,0,\widetilde{f})^\top\in\ker(H_\tau-\lambda)$. Hence, $\sigma(\mathcal{C}_q)\subset\sigma_p(H_\tau)$ and by the symmetry of spectra $\sigma(\mathcal{C}_{q^\sharp})\subset\sigma_p(H_\tau)$, so that $\sigma(\mathcal{C})\subset\sigma_p(H_\tau)$.

Now, let $\lambda\in\rho(H_\tau)$ and let $g=\widetilde{g}=0$. Then $\lambda\in\rho(\mathcal{C})$ and it follows from the second and the third equalities in (5.5) that

$$P\Gamma_0^0\widehat{f}_0=(I-P)\Gamma_1^0\widehat{f}_0=0.$$

Therefore, $\widehat{f}_0\in A_\tau$ and the first equality in (5.5) means that

$$\{f_0,g_0\}\in A_\tau-\lambda.$$

By assumption $\lambda\in\rho(H_\tau)$ and since $g_0\in\mathfrak{H}_0$ is arbitrary it follows that $\lambda\in\rho(A_\tau)$. Therefore, $\rho(H_\tau)\subset\rho(A_\tau)\cap\rho(\mathcal{C})$.

(ii) The statement follows from the identity (with $\lambda\in\rho(H_\tau)$)

$$(H_\tau-\lambda)^{-1}(g_0,0,0)^\top=\Big((A_\tau-\lambda)^{-1}g_0,\Lambda^\top\Gamma_1^0\widehat{f}_0,-(\mathcal{C}_q-\lambda)^{-1}(\Gamma_0^0\widehat{f}_0\otimes e_n)\Big)^\top.$$

(iii) Clearly, $\mathcal{L}$ is a neutral subspace of $\mathfrak{H}_0\oplus\mathfrak{H}_Q$ with dimension dn, so that it is maximal neutral, cf. [5]. Moreover,

$$(H_\tau-\lambda)^{-1}(0,Pg,(I-P)\widetilde{g})^\top=(0,X_nPg,(\mathcal{C}_q-\lambda)^{-1}(I-P)\widetilde{g}),$$

where X_n stands for

$$X_n=\begin{pmatrix}0&0&\dots&0\\ I&0&\ddots&\vdots\\ \vdots&\ddots&\ddots&0\\ \lambda^{n-2}&\dots&I&0\end{pmatrix}.$$

This implies (5.2).

(iv) The transform of the boundary mappings in (5.4) corresponds to the following transform of the Weyl function $M = q^\sharp M_0 q$ (cf. [13]):

$$\begin{aligned}
\widetilde{M}_\tau &= [(I-P)+PM][P-(I-P)M]^{-1} \\
&= \begin{pmatrix} \widetilde{q}^\sharp M_{11}\widetilde{q} & \widetilde{q}^\sharp M_{12}\widetilde{q} \\ 0 & I \end{pmatrix} \begin{pmatrix} I & 0 \\ -\widetilde{q}^{-1}M_{22}^{-1}M_{21}\widetilde{q} & -\widetilde{q}^{-1}M_{22}^{-1}\widetilde{q}^{-\sharp} \end{pmatrix} \\
&= \begin{pmatrix} \widetilde{q}^\sharp (M_{11}-M_{12}M_{22}^{-1}M_{21})\widetilde{q} & -\widetilde{q}^\sharp M_{12}M_{22}^{-1}\widetilde{q}^{-\sharp} \\ -\widetilde{q}^{-1}M_{22}^{-1}M_{21}\widetilde{q} & -\widetilde{q}^{-1}M_{22}^{-1}\widetilde{q}^{-\sharp} \end{pmatrix},
\end{aligned}$$

from which the statement follows. □

Recall that $\rho(A_\tau) \subset \rho(\widetilde{M}_{0,\tau})$ and $\rho(H_\tau) \subset \rho(\widetilde{M}_\tau)$, and that the inclusions are equalities if S_0 and S are simple. These properties are also reflected in (i) and (iv) of Theorem 5.1.

The proof of Theorem 5.1 gives also the following result, which shows the difference between the cases $\tau = 0$ and $\tau \neq 0$, cf. Proposition 3.5.

Corollary 5.2. *Let the assumptions be as in Theorem* 5.1 *and let* τ *be given by* $\tau = \{\{Ph, (I-P)h\} : h \in \mathcal{H}\}$ *for some orthogonal projection* P *in* $\mathcal{H}$. *If* $\tau \neq 0$ *(i.e.,* $P \neq I$*) then* $\sigma(q^\sharp q) \subset \sigma_p(H_\tau)$ *and* $\sigma(H_\tau) = \sigma(A_\tau) \cup \sigma(q^\sharp q)$.

Acknowledgment

This work has been supported by the Research Institute for Technology at the University of Vaasa, the Academy of Finland (project 79774), and the Dutch Association for Mathematical Physics (MF00/34).

References

[1] S. Albeverio, F. Gesztesy, R. Høegh-Krohn, and H. Holden, *Solvable models in quantum mechanics*, Springer-Verlag, New York, 1988.

[2] S. Albeverio and P. Kurasov, "Finite rank perturbations and distribution theory", Proc. Amer. Math. Soc., 127 (1999), 1151–1161.

[3] S. Albeverio and P. Kurasov, *Singular perturbations of differential operators*, London Math. Soc., Lecture Notes Series 271, Cambridge University Press, 1999.

[4] Yu.M. Arlinskiĭ, S. Hassi, Z. Sebestyén, and H.S.V. de Snoo, "On the class of extremal extensions of a nonnegative operator", Béla Szőkefalvi-Nagy memorial volume, Oper. Theory Adv. Appl., 127 (2001), 41–81.

[5] T.Ya. Azizov and I.S. Iokhvidov, *Foundations of the theory of linear operators in spaces with an indefinite metric*, Nauka, Moscow, 1986 (Russian) (English translation: Wiley, New York, 1989).

[6] Yu.M. Berezanskiĭ, *Expansions in eigenfunctions of self-adjoint operators*, Naukova Dumka, Kiev, 1965 (Russian) (English translation: Translations of Mathematical Monographs, Volume 17, American Mathematical Society, 1968).

[7] V.A. Derkach, "On generalized resolvents of Hermitian relations", J. Math. Sciences, 97 (1999), 4420–4460.

[8] V.A. Derkach and S. Hassi, "A reproducing kernel space model for $\mathbf{N}_\kappa$-functions", Proc. Amer. Math. Soc., 131 (2003), 3795–3806.

[9] V.A. Derkach, S. Hassi, and H.S.V. de Snoo, "Operator models associated with singular perturbations", Methods of Functional Analysis and Topology, 7 (2001), 1–21.

[10] V.A. Derkach, S. Hassi, and H.S.V. de Snoo, "Rank one perturbations in a Pontryagin space with one negative square", J. Funct. Anal., 188 (2002), 317–349.

[11] V.A. Derkach, S. Hassi, and H.S.V. de Snoo, "Singular perturbations of self-adjoint operators", Mathematical Physics, Analysis and Geometry, 6 (2003), 349–384.

[12] V.A. Derkach and M.M. Malamud, "Generalized resolvents and the boundary value problems for hermitian operators with gaps", J. Funct. Anal., 95 (1991), 1–95.

[13] V.A. Derkach and M.M. Malamud, "The extension theory of hermitian operators and the moment problem", J. Math. Sciences, 73 (1995), 141–242.

[14] J.F. van Diejen and A. Tip, "Scattering from generalized point interaction using self-adjoint extensions in Pontryagin spaces", J. Math. Phys., 32 (3), (1991), 631–641.

[15] A. Dijksma, H. Langer, Yu.G. Shondin, and C. Zeinstra, "Self-adjoint operators with inner singularities and Pontryagin spaces", Oper. Theory Adv. Appl., 118 (2000), 105–175.

[16] V.I. Gorbachuk and M.L. Gorbachuk, *Boundary value problems for operator differential equations*, Naukova Dumka, Kiev, 1984 (Russian) (English translation: Kluwer Academic Publishers, Dordrecht, Boston, London, 1990).

[17] S. Hassi, M. Kaltenbäck, and H.S.V. de Snoo, "Triplets of Hilbert spaces and Friedrichs extensions associated with the subclass $\mathbf{N}_1$ of Nevanlinna functions", J. Operator Theory, 37 (1997), 155–181.

[18] S. Hassi, H. Langer, and H.S.V. de Snoo, "Self-adjoint extensions for a class of symmetric operators with defect numbers $(1,1)$", 15th OT Conference Proceedings, (1995), 115–145.

[19] S. Hassi and H.S.V. de Snoo, "One-dimensional graph perturbations of self-adjoint relations", Ann. Acad. Sci. Fenn. A.I. Math., 22 (1997), 123–164.

[20] S. Hassi and H.S.V. de Snoo, "Nevanlinna functions, perturbation formulas and triplets of Hilbert spaces", Math. Nachr., 195 (1998), 115–138.

[21] M.G. Kreĭn and H. Langer, "Über die Q-Funktion eines π-hermiteschen Operators im Raume Π_κ", Acta Sci. Math. (Szeged), 34 (1973), 191–230.

[22] P. Kurasov, "$\mathfrak{H}_{-n}$-perturbations of self-adjoint operators and Kreĭn's resolvent formula", Integr. Equ. Oper. Theory, 45 (2003), 437–460.

[23] Yu.G. Shondin, "Quantum-mechanical models in $\mathbb{R}_n$ associated with extensions of the energy operator in Pontryagin space", Teor. Mat. Fiz., 74 (1988), 331–344) (Russian) (English translation: Theor. Math. Phys., 74 (1988), 220–230).

[24] A.V. Štraus, "Extensions and generalized resolvents of a symmetric operator which is not densely defined", Izv. Akad. Nauk SSSR, Ser. Mat., 34 (1970), 175–202 (Russian) (English translation: Math. USSR-Izvestija, 4 (1970), 179–208).

[25] E.R. Tsekanovskiĭ and Yu. L. Shmulyan, "The theory of bi-extensions of operators on rigged Hilbert spaces. Unbounded operator colligations and characteristic functions", Uspekhi Mat. Nauk, 32:5 (1977), 69–124 (Russian) (English translation: Russian Math. Surveys, 32:5 (1977), 73–131).

Vladimir Derkach
Department of Mathematics
Donetsk National University
Universitetskaya str. 24
83055 Donetsk
Ukraine
e-mail: derkach@univ.donetsk.ua

Seppo Hassi
Department of Mathematics and Statistics
University of Vaasa
P.O. Box 700, 65101 Vaasa
Finland
e-mail: sha@uwasa.fi

Henk de Snoo
Department of Mathematics and Computing Science
University of Groningen
P.O. Box 800, 9700 AV Groningen
Nederland
e-mail: desnoo@math.rug.nl

Operator Theory:
Advances and Applications, Vol. 154, 69–90

Minimal Realizations of Scalar Generalized Nevanlinna Functions Related to Their Basic Factorization

Aad Dijksma, Heinz Langer, Annemarie Luger and Yuri Shondin

Abstract. In this paper we present a minimal realization of a scalar generalized Nevanlinna function q which corresponds to the basic factorization of q as a product of a Nevanlinna function q_0 and of a rational function $r^{\#}r$, which collects the generalized poles and generalized zeros of q that are not of positive type. The key tool are reproducing kernel Pontryagin spaces.

Mathematics Subject Classification (2000). Primary 47A06, 47A50, 47B32, 47B50 .

Keywords. Generalized Nevanlinna function, generalized pole, generalized zero, reproducing kernel Pontryagin space, symmetric operator, self-adjoint extension.

1. Introduction

In [DLLSh] (see also [DeHS1]) it was shown that a generalized Nevanlinna function $q \in \mathcal{N}_\kappa$ admits a unique *basic factorization*

$$q(z) = r^{\#}(z)q_0(z)r(z) \tag{1.1}$$

with a Nevanlinna function $q_0 \in \mathcal{N}_0$ and a rational function r, whose zeros (poles, respectively) are the generalized zeros (poles, respectively) of q in $\mathbb{C}^+ \cup \mathbb{R}$ ($\mathbb{C}^- \cup \mathbb{R}$, respectively) which are not of positive type. Here and in the following, for a vector function f, by $f^{\#}$ we denote the function $f^{\#}(z) := f(z^*)^*$; for the definition of r and of generalized poles and zeros we refer to Section 3.

The research for this paper was supported by the Netherlands Organization for Scientific Research NWO (grant 047-008-008), the Russian Foundation for Basic Research RFBR (grant N 0001-00544), the Research Training Network HPRN-CT-2000-00116 of the European Union, and the "Fond zur Förderung der wissenschaftlichen Forschung" (FWF, Austria) grant number P15540-N05.

It is well known that every generalized Nevanlinna function $q \in \mathcal{N}_\kappa$ admits a minimal realization in some Pontryagin space of negative index κ. In the present paper we construct a minimal realization of $q \in \mathcal{N}_\kappa$, $\kappa > 0$, which corresponds to the basic factorization (1.1) of q. This means, we construct a realization of q in terms of realizations of the function q_0 and of the matrix generalized Nevanlinna function

$$\mathcal{M}_r(z) = \begin{pmatrix} 0 & r^{\#}(z) \\ r(z) & 0 \end{pmatrix}.$$

Recall that a *realization* (A, Γ_z) for an $n \times n$ matrix generalized Nevanlinna function Q is given by a self-adjoint relation A in some Pontryagin space $\mathcal{P}$ and a corresponding Γ-*field*, that is, a family of mappings $\Gamma_z \in \mathcal{L}(\mathbb{C}^n, \mathcal{P})$, $z \in \rho(A)$, which satisfy

$$\Gamma_w = (I + (w - z)(A - w)^{-1})\Gamma_z, \qquad z, w \in \rho(A), \tag{1.2}$$

and

$$\frac{Q(z) - Q(w)^*}{z - w^*} = \Gamma_w^* \Gamma_z, \quad z, w \in \rho(A), \ z \neq w^*. \tag{1.3}$$

If a point $z_0 \in \rho(A)$ is fixed this implies the following representation of Q:

$$Q(z) = Q(z_0)^* + (z - z_0^*)\Gamma_{z_0}^*(I + (z - z_0)(A - z)^{-1})\Gamma_{z_0}, \qquad z \in \mathcal{D}(Q).$$

Here $\mathcal{D}(Q)$ denotes the domain of holomorphy of the function Q. Note that the function Q is determined by A and the Γ-field Γ_z up to an additive constant self-adjoint $n \times n$ matrix. The space $\mathcal{P}$ is called the *state space* of the realization (A, Γ_z). The realization (A, Γ_z) can always be chosen *minimal* which means that

$$\mathcal{P} = \overline{\operatorname{span}}\, \{\Gamma_z \mathbf{c} \mid z \in \varrho(A), \mathbf{c} \in \mathbb{C}^n\}.$$

In this case two minimal realizations of Q are unitarily equivalent and, moreover, $\mathcal{D}(Q) = \rho(A)$; see [DLS2, Theorem 1.1]. In a minimal realization of Q the negative index of the state space $\mathcal{P}$ is equal to the number of negative squares of the kernel

$$K_Q(z, w) := \frac{Q(z) - Q(w)^*}{z - w^*}, \qquad z, w \in \mathcal{D}(Q),$$

where the expression on the right-hand side for $w = z^*$ is to be understood as $Q'(z)$.

For $\kappa = 0$ and $n = 1$ (and under some additional growth condition on Q) the self-adjoint operator of a realization of Q is given by the operator of multiplication in the space L^2_σ, where σ is the measure in an integral representation of Q, see, for example, [LLu].

With a realization (A, Γ_z) often a symmetric restriction S of the relation A is defined by

$$S := \{\{f, g\} \in A \,|\, \Gamma_{z_0}^*(g - z_0^* f) = 0\}.$$

This definition is independent of $z_0 \in \mathcal{D}(Q)$ and Γ_z maps $\mathbb{C}^n$ onto the defect subspace $(\operatorname{ran}(S - z^*))^\perp$ of S. The triplet (A, Γ_z, S) is sometimes called a *model* for the realization of Q or, for short, a model for the function Q.

In Section 2, Theorem 2.1, we construct the canonical model for a realization of a generalized Nevanlinna function $Q \in \mathcal{N}_\kappa^{n\times n}$. Here by *canonical* we mean that the model acts in the reproducing kernel Pontryagin space $\mathcal{L}(Q)$. To emphasize the dependence of the canonical model on the function Q we sometimes write A_Q, S_Q etc. instead of A, S, etc. As a byproduct of this construction, in $\mathcal{L}(Q)$ a simple characterization of the Jordan chains of the operator A_Q is given. Finally, in Theorem 2.4 we describe all self-adjoint extensions in $\mathcal{L}(Q)$ of the symmetric operator S_Q.

In Section 3 with the factorization (1.1) the matrix function

$$\mathcal{Q}(z) = \begin{pmatrix} q_0(z) & 0 & 0 \\ 0 & 0 & r^{\#}(z) \\ 0 & r(z) & 0 \end{pmatrix},$$

is defined and it is shown that a realization of q can be derived from a realization of $\mathcal{Q}$. The latter is evidently the orthogonal sum of realizations of $q_0 \in \mathcal{N}_0$ and of $\mathcal{M}_r(z) = \begin{pmatrix} 0 & r^{\#}(z) \\ r(z) & 0 \end{pmatrix}$. Then the self-adjoint relation A of a minimal realization of q can, for example, be obtained as the compression of a certain self-adjoint extension of $S_{\mathcal{Q}}$ to the orthogonal complement of a finite-dimensional Hilbert subspace $\mathcal{H}_0$ of $\mathcal{L}(\mathcal{Q}) = \mathcal{L}(q_0) \oplus \mathcal{L}(\mathcal{M}_r)$. Moreover, in Theorem 3.4 the finite-dimensional space $\mathcal{L}(\mathcal{M}_r)$ is described explicitly.

In Section 4 the Hilbert subspace $\mathcal{H}_0$ of $\mathcal{L}(q_0) \oplus \mathcal{L}(\mathcal{M}_r)$ is characterized in terms of eigenfunctions of A_{q_0} corresponding to those generalized zeros of q which belong to the point spectrum $\sigma_p(A_{q_0})$ of A_{q_0} and of eigenfunctions of A_{-1/q_0} corresponding to those generalized poles of q which belong to $\sigma_p(A_{-1/q_0})$. In particular, the dimension of the Hilbert space $\mathcal{H}_0$ is determined.

Recently, V. Derkach and S. Hassi in [DeH] constructed a model for q by a certain coupling of the self-adjoint relations A_{q_0} and $A_{\mathcal{M}_r}$ in terms of boundary triples. Earlier, jointly with H. de Snoo in [DeHS2, DeHS3] they also used the coupling method to construct models for matrix-valued generalized Nevanlinna functions q with a matrix polynomial r. In all three papers necessary and sufficient conditions for minimality are formulated. As is well known minimal models are unitarily equivalent, which implies that the models are essentially the same. In the present paper we obtain a minimal model in a more direct way. Minimality is described analytically in Theorem 4.4. The fact that we use canonical models for the factors q_0 and r makes the model in this paper rather explicit and in fact unique from the point of view of reproducing kernel Pontryagin spaces.

We assume that the reader is familiar with operator theory in spaces with an indefinite metric such as in [AzI] and [IKL] and with the theory of reproducing kernel spaces such as in [ADy] and [ADRS]. For linear relations in Kreĭn spaces we refer to, for example, [DS] and for canonical models to [dB], [ADRS], [DLS2], and [DeHS1].

2. The canonical model for a matrix generalized Nevanlinna function

1. The *generalized Nevanlinna class* $\mathcal{N}_\kappa^{n\times n}$ ($\mathcal{N}_\kappa := \mathcal{N}_\kappa^{1\times 1}$) consists of all $n \times n$ matrix functions Q which are meromorphic in $\mathbb{C} \setminus \mathbb{R}$, satisfy $Q(z^*)^* = Q(z)$ for $z \in \mathcal{D}(Q)$, and for which the kernel

$$K_Q(z,w) = \frac{Q(z) - Q(w)^*}{z - w^*}, \quad z, w \in \mathcal{D}(Q), \ z \neq w^*,$$

has κ negative squares. We denote by $\mathcal{L}(Q)$ the corresponding reproducing kernel Pontryagin space: This is the completion of the linear span of all functions of the form $K_Q(\,\cdot\,, w)\mathbf{c}$, $\mathbf{c} \in \mathbb{C}^n$, $w \in \mathcal{D}(Q)$, with respect to the norm generated by the inner product

$$\langle K_Q(\,\cdot\,,w)\mathbf{c}, K_Q(\,\cdot\,,z)\mathbf{d}\rangle_{\mathcal{L}(Q)} := \mathbf{d}^* K_Q(z,w)\mathbf{c},$$

which has κ negative squares. The elements of $\mathcal{L}(Q)$ are holomorphic n-vector functions f on $\mathcal{D}(Q)$. Sometimes the function f is also denoted by $f(\zeta)$. The reproducing property of the kernel $K_Q(z,w)$ is expressed by the inner product formula

$$\langle f(\cdot), K_Q(\,\cdot\,,w)\mathbf{c}\rangle_{\mathcal{L}(Q)} = (f(w), \mathbf{c})_{\mathbb{C}^n} = \mathbf{c}^* f(w).$$

The following theorem describes a realization of the function Q in the space $\mathcal{L}(Q)$. The origins of this theorem go back to M.G. Kreĭn for the case $\kappa = 0$ and to [KL1, KL2, KL3] for the general case, see also [DLS1]. In these papers the so-called ε-method or formal sum method was used to obtain a realization of Q. Alpay showed in [A] that the space $\mathcal{L}(Q)$ is invariant under the difference-quotient operator

$$R_w(f)(z) = \frac{f(z) - f(w)}{z - w} \tag{2.1}$$

and that this operator satisfies the resolvent identity, see also [dB]. In turn this result was used in [DLS2, Section 3], see also [ABDS], to obtain the realization below. Here we make use of reproducing kernel space methods only. To keep this paper reasonably self-contained and for the convenience of the reader, we prove most of the statements.

The definition of a matrix generalized Nevanlinna function can easily be extended to functions whose values are bounded linear operators in a Hilbert space. As can be seen from its proof, the following theorem (except for the formula for the defect indices in statement) also holds in that case. In the sequel we use the theorem for the cases $n = 1, 2$, and 3 only.

Theorem 2.1. *Let $Q \in \mathcal{N}_\kappa^{n\times n}$ be given. Then:*

(i) $A := \{\{f,g\} \in \mathcal{L}(Q)^2 \,|\, \exists\, \mathbf{c} \in \mathbb{C}^n : \ g(\zeta) - \zeta f(\zeta) \equiv \mathbf{c}\}$ *is a self-adjoint linear relation in $\mathcal{L}(Q)$ with $\rho(A) \neq \emptyset$, and*

$$(\Gamma_z \mathbf{c})(\zeta) = K_Q(\zeta, z^*)\mathbf{c} = \frac{Q(\zeta) - Q(z)}{\zeta - z}\mathbf{c}, \quad \mathbf{c} \in \mathbb{C}^n,$$

is a corresponding Γ-field. The pair (A, Γ_z) is a minimal realization of Q.

(ii) *The resolvent of A is the difference-quotient operator in $\mathcal{L}(Q)$:*

$$(A-w)^{-1}=R_w, \quad w\in\rho(A).$$

(iii) $S:=\{\{f,g\}\in\mathcal{L}(Q)^2\,|\,g(\zeta)-\zeta f(\zeta)\equiv 0\}$ *is a symmetric operator in* $\mathcal{L}(Q)$ *with equal defect indices* $n-d$, *where* $d=\dim\ker\Gamma_z$. *Moreover,* $\sigma_p(S)=\emptyset$ *and the adjoint of S is given by*

$$\begin{aligned}S^* &= \overline{\operatorname{span}}\,\{\{\Gamma_z\mathbf{h}, z\Gamma_z\mathbf{h}\}\mid \mathbf{h}\in\mathbb{C}^n, z\in\mathcal{D}(Q)\}\\ &=\{\{f,g\}\in\mathcal{L}(Q)^2\,|\,\exists\,\mathbf{c},\mathbf{d}\in\mathbb{C}^n:\ g(\zeta)-\zeta f(\zeta)\equiv\mathbf{c}-Q(\zeta)\mathbf{d}\}.\end{aligned}$$

Recall that a nonself-adjoint symmetric operator is called *simple* if it has no non-real eigenvalues and the defect elements to the nonreal spectral points form a total subset of the whole space. Hence the symmetric operator S in the theorem is simple. If $n=1$ the function

$$\varphi(\,\cdot\,,z):=\Gamma_z 1=(I+(z-w)(A-z)^{-1})\varphi(\,\cdot\,,w),$$

called a *defect function* for S and A, spans the defect subspace of S at z and the representation of Q takes the form

$$Q(z)=Q(w)^*+(z-w^*)\langle\varphi(\,\cdot\,,z),\varphi(\,\cdot\,,w)\rangle_{\mathcal{L}(Q)}.$$

Proof of Theorem 2.1. Consider the linear relations A_0 and U_0 in $\mathcal{L}(Q)$ defined by

$$A_0:=\Big\{\{\sum_w K_Q(\,\cdot\,,w)\mathbf{c}_w,\sum_w w^*K_Q(\,\cdot\,,w)\mathbf{c}_w\}\,\Big|\,\mathbf{c}_w\in\mathbb{C}^n,\sum_w\mathbf{c}_w=0\Big\}$$

and

$$U_0:=\Big\{\{\sum_w(w^*-\mu)K_Q(\,\cdot\,,w)\mathbf{c}_w,\sum_w(w^*-\mu^*)K_Q(\,\cdot\,,w)\mathbf{c}_w\}\Big|\mathbf{c}_w\in\mathbb{C}^n,\sum_w\mathbf{c}_w=0\Big\}.$$

Here $\sum_w$ stands for sums over finite sets of points $w\in\mathcal{D}(Q)$ and μ is a fixed non-real point in $\mathcal{D}(Q)$. Then $A_0\subset A_0^*$ and U_0 is isometric. The equalities

$$\begin{aligned}\operatorname{dom}U_0 &= \operatorname{span}\{K_Q(\,\cdot\,,w)\mathbf{c}\,|\,w\neq\mu^*,\ \mathbf{c}\in\mathbb{C}^n\},\\ \operatorname{ran}U_0 &= \operatorname{span}\{K_Q(\,\cdot\,,w)\mathbf{c}\,|\,w\neq\mu,\ \mathbf{c}\in\mathbb{C}^n\}\end{aligned}$$

imply that the domain and the range of U_0 are dense in $\mathcal{L}(Q)$. Hence (see [ADRS, Theorem 1.4.2]) the closure U of U_0 is a (bounded) unitary operator on $\mathcal{L}(Q)$ and the closure A of A_0 is a self-adjoint relation in $\mathcal{L}(Q)$. Moreover, we have

$$A=\{\{(U-1)u,(\mu U-\mu^*)u\}\,|\,u\in\mathcal{L}(Q)\}.$$

From the relation

$$(A-\mu^*)^{-1}=\frac{1}{\mu^*-\mu}(U-1)$$

it follows that $\mu^*\in\rho(A)$, that is, $\mathcal{D}(Q)\cap(\mathbb{C}\setminus\mathbb{R})\subset\rho(A)$. Note that for arbitrary $\mathbf{c}\in\mathbb{C}^n$ and $\lambda,w\in\mathcal{D}(Q)$

$$\{K_Q(\,\cdot\,,\lambda)\mathbf{c}-K_Q(\,\cdot\,,w)\mathbf{c},\ \lambda^*K_Q(\,\cdot\,,\lambda)\mathbf{c}-w^*K_Q(\,\cdot\,,w)\mathbf{c}\}\in A_0\subset A$$

and hence

$$(A - w^*)^{-1} K_Q(\zeta, \lambda)\mathbf{c} = \frac{K_Q(\zeta, \lambda) - K_Q(\zeta, w)}{\lambda^* - w^*}\,\mathbf{c}.$$

Taking the $\mathcal{L}(Q)$ inner product of $f \in \mathcal{L}(Q)$ with the elements on both sides of this equality we find

$$(A - w)^{-1} f(\lambda) = \frac{f(\lambda) - f(w)}{\lambda - w} = R_w f(\lambda), \tag{2.2}$$

or, in words, the difference-quotient operator is a bounded operator on $\mathcal{L}(Q)$ and coincides with the resolvent operator of A. This implies that

$$\begin{aligned} A &= \{\{R_w h, h + w R_w h\} \mid h \in \mathcal{L}(Q)\} \\ &= \{\{f, g\} \in \mathcal{L}(Q)^2 \mid \exists \mathbf{c} \in \mathbb{C}^n : \; g(\zeta) - \zeta f(\zeta) \equiv \mathbf{c}\}. \end{aligned}$$

From formula (2.2) it follows that the family Γ_z given in (i) is a Γ-field for A. We find that $\Gamma_w^* =: E_{w^*}$ is the operator of point evaluation at w^* and hence the representation for Q holds. This realization is minimal, since $\mathcal{L}(Q)$, by definition a reproducing kernel space, is generated by the functions $K_Q(\,\cdot\,, w)$. This completes the proofs of (i) and (ii).

From $S \subset A = A^*$ it follows that S is a symmetric operator with equal defect indices. The point spectrum $\sigma_p(S)$ of S is empty and by the reproducing property of the kernel $K_Q(z, w)$ the first equality for S^* in (iii) holds true. Now we use that S can be written in the form

$$S = \{\{f, g\} \in A \mid \Gamma_\mu^*(g - \mu^* f) = 0\},$$

where the right-hand side does not depend on the choice of the non-real point $\mu \in \mathcal{D}(Q)$. On the one hand this implies

$$\begin{aligned} S^* &= A + \{\{\Gamma_\mu \mathbf{d}, \mu \Gamma_\mu \mathbf{d}\} \mid \mathbf{d} \in \mathbb{C}^n\} \\ &= \{\{f, g\} \in \mathcal{L}(Q)^2 \mid \exists \mathbf{c}, \mathbf{d} \in \mathbb{C}^n : \; g(\zeta) - \zeta f(\zeta) \equiv \mathbf{c} - Q(\zeta)\mathbf{d}\}. \end{aligned}$$

On the other hand it also implies that $\operatorname{ran}(S - \mu^*) = \ker \Gamma_\mu^* = (\operatorname{ran} \Gamma_\mu)^\perp$ and hence

$$\dim \ker (S^* - \mu) = \dim (\operatorname{ran}(S - \mu^*))^\perp = \dim \operatorname{ran} \Gamma_\mu = n - \dim \ker \Gamma_\mu,$$

that is, the defect indices of S are $n - d$ with d as in the theorem. □

We call the realization of Q in Theorem 2.1 the *canonical realization* for the function Q and the corresponding model (A, Γ_z, S) the *canonical model.* As mentioned in the introduction, if need be, to denote the dependence of the canonical model on Q we write A_Q, Γ_{Qz}, and S_Q.

Recall that a point $\alpha \in \mathbb{C} \cup \{\infty\}$ is called a *generalized pole* of $Q \in N_\kappa$ if it is an eigenvalue of the relation A_Q. A *generalized zero* of Q is by definition a generalized pole of the function $-Q(z)^{-1}$. As a direct consequence of Theorem 2.1 we find the following description of the algebraic eigenspaces of the relation A_Q.

Corollary 2.2. *Let $Q \in \mathcal{N}_\kappa^{n\times n}$ be given. Then:*

(i) *The point $\alpha \in \mathbb{C}$ is a generalized pole of Q if and only if for some non-zero vector $\mathbf{c}_1 \in \mathbb{C}^n$*
$$\frac{\mathbf{c}_1}{\zeta-\alpha} \in \mathcal{L}(Q).$$
In this case, the Jordan chains of length k at the eigenvalue α of A_Q are the sequences in $\mathcal{L}(Q)$ of the form
$$\frac{\mathbf{c}_1}{\zeta-\alpha},\ \frac{\mathbf{c}_1}{(\zeta-\alpha)^2}+\frac{\mathbf{c}_2}{\zeta-\alpha},\ \ldots,\ \frac{\mathbf{c}_1}{(\zeta-\alpha)^k}+\frac{\mathbf{c}_2}{(\zeta-\alpha)^{k-1}}+\cdots+\frac{\mathbf{c}_k}{\zeta-\alpha}$$
with some vectors $\mathbf{c}_i \in \mathbb{C}^n$, $i=1,2,\ldots,k$.

(ii) *The point ∞ is a generalized pole of Q if and only if for some non-zero vector $\mathbf{c}_1 \in \mathbb{C}^n$ the constant function $\mathbf{c}_1$ belongs to $\mathcal{L}(Q)$. In this case, the Jordan chains of length k at the eigenvalue ∞ of A_Q are the sequences in $\mathcal{L}(Q)$ of the form*
$$\mathbf{c}_1,\ \zeta\mathbf{c}_1+\mathbf{c}_2,\ \ldots,\ \zeta^{k-1}\mathbf{c}_1+\zeta^{k-2}\mathbf{c}_2+\cdots+\mathbf{c}_k$$
with some vectors $\mathbf{c}_i \in \mathbb{C}^n$, $i=1,2,\ldots,k$.

The following corollary gives a connection between the spaces $\mathcal{L}(Q)$ and $\mathcal{L}\left(-Q^{-1}\right)$ and a description of the generalized zeros of Q.

Corollary 2.3. *Let $Q \in \mathcal{N}_\kappa^{n\times n}$ and assume that $Q(\mu)$ is invertible for some non-real point $\mu \in \mathcal{D}(Q)$. Then:*

(i) *The mapping $f(\zeta) \longmapsto Q(\zeta)f(\zeta)$ from $\mathcal{L}\left(-Q^{-1}\right)$ onto $\mathcal{L}(Q)$ is unitary, and under this mapping the self-adjoint relation $A_{-Q^{-1}}$ in $\mathcal{L}\left(-Q^{-1}\right)$ is isomorphic to the self-adjoint relation*
$$\widehat{A}_Q := \left\{\{f,g\} \in \mathcal{L}(Q)^2 \,\middle|\, \exists\, \mathbf{d} \in \mathbb{C}^n : g(\zeta)-\zeta f(\zeta) \equiv Q(\zeta)\mathbf{d}\right\}$$
in $\mathcal{L}(Q)$. In particular, $\rho(\widehat{A}_Q) \neq \emptyset$.

(ii) *The point $\beta \in \mathbb{C}$ is a generalized zero of Q if and only if for some non-zero vector $\mathbf{d}_1 \in \mathbb{C}^n$,*
$$\frac{Q(\zeta)\mathbf{d}_1}{\zeta-\beta} \in \mathcal{L}(Q).$$
In this case, the Jordan chains of length k at the eigenvalue β of $\widehat{A}_Q$ are the sequences in $\mathcal{L}(Q)$ of the form
$$\frac{Q(\zeta)\mathbf{d}_1}{\zeta-\beta},\ \frac{Q(\zeta)\mathbf{d}_1}{(\zeta-\beta)^2}+\frac{Q(\zeta)\mathbf{d}_2}{\zeta-\beta},\ \ldots,\ \frac{Q(\zeta)\mathbf{d}_1}{(\zeta-\beta)^k}+\frac{Q(\zeta)\mathbf{d}_2}{(\zeta-\beta)^{k-1}}+\cdots+\frac{Q(\zeta)\mathbf{d}_k}{\zeta-\beta}$$
with some vectors $\mathbf{d}_i \in \mathbb{C}^n$, $i=1,2,\ldots,k$.

(iii) *The point ∞ is a generalized zero of Q if and only if for some non-zero vector $\mathbf{d}_1 \in \mathbb{C}^n$ the function $Q(\zeta)\mathbf{d}_1$ belongs to the space $\in \mathcal{L}(Q)$. In this case, the Jordan chains of length k at the eigenvalue ∞ of $\widehat{A}_Q$ are the sequences in*

$\mathcal{L}(Q)$ of the form

$$Q(\zeta)\mathbf{d}_1,\ \zeta Q(\zeta)\mathbf{d}_1+Q(\zeta)\mathbf{d}_2,\ \ldots,\ \zeta^{k-1}Q(\zeta)\mathbf{d}_1+\zeta^{k-2}Q(\zeta)\mathbf{d}_2+\cdots+Q(\zeta)\mathbf{d}_k$$

with some vectors $\mathbf{d}_i \in \mathbb{C}^n$, $i = 1, 2, \ldots, k$.

Proof. The first statement is implied by the following relation for the kernels:

$$K_{-Q^{-1}}(\zeta, z) = Q(\zeta)^{-1} K_Q(\zeta, z) Q(z^*)^{-1}$$

and [ADRS, Theorem 1.5.7]. The rest is an immediate consequence of the previous corollary. □

2. A function $Q \in \mathcal{N}_\kappa^{n\times n}$ is called *strict* if for some non-real point $\mu \in \mathcal{D}(Q)$ it holds

$$\bigcap_{z\in\mathcal{D}(Q)} \ker K_Q(z,\mu) = \{0\}.$$

For a minimal realization (A, Γ_z) this means that the mappings Γ_z are for some and hence for every $z \in \rho(A_Q)$ injective. In this case the following theorem describes all canonical self-adjoint extensions of S. This theorem can be derived from [De, Corollary 2.4], but we give a direct proof using the tools developed so far. The formula (2.3) can be seen as Kreĭn's resolvent formula.

Theorem 2.4. *Suppose that the generalized Nevanlinna function $Q \in \mathcal{N}_\kappa^{n\times n}$ is strict and let $(A,\ \Gamma_z,\ S)$ be the canonical model for Q. Then:*

(i) *A linear relation is a canonical self-adjoint extension of S if and only if it is of the form*

$$A_{\mathcal{A},\mathcal{B}} := \{\{f,g\} \in \mathcal{L}(Q)^2 \,|\, \exists\, \mathbf{h} \in \mathbb{C}^n :\ g(\zeta) - \zeta f(\zeta) \equiv (\mathcal{A} + Q(\zeta)\mathcal{B})\mathbf{h}\}$$

with $n \times n$ matrices $\mathcal{A}$ and $\mathcal{B}$ satisfying the relations

$$\operatorname{rank}\begin{pmatrix}\mathcal{A}\\ \mathcal{B}\end{pmatrix} = n, \qquad \mathcal{A}^*\mathcal{B} - \mathcal{B}^*\mathcal{A} = 0.$$

If $A_{\mathcal{A},\mathcal{B}}$ and $A_{\mathcal{A}',\mathcal{B}'}$ are two such canonical self-adjoint extensions of S then $A_{\mathcal{A}',\mathcal{B}'} = A_{\mathcal{A},\mathcal{B}}$ if and only if $\mathcal{A}' = \mathcal{A}\mathcal{C}$ and $\mathcal{B}' = \mathcal{B}\mathcal{C}$ for some invertible $n\times n$ matrix $\mathcal{C}$.

(ii) *$\rho(A_{\mathcal{A},\mathcal{B}}) \neq \emptyset$ if and only if for some non-real point $w \in \mathcal{D}(Q)$ the matrices $\mathcal{A}+Q(w)\mathcal{B}$ and $\mathcal{A}+Q(w^*)\mathcal{B}$ are invertible. In this case for $z \in \rho(A_{\mathcal{A},\mathcal{B}}) \cap \rho(A)$:*

$$(A_{\mathcal{A},\mathcal{B}} - z)^{-1} = (A-z)^{-1} - \Gamma_z \mathcal{B}(\mathcal{A} + Q(z)\mathcal{B})^{-1}\Gamma_{z^*}^*. \tag{2.3}$$

Proof. (i) With $\mu \in \mathcal{D}(Q) \cap \mathbb{C}^+$ we have the direct decomposition

$$S^* = A \dotplus \left\{\{\Gamma_\mu \mathbf{x}, \mu\Gamma_\mu\mathbf{x}\} \,\middle|\, \mathbf{x} \in \mathbb{C}^n\right\}.$$

Hence each element $\{f,g\} \in S^*$ can be written as

$$\{f,g\} = \{f_A + \Gamma_\mu\mathbf{x}, g_A + \mu\Gamma_\mu\mathbf{x}\}$$

with $\{f_A, g_A\} \in A$ and a vector $\mathbf{x} \in \mathbb{C}^n$, which is uniquely determined since Γ_μ is by assumption injective. Define the mapping $\mathbf{b} : S^* \to \mathbb{C}^{2n}$ by

$$\mathbf{b}(\{f,g\}) := \begin{pmatrix} B_0(\{f,g\}) \\ B_1(\{f,g\}) \end{pmatrix} := \begin{pmatrix} \Gamma_\mu^* (g_A - \mu^* f_A) + Q(\mu)\mathbf{x} \\ \mathbf{x} \end{pmatrix}.$$

Since A is self-adjoint and $\mu \in \mathcal{D}(Q)$ the elements $g_A - \mu^* f_A$ run through $\mathcal{L}(Q)$ when $\{f_A, g_A\}$ runs through A. From the injectivity of Γ_μ it follows that Γ_μ^* and hence also $\mathbf{b}$ are surjective. Also, evidently, $S \subset \ker \mathbf{b}$. Moreover, Greens identity holds: If $\{f,g\} = \{f_A + \Gamma_\mu \mathbf{x}, g_A + \mu\Gamma_\mu \mathbf{x}\}$ and $\{h,k\} = \{h_A + \Gamma_\mu \mathbf{y}, k_A + \mu\Gamma_\mu \mathbf{y}\}$ are arbitrary elements of S^*, then

$$\begin{aligned} &\langle k, f\rangle_{\mathcal{L}(Q)} - \langle h, g\rangle_{\mathcal{L}(Q)} \\ &= \langle k_A - \mu^* h_A, \Gamma_\mu \mathbf{x}\rangle_{\mathcal{L}(Q)} - \langle \Gamma_\mu \mathbf{y}, g_A - \mu^* f_A\rangle_{\mathcal{L}(Q)} + (\mu - \mu^*) \langle \Gamma_\mu \mathbf{y}, \Gamma_\mu \mathbf{x}\rangle_{\mathcal{L}(Q)} \\ &= \langle B_0\{h,k\}, B_1\{f,g\}\rangle_{\mathbb{C}^n} - \langle B_1\{h,k\}, B_0\{f,g\}\rangle_{\mathbb{C}^n}. \end{aligned}$$

Hence $\mathbf{b}$ is a boundary mapping as in [M] and [De], where boundary triples are considered, and as in [DLS1, Section 3] and [AzCD].

From Greens formula it follows that the self-adjoint extensions of S can be parametrized as restrictions of S^*:

$$A_\tau = \left\{ \{f,g\} \in S^* \middle| \{B_1(\{f,g\}), B_0(\{f,g\})\} \in \tau \right\},$$

where the relation

$$\tau = \left\{ \{B_1(\{f,g\}), B_0(\{f,g\})\} \middle| \{f,g\} \in A_\tau \right\} \subset (\mathbb{C}^n)^2$$

is self-adjoint. Choose $z_0 \in \mathbb{C}^+$ and define $\mathcal{A}$ and $\mathcal{B}$ as the matrix representations of the operators $-(I + z_0(\tau - z_0)^{-1})$ and $(\tau - z_0)^{-1}$, respectively. They satisfy the relations in (i) and we have

$$\tau = \left\{ \{-\mathcal{B}\mathbf{h}, \mathcal{A}\mathbf{h}\} \middle| \mathbf{h} \in \mathbb{C}^n \right\}.$$

Since S^* in $\mathcal{L}(Q)$ is given by

$$\begin{aligned} S^* &= A \dotplus \left\{ \{\Gamma_\mu \mathbf{d}, \mu\Gamma_\mu \mathbf{d}\} \,\middle|\, \mathbf{d} \in \mathbb{C}^n \right\} \\ &= \left\{ \{f,g\} \in \mathcal{L}(Q)^2 \,\middle|\, \exists\, \mathbf{c}, \mathbf{d} \in \mathbb{C}^n : \ g(\zeta) - \zeta f(\zeta) \equiv \mathbf{c} - Q(\zeta)\mathbf{d} \right\}, \end{aligned}$$

the boundary mapping $\mathbf{b} = \begin{pmatrix} B_0 \\ B_1 \end{pmatrix}$ can be written as

$$B_0(\{f,g\}) = \mathbf{c}, \quad B_1(\{f,g\}) = \mathbf{d}.$$

Hence the relation $A_\tau =: A_{\mathcal{A},\mathcal{B}}$ is described by the formula

$$A_{\mathcal{A},\mathcal{B}} = \left\{ \{f,g\} \in \mathcal{L}(Q)^2 \,\middle|\, \exists\, \mathbf{h} \in \mathbb{C}^n : \ g(\zeta) - \zeta f(\zeta) \equiv (\mathcal{A} + Q(\zeta)\mathcal{B})\mathbf{h} \right\}.$$

With the above descriptions of A_τ and $A_{\mathcal{A},\mathcal{B}}$ Kreĭn's formula reads as

$$(A_\tau - z)^{-1} = (A - z)^{-1} - \Gamma_z (Q(z) - \tau)^{-1} \Gamma_{z^*}^*, \ z \in \rho(A_\tau) \cap \rho(A),$$

or, equivalently,

$$(A_{\mathcal{A},\mathcal{B}} - z)^{-1} = (A - z)^{-1} - \Gamma_z \mathcal{B}(Q(z)\mathcal{B} + \mathcal{A})^{-1}\Gamma_{z^*}^*, \; z \in \rho(A_{\mathcal{A},\mathcal{B}}) \cap \rho(A).$$

The proofs of the other statements in (i) are left to the reader.
(ii) We first assume that $\rho\left(A_{\mathcal{A},\mathcal{B}}\right) \neq \emptyset$. Then, since $A_{\mathcal{A},\mathcal{B}}$ is self-adjoint and $\Gamma_{w^*}^* = E_w$ is surjective, we find

$$\left\{E_w(g - wf) \,\middle|\, \{f,g\} \in A_{\mathcal{A},\mathcal{B}}\right\} = \mathbb{C}^n, \quad w \in \rho(A_{\mathcal{A},\mathcal{B}}).$$

Therefore, by the definition of $A_{\mathcal{A},\mathcal{B}}$, the matrix $\mathcal{A} + Q(w)\mathcal{B}$ is invertible.

To prove the converse denote the operator on the right-hand side of (2.3) by $K(z)$. Using

$$\Gamma_z - \Gamma_w = (z - w)(A - z)^{-1}\Gamma_w, \quad (A - w)^{-1}\Gamma_z = (A - z)^{-1}\Gamma_w,$$

and

$$\mathcal{B}(\mathcal{A} + Q(z)\mathcal{B})^{-1} = (\mathcal{A}^* + \mathcal{B}^* Q(z))^{-1}\mathcal{B}^*$$

we find that $K(z)$ has all the properties of a resolvent operator of a canonical self-adjoint extension of S with a nonempty resolvent set: $\mathcal{D}(K)$, the domain of holomorphy of K, is symmetric with respect to the real axis and

$$K(z^*)^* = K(z), \quad K(z) - K(w) = (z - w)K(z)K(w), \quad K(z)(S - z) \subset I.$$

If we denote this self-adjoint extension by A_K, then

$$A_K = \left\{\{K(z)h, h + zK(z)h\} \,\middle|\, h \in \mathcal{L}(Q)\right\}.$$

Using again $\Gamma_{z^*}^* = E_z$ we see that for $h \in \mathcal{L}(Q)$,

$$h(\zeta) + (z - \zeta)K(z)h(\zeta) \equiv (\mathcal{A} + Q(\zeta)\mathcal{B})\mathbf{c}, \quad \mathbf{c} = (\mathcal{A} + Q(z)\mathcal{B})^{-1}h(z).$$

Hence $A_K \subset A_{\mathcal{A},\mathcal{B}}$. Since both relations are self-adjoint, equality prevails. □

Remark 2.5. Note that according to (2.3) the resolvent of $\widehat{A}_Q$ in Corollary 2.3 is an at most n-dimensional perturbation of the resolvent of A_Q.

3. A minimal model related to the basic factorization

1. Recall that a generalized pole α of $q \in \mathcal{N}_\kappa$ is by definition an eigenvalue of the relation A_q. A generalized pole α is called *of positive type* if the corresponding eigenspace of A_q is positive. If the generalized pole α is not of positive type, its *degree of non-positivity* is by definition the dimension of a maximal non-positive invariant subspace of the algebraic eigenspace of A_q at α. A generalized zero β of q, which is by definition a generalized pole of $-q(z)^{-1}$, is called of *positive type* if it is a pole of positive type of $-q(z)^{-1}$, and if β is not of positive type its *degree of non-positivity* is the degree of non-positivity of β as a generalized pole of $-q(z)^{-1}$.

If $q \in \mathcal{N}_\kappa$, its *basic factorization* (1.1) is defined as follows. Let the points α_i, $i = 1, 2, \ldots, \ell$, (β_j, $j = 1, 2, \ldots, k$, respectively) be the generalized poles (zeros, respectively) of q in $\mathbb{C}^+ \cup \mathbb{R}$ that are not of positive type, denote by ν_i (κ_j, respectively) the degrees of non-positivity of α_i (β_j, respectively), and define

$$r(z) := \frac{(z-\beta_1)^{\kappa_1} \ldots (z-\beta_k)^{\kappa_k}}{(z-\alpha_1^*)^{\nu_1} \ldots (z-\alpha_\ell^*)^{\nu_\ell}}. \tag{3.1}$$

Then there exists a function $q_0 \in \mathcal{N}_0$ such that

$$q(z) = r^{\#}(z) q_0(z) r(z), \tag{3.2}$$

see [DLLSh], [DeHS1]. Note that, if

$$\tau := \kappa_1 + \cdots + \kappa_k - (\nu_1 + \cdots + \nu_\ell) \tag{3.3}$$

is positive (negative, respectively), then ∞ is a generalized pole (zero, respectively) of q which is not of positive type and with degree of non-positivity $|\tau|$. Since ∞ cannot be a generalized zero and a generalized pole at the same time, we have

$$\kappa = \max\{\kappa_1 + \cdots + \kappa_k, \nu_1 + \cdots + \nu_\ell\}.$$

By $\mathcal{M}_r(z)$ and $\mathcal{Q}(z)$ we denote the following matrix functions:

$$\mathcal{M}_r(z) = \begin{pmatrix} 0 & r^{\#}(z) \\ r(z) & 0 \end{pmatrix}, \quad \mathcal{Q}(z) = \begin{pmatrix} q_0(z) & 0 & 0 \\ 0 & 0 & r^{\#}(z) \\ 0 & r(z) & 0 \end{pmatrix},$$

which belong to the classes $\mathcal{N}_\kappa^{2\times 2}$ and $\mathcal{N}_\kappa^{3\times 3}$, respectively (see Theorem 3.4 below). Note that the elements of $\mathcal{L}(q_0)$ are scalar functions, the elements of $\mathcal{L}(\mathcal{M}_r)$ are 2-vector functions, and the elements of $\mathcal{L}(\mathcal{Q}) = \mathcal{L}(q_0) \oplus \mathcal{L}(\mathcal{M}_r)$ are 3-vector functions. By $\mathbf{v}(\zeta)$ we denote the vector function

$$\mathbf{v}(\zeta) := \begin{pmatrix} r(\zeta) \\ 1 \\ r(\zeta) q_0(\zeta) \end{pmatrix},$$

which is not necessarily an element of $\mathcal{L}(\mathcal{Q})$. The first result concerns the state space $\mathcal{L}(q)$, which will be described in terms of a unitary mapping from a subspace of $\mathcal{L}(\mathcal{Q})$ to $\mathcal{L}(q)$.

Lemma 3.1.

(i) *The bounded linear operator* $\mathbf{T} : \mathcal{L}(\mathcal{Q}) \longmapsto \mathcal{L}(q)$ *defined by*

$$(\mathbf{T}\widetilde{f})(\zeta) := \mathbf{v}^{\#}(\zeta) \widetilde{f}(\zeta)$$

is a partial isometry with $\operatorname{ran} \mathbf{T} = \mathcal{L}(q)$.

(ii) *The linear manifold*

$$\mathcal{H}_0 := \ker \mathbf{T} = \left\{ \widetilde{f} \in \mathcal{L}(\mathcal{Q}) \,\middle|\, \mathbf{v}^{\#}(\zeta) \widetilde{f}(\zeta) \equiv 0 \right\}$$

is a finite-dimensional Hilbert subspace of $\mathcal{L}(\mathcal{Q})$.

(iii) *With* $\mathcal{P} := \mathcal{H}_0^\perp$ *the restriction* $T := \mathbf{T}\,|_{\mathcal{P}}$ *is a unitary mapping from* $\mathcal{P}$ *onto* $\mathcal{L}(q)$.

Below in Corollary 4.3 we give a formula for the dimension of the space $\mathcal{H}_0$.

Proof of Lemma 3.1. A straightforward calculation shows that the two kernels $K_q(z,w)$ and $K_{\mathcal{Q}}(z,w)$ are related by the formula

$$K_q(\zeta,z) = \mathbf{v}^{\#}(\zeta)K_{\mathcal{Q}}(\zeta,z)\mathbf{v}(z^*).$$

Since the numbers of negative squares of these kernels coincide, according to [ADRS, Theorem 1.5.7.] the mapping $\mathbf{T}$ is a surjective partial isometry and its kernel $\mathcal{H}_0$ is a Hilbert space. Write

$$\mathbf{T} = \begin{pmatrix} 0 & T \end{pmatrix} : \begin{pmatrix} \mathcal{H}_0 \\ \mathcal{P} \end{pmatrix} \to \mathcal{L}(q).$$

Then $T : \mathcal{P} \to \mathcal{L}(q)$ satisfies (iii) of the theorem. From the defining equation for $\mathcal{H}_0$ we find $\dim \mathcal{H}_0 \le \dim \mathcal{L}(\mathcal{M}_r)$. It is well known (or follows from Theorem 3.4 below) that $\dim \mathcal{L}(\mathcal{M}_r) = 2\kappa$. □

Theorem 3.2. *Let $q \in \mathcal{N}_\kappa$ be given. Then:*

(i) *The self-adjoint relation*

$$A := \left\{ \{\widetilde{f},\widetilde{g}\} \in \mathcal{P}^2 \,|\, \exists c \in \mathbb{C} : \mathbf{v}^{\#}(\zeta)(\widetilde{g}(\zeta) - \zeta\widetilde{f}(\zeta)) \equiv c \right\}$$

and the corresponding Γ-field $\Gamma_z 1$ with

$$\Gamma_z 1(\zeta) := \varphi(\zeta,z) := K_{\mathcal{Q}}(\zeta,z^*)\mathbf{v}(z)$$
$$= \frac{\mathcal{Q}(\zeta)-\mathcal{Q}(z)}{\zeta - z}\mathbf{v}(z) = (\Gamma_{\mathcal{Q}z}\mathbf{v}(z))(\zeta), \quad z \in \mathcal{D}(q),$$

form a minimal realization of q in the state space $\mathcal{P}$.

(ii) *Under the mapping T this realization is unitarily equivalent to the canonical model for q.*

(iii) *The corresponding symmetric operator S is given by*

$$S := \left\{ \{\widetilde{f},\widetilde{g}\} \in \mathcal{P}^2 \,\middle|\, \mathbf{v}^{\#}(\zeta)(\widetilde{g}(\zeta) - \zeta\widetilde{f}(\zeta)) \equiv 0 \right\}$$

and its adjoint by

$$S^* = \left\{ \{\widetilde{f},\widetilde{g}\} \in \mathcal{P}^2 \,\middle|\, \exists c,d \in \mathbb{C} : \mathbf{v}^{\#}(\zeta)(\widetilde{g}(\zeta) - \zeta\widetilde{f}(\zeta)) \equiv c - q(\zeta)d \right\}.$$

Proof. Theorem 2.1 and the relation

$$A = \left\{ \{\widetilde{f},\widetilde{g}\} \,\middle|\, \{T\widetilde{f},T\widetilde{g}\} \in A_q \right\}$$

imply that A is self-adjoint. Likewise, with $S = T^{-1}S_qT$ and

$$S^* = \left\{ \{\widetilde{f},\widetilde{g}\} \,\middle|\, \{T\widetilde{f},T\widetilde{g}\} \in S_q^* \right\}$$

the statement (iii) follows. The reproducing property of $K_{\mathcal{Q}}(\zeta,z)$ implies for $\widetilde{f} \in \mathcal{H}_0$ and $z \in \mathcal{D}(q)$

$$\left\langle \widetilde{f}(\,\cdot\,), K_{\mathcal{Q}}(\,\cdot\,,z^*)\mathbf{v}(z)\right\rangle_{\mathcal{L}(\mathcal{Q})} = \mathbf{v}(z)^*\widetilde{f}(z^*) = \mathbf{v}^{\#}(z^*)\widetilde{f}(z^*) = 0,$$

and hence $\varphi(\,\cdot\,,z) \in \mathcal{P}$. From the relation

$$T\varphi(\,\cdot\,,z) = \mathbf{v}^{\#}(\,\cdot\,)K_{\mathcal{Q}}(\,\cdot\,,z^*)\mathbf{v}(z) = K_q(\,\cdot\,,z^*) = \varphi_q(\,\cdot\,,z)$$

we see that $\varphi(\,\cdot\,,z)$ is a Γ-field corresponding to A and, moreover, since T is unitary, we have the representation

$$\frac{q(z)-q(z_0^*)}{z-z_0^*} = \langle \varphi_q(\,\cdot\,,z),\varphi_q(\,\cdot\,,z_0)\rangle_{\mathcal{L}(q)} = \langle \varphi(\,\cdot\,,z),\varphi(\,\cdot\,,z_0)\rangle_{\mathcal{P}}.$$

The equality

$$\left\langle \widetilde{f}(\cdot), K_{\mathcal{Q}}(\,\cdot\,,z)\mathbf{v}(z^*)\right\rangle_{\mathcal{L}(\mathcal{Q})} = \mathbf{v}^{\#}(z)\widetilde{f}(z) = (\mathbf{T}\widetilde{f})(z), \quad \widetilde{f} \in \mathcal{L}(\mathcal{Q}),\ z \in \mathcal{D}(q),$$

where $\mathbf{T}$ is the partial isometry defined in Lemma 3.1, and the fact that $\mathcal{H}_0 = \ker \mathbf{T}$ imply

$$\mathcal{P} = \overline{\operatorname{span}}\{\varphi(\,\cdot\,,z) \mid z \in \mathcal{D}(q)\},$$

that is, the realization is minimal. □

2. In the following we give another description of the self-adjoint relation A in $\mathcal{P}$. The symmetric operator $S_{\mathcal{Q}}$ in the canonical model for $\mathcal{Q}$ can be decomposed as $S_{\mathcal{Q}} = S_{q_0} \oplus S_{\mathcal{M}_r}$ and it has defect indices $(3,3)$. The self-adjoint linear relation $A_{\mathcal{Q}} = A_{q_0} \oplus A_{\mathcal{M}_r}$ is a canonical self-adjoint extension of $S_{\mathcal{Q}}$. By Theorem 2.4, another canonical self-adjoint extension of $S_{\mathcal{Q}}$ is given by

$$\widetilde{A} := \left\{\{\widetilde{f},\widetilde{g}\} \in (\mathcal{L}(\mathcal{Q}))^2 \,\middle|\, \exists\, \mathbf{h} \in \mathbb{C}^3 :\ \widetilde{g}(\zeta) - \zeta\widetilde{f}(\zeta) \equiv (\mathcal{I} + \mathcal{Q}(\zeta)\mathcal{B})\mathbf{h}\right\},$$

where $\mathcal{I}$ is the 3×3 unit matrix and

$$\mathcal{B} = -\begin{pmatrix} 0 & 0 & 1 \\ 0 & 0 & 0 \\ 1 & 0 & 0 \end{pmatrix}.$$

It has a nonempty resolvent set since the matrix

$$\mathcal{I} + \mathcal{Q}(z)\mathcal{B} = \begin{pmatrix} 1 & 0 & -q_0(z) \\ -r^{\#}(z) & 1 & 0 \\ 0 & 0 & 1 \end{pmatrix}$$

is invertible. Furthermore,

$$(\widetilde{A} - z)^{-1} = (A_{\mathcal{Q}} - z)^{-1} - \Gamma_{\mathcal{Q}z}\mathcal{B}\left(\mathcal{I} + \mathcal{Q}(z)\mathcal{B}\right)^{-1}\Gamma_{\mathcal{Q}z^*}^*. \tag{3.4}$$

The next theorem implies in particular that in the state space $\mathcal{L}(Q)$ also the pair $(\widetilde{A},\varphi(\,\cdot\,,z))$ is a (not necessarily minimal) realization of q.

Theorem 3.3. *With respect to the decomposition $\mathcal{L}(Q) = \mathcal{P} \oplus \mathcal{H}_0$ we have*

$$(\widetilde{A} - z)^{-1} = (A - z)^{-1} \oplus (A_0 - z)^{-1}, \tag{3.5}$$

where A is the self-adjoint relation in $\mathcal{P}$ as in Theorem 3.2 and A_0 is a self-adjoint relation in $\mathcal{H}_0$.

It follows that $A = \widetilde{A} \cap \mathcal{P}^2$, $A_0 = \widetilde{A} \cap \mathcal{H}_0^2$, and the relation (3.5) can be written as

$$\widetilde{A} = \{\{f+h, g+k\} \mid \{f,g\} \in A, \{h,k\} \in A_0\} = A \oplus A_0.$$

Proof of Theorem 3.3. We show first that for $z, w \in \rho(\widetilde{A})$,

$$(z-w)(\widetilde{A}-z)^{-1}\varphi(\,\cdot\,,w) = \varphi(\,\cdot\,,z) - \varphi(\,\cdot\,,w). \tag{3.6}$$

For this we apply (3.4) with

$$\mathcal{B}(\mathcal{I}+\mathcal{Q}(z)\mathcal{B})^{-1} = -\begin{pmatrix} 0 & 0 & 1 \\ 0 & 0 & 0 \\ 1 & 0 & q_0(z) \end{pmatrix}$$

to the Γ-field Γ_{Qw}. Using the defining relations (1.2) and (1.3) for a Γ-field, we obtain

$$\begin{aligned} (z-w)(\widetilde{A}-z)^{-1}\Gamma_{Qw} &= \Gamma_{Qz} - \Gamma_{Qw} \\ &\quad + \Gamma_{Qz}\begin{pmatrix} 0 & 0 & 1 \\ 0 & 0 & 0 \\ 1 & 0 & q_0(z) \end{pmatrix}\begin{pmatrix} q_0(z)-q_0(w) & 0 & 0 \\ 0 & 0 & r^{\#}(z)-r^{\#}(w) \\ 0 & r(z)-r(w) & 0 \end{pmatrix} \\ &= \Gamma_{Qz} - \Gamma_{Qw} + \Gamma_{Qz}\begin{pmatrix} 0 & r(z)-r(w) & 0 \\ 0 & 0 & 0 \\ q_0(z)-q_0(w) & q_0(z)(r(z)-r(w)) & 0 \end{pmatrix}. \end{aligned}$$

Multiplying both sides of this equality from the right by the column vector $\mathbf{v}(w)$ and observing that $\Gamma_{Qw}\mathbf{v}(w) = \varphi(\,\cdot\,,w)$ we find

$$\begin{aligned} &(z-w)(\widetilde{A}-z)^{-1}\varphi(\,\cdot\,,w) \\ &= \Gamma_{Qz}\mathbf{v}(w) - \varphi(\,\cdot\,,w) + \Gamma_{Qz}\begin{pmatrix} r(z)-r(w) \\ 0 \\ q_0(z)r(z)-q_0(w)r(w) \end{pmatrix} \\ &= \Gamma_{Qz}\mathbf{v}(w) - \varphi(\,\cdot\,,w) + \Gamma_{Qz}(\mathbf{v}(z)-\mathbf{v}(w)) \\ &= \varphi(\,\cdot\,,z) - \varphi(\,\cdot\,,w). \end{aligned}$$

This implies (3.6). From Theorem 3.2 we have that

$$(z-w)(A-z)^{-1}\varphi(\,\cdot\,,w) = \varphi(\,\cdot\,,z) - \varphi(\,\cdot\,,w),$$

and comparing this with (3.6) we obtain

$$(\widetilde{A}-z)^{-1}\varphi(\,\cdot\,,w) = (A-z)^{-1}\varphi(\,\cdot\,,w).$$

Since the elements $\varphi(\,\cdot\,,w)$, $w \in \rho(A)$, form a total set in $\mathcal{P}$ we have in fact that

$$(\widetilde{A}-z)^{-1}h = (A-z)^{-1}h, \quad h \in \mathcal{P}.$$

It follows that $\mathcal{P}$ is invariant under $(\widetilde{A}-z)^{-1}$ for all $z \in \rho(\widetilde{A})$, and by taking adjoints we see that also $\mathcal{H}_0$ is invariant under this resolvent operator. Thus the decomposition of $(\widetilde{A} - z)^{-1}$ with respect to the decomposition $\mathcal{L}(Q) = \mathcal{P} \oplus \mathcal{H}_0$ is a diagonal block operator matrix with entries $(A - z)^{-1}$ and an operator $R_0(z)$, say, on $\mathcal{H}_0$. Since $R_0(z)$ satisfies the resolvent identity and $R_0(z)^* = R_0(z^*)$, it is the resolvent of a self-adjoint relation A_0 in $\mathcal{H}_0$. □

3. Now we give a description of $\mathcal{L}(\mathcal{M}_r)$. This space is also considered in [DeH], but here we emphasize the structure of the space via the spectral decomposition of the self-adjoint relation $A_{\mathcal{M}_r}$ in the model for the function $\mathcal{M}_r$. Recall that the function r is given by (3.1). We write

$$r(z) = r_0(z) + r_1(z) + \cdots + r_\ell(z),$$

with

$$r_0(z) := \sigma_{\nu_0} z^{\nu_0} + \sigma_{\nu_0 - 1} z^{\nu_0 - 1} + \cdots + \sigma_1 z + \sigma_0,$$

$$r_i(z) := \sum_{j=1}^{\nu_i} \frac{-\sigma_{ij}}{(z - \alpha_i^*)^j}, \quad i = 1, 2, \ldots, \ell,$$

where all the coefficients are complex numbers, $\sigma_{\nu_0} \neq 0$, and $\sigma_{i\nu_i} \neq 0$ for $i = 1, 2, \ldots, \ell$. Set $\mathcal{M}_i(z) := \mathcal{M}_{r_i}(z)$, $A_i := A_{\mathcal{M}_{r_i}}$, $i = 0, 1, 2, \ldots, \ell$.

Theorem 3.4. *We have*

$$\mathcal{L}(\mathcal{M}_r) = \mathcal{L}(\mathcal{M}_0) \oplus \mathcal{L}(\mathcal{M}_1) \oplus \cdots \oplus \mathcal{L}(\mathcal{M}_\ell) \tag{3.7}$$

and this orthogonal decomposition is the spectral decomposition for the self-adjoint relation $A_{\mathcal{M}_r} = A_0 \oplus A_1 \oplus \cdots \oplus A_\ell$ in the model for $\mathcal{M}_r$:

(a) *For $i = 1, 2, \ldots, \ell$, the space $\mathcal{L}(\mathcal{M}_i)$ is spanned by two Jordan chains, each of length ν_i, corresponding to the eigenvalues α_i and α_i^* of A_i in $\mathcal{L}(\mathcal{M}_i)$:*

$$\begin{pmatrix} \frac{1}{\zeta - \alpha_i} \\ 0 \end{pmatrix}, \begin{pmatrix} \frac{1}{(\zeta - \alpha_i)^2} \\ 0 \end{pmatrix}, \ldots, \begin{pmatrix} \frac{1}{(\zeta - \alpha_i)^{\nu_i}} \\ 0 \end{pmatrix} \text{ and } \begin{pmatrix} 0 \\ \frac{1}{\zeta - \alpha_i^*} \end{pmatrix}, \begin{pmatrix} 0 \\ \frac{1}{(\zeta - \alpha_i^*)^2} \end{pmatrix}, \ldots, \begin{pmatrix} 0 \\ \frac{1}{(\zeta - \alpha_i^*)^{\nu_i}} \end{pmatrix}$$

The Gram matrix $\widetilde{\mathcal{G}}_i$ for this basis is given by

$$\widetilde{\mathcal{G}}_i := \begin{pmatrix} 0 & \mathcal{G}_i \\ \mathcal{G}_i^* & 0 \end{pmatrix}, \qquad \mathcal{G}_i := \begin{pmatrix} \sigma_{i1} & \sigma_{i2} & \cdots & \sigma_{i\nu_i} \\ \sigma_{i2} & & \cdots & 0 \\ \vdots & \vdots & \vdots & \vdots \\ \sigma_{i\nu_i} & 0 & \cdots & 0 \end{pmatrix}^{-1};$$

in particular the linear spans of the two chains are skewly-linked.

(b) *If the polynomial r_0 is not a constant, the space $\mathcal{L}(\mathcal{M}_0)$ is spanned by two chains of length ν_0, corresponding to the eigenvalue ∞ of A_0 in $\mathcal{L}(\mathcal{M}_0)$:*

$$\begin{pmatrix} 1 \\ 0 \end{pmatrix}, \begin{pmatrix} \zeta \\ 0 \end{pmatrix}, \ldots, \begin{pmatrix} \zeta^{\nu_0 - 1} \\ 0 \end{pmatrix} \text{ and } \begin{pmatrix} 0 \\ 1 \end{pmatrix}, \begin{pmatrix} 0 \\ \zeta \end{pmatrix}, \ldots, \begin{pmatrix} 0 \\ \zeta^{\nu_0 - 1} \end{pmatrix}.$$

The Gram matrix $\widetilde{\mathcal{G}}_0$ for this basis is given by

$$\widetilde{\mathcal{G}}_0 := \begin{pmatrix} 0 & \mathcal{G}_0 \\ \mathcal{G}_0^* & 0 \end{pmatrix}, \qquad \mathcal{G}_0 := \begin{pmatrix} \sigma_1 & \sigma_2 & \dots & \sigma_{\nu_0} \\ \sigma_2 & \dots & \dots & 0 \\ \vdots & \vdots & \vdots & \vdots \\ \sigma_{\nu_0} & 0 & \dots & 0 \end{pmatrix}^{-1};$$

in particular the spans of the chains are skewly-linked.

Proof. We write $K_i(\,\cdot\,,w)$ for $K_{\mathcal{M}_i}(\,\cdot\,,w)$, $i = 0,1,2,\dots,\ell$. From $\mathcal{M}_r(z) = \mathcal{M}_0(z)+\mathcal{M}_1(z)+\cdots+\mathcal{M}_\ell(z)$, we obtain

$$K_{\mathcal{M}_r}(\,\cdot\,,w) = K_0(\,\cdot\,,w) + K_1(\,\cdot\,,w) + \cdots + K_\ell(\,\cdot\,,w),$$

whence (3.7), where, by [ADRS, Section 1.5], the sum is orthogonal because it is direct. The latter follows from the description of these spaces which we give now; in particular, elements from different summands have different singularities.

(a) For $\begin{pmatrix} a \\ b \end{pmatrix}, \begin{pmatrix} c \\ d \end{pmatrix} \in \mathbb{C}^2$ we have

$$\left\langle K_i(\cdot,w)\begin{pmatrix} a \\ b \end{pmatrix}, K_i(\cdot,v)\begin{pmatrix} c \\ d \end{pmatrix}\right\rangle_{\mathcal{L}(M_i)} = \begin{pmatrix} c^* & d^* \end{pmatrix} K_i(v,w)\begin{pmatrix} a \\ b \end{pmatrix}$$

$$= c^*\frac{r_i^{\#}(v) - r_i^{\#}(w^*)}{v-w^*}b + d^*\frac{r_i(v)-r_i(w^*)}{v-w^*}a.$$

Choosing in particular $a = c = 0$ or $b = d = 0$ we find that the subspaces

$$\left\{\begin{pmatrix} 0 \\ g \end{pmatrix} \in \mathcal{L}(\mathcal{M}_i)\right\}, \quad \left\{\begin{pmatrix} f \\ 0 \end{pmatrix} \in \mathcal{L}(\mathcal{M}_i)\right\}$$

of $\mathcal{L}(\mathcal{M}_i)$ are neutral subspaces. This accounts for the zero entries on the main diagonal of the Gram matrix $\widetilde{\mathcal{G}}_i$. For $a = d = 1$ and $b = c = 0$, the above formula reads as

$$\left\langle \begin{pmatrix} 0 \\ \frac{r_i(\zeta)-r_i(w^*)}{\zeta-w^*} \end{pmatrix}, \begin{pmatrix} \frac{r_i^{\#}(\zeta)-r_i^{\#}(v^*)}{\zeta-v^*} \\ 0 \end{pmatrix}\right\rangle_{\mathcal{L}(\mathcal{M}_i)} = \frac{r_i(v)-r_i(w^*)}{v-w^*}. \tag{3.8}$$

Now we use that

$$\frac{r_i(\zeta)-r_i(w^*)}{\zeta-w^*} = \sum_{j=1}^{\nu_i}\sum_{s=0}^{j-1}\frac{\sigma_{ij}}{(\zeta-\alpha_i^*)^{j-s}(w^*-\alpha_i^*)^{s+1}}$$

and

$$\frac{r_i^{\#}(\zeta)-r_i^{\#}(v^*)}{\zeta-v^*} = \sum_{j=1}^{\nu_i}\sum_{s=0}^{j-1}\frac{\sigma_{ij}^*}{(\zeta-\alpha_i)^{j-s}(v^*-\alpha_i)^{s+1}}.$$

We multiply both sides of (3.8) by $(w^*-\alpha_i^*)^n(v-\alpha_i^*)^m$, integrate over a small circle around α_i^* with respect to w^* and likewise with respect to v, and obtain for

$n, m = 0, 1, \ldots, \nu_i - 1,$

$$\left\langle \begin{pmatrix} 0 \\ \sum_{j=n+1}^{\nu} \frac{\sigma_{ij}}{(\zeta-\alpha_i^*)^{j-n}} \end{pmatrix}, \begin{pmatrix} \sum_{j=m+1}^{\nu} \frac{\sigma_{ij}^*}{(\zeta-\alpha_i)^{j-m}} \\ 0 \end{pmatrix} \right\rangle_{\mathcal{L}(\mathcal{M}_i)} = \sigma_{i,n+m+1},$$

where we set $\sigma_{i,j} = 0$ if $j \geq \nu_i + 1$. These equalities can be written in matrix form $\mathcal{G}_i^{-1}\mathcal{H}_i\mathcal{G}_i^{-1} = \mathcal{G}_i^{-1}$ where $\mathcal{H}_i = (h_{l,m})_{l,m=1}^{\nu_i}$ with

$$h_{l,m} = \left\langle \begin{pmatrix} 0 \\ \frac{1}{(\zeta-\alpha_i^*)^m} \end{pmatrix}, \begin{pmatrix} \frac{1}{(\zeta-\alpha_i)^l} \\ 0 \end{pmatrix} \right\rangle_{\mathcal{L}(\mathcal{M}_i)}$$

From this relation we see that $\mathcal{H}_i = \mathcal{G}_i$, and the formula for the Gram matrix $\widetilde{\mathcal{G}}_i$ now readily follows. The rest of the proof of part (a) is left to the reader.
(b) The proof is similar to the previous one. The analog of (3.8) is

$$\left\langle \begin{pmatrix} 0 \\ \sum_{j=1}^{\nu_0} \sum_{s=0}^{j-1} \sigma_j \zeta^{j-1-s} w^{*s} \end{pmatrix}, \begin{pmatrix} \sum_{j=1}^{\nu_0} \sum_{s=0}^{j-1} \sigma_j^* \zeta^{j-1-s} v^{*s} \\ 0 \end{pmatrix} \right\rangle_{\mathcal{L}(M_0)} = \sum_{j=1}^{\nu_0} \sum_{s=0}^{j-1} \sigma_j v^{j-1-s} w^{*s}.$$

If we multiply with $w^{*(-n-1)}v^{-m-1}$ and integrate with respect to w^* and v over small circles around the origin, we obtain for $n, m = 0, 1, \ldots, \nu_0 - 1$,

$$\left\langle \begin{pmatrix} 0 \\ \sum_{j=n+1}^{\nu_0} \sigma_j \zeta^{j-n-1} \end{pmatrix}, \begin{pmatrix} \sum_{j=m+1}^{\nu_0} \sigma_j^* \zeta^{j-m-1} \\ 0 \end{pmatrix} \right\rangle_{\mathcal{L}(\mathcal{M}_0)} = \sigma_{n+m+1},$$

where $\sigma_j = 0$ if $j \geq \nu_0 + 1$. This readily leads to the formula for the Gram matrix $\widetilde{\mathcal{G}}_0$ in this case. □

4. The space $\mathcal{H}_0$

It may happen that in the factorization (3.2) a real generalized zero β (generalized pole α, respectively) of the function q is a generalized pole (generalized zero, respectively) of q_0 as well.

Example. The function $q_0(z) = 1 - \frac{1}{z} \in \mathcal{N}_0$ has a generalized pole at $\beta = 0$, which is of positive type. If $q(z) = 1 + \frac{1}{z-1}$, then $q \in \mathcal{N}_1$ and $q(z) = r^{\#}(z)q_0(z)r(z)$ with $r(z) = \frac{z}{z-1}$.

In this example the factor $z - \beta$ cancels and the function $q \in \mathcal{N}_1$ does not have a pole at $z = 0$. In the more general situation this "cancellation" corresponds to the fact that the model $(\widetilde{A}, \varphi(\,\cdot\,, z))$ is not minimal in the state space $\mathcal{L}(Q)$. In this section we describe the space $\mathcal{H}_0$, which is nontrivial in exactly this case.

Lemma 4.1. *Let $q_0 \in \mathcal{N}_0$ be given.*

(i) *Assume either $\beta \in \mathbb{C}$ and $n > 1$ or $n = 1$ and the point $\beta \in \mathbb{C}$ is not a generalized pole of q_0. If*

$$\lim_{z \hat{\to} \beta} (z - \beta)^n q_0(z) = 0,$$

then every $f \in \mathcal{L}(q_0)$ has the same behavior at β as q_0, that is,

$$\lim_{\zeta \hat{\to} \beta} (\zeta - \beta)^n f(\zeta) = 0.$$

(ii) *Assume $n > 1$ or $n = 1$ and ∞ is not a generalized pole of q_0. If $\lim_{z \hat{\to} \infty} \frac{q_0(z)}{z^n} = 0$ then also $\lim_{\zeta \hat{\to} \infty} \frac{f(\zeta)}{\zeta^n} = 0$ for every $f \in \mathcal{L}(q_0)$.*

Proof. We only prove statement (i), the other case can be proven in a similar way. With the abbreviation $x(\,\cdot\,, \zeta^*) := (\zeta^* - \beta^*)^n K_{q_0}(\,\cdot\,, \zeta)$ we find under the assumptions of the lemma and since $q_0(z) \in \mathcal{N}_0$ that for $w \in \mathcal{D}(q_0)$

$$\lim_{\zeta \hat{\to} \beta} \langle x(\,\cdot\,, \zeta^*), K_{q_0}(\,\cdot\,, w)\rangle_{\mathcal{L}(q_0)} = \lim_{\zeta \hat{\to} \beta} x(w, \zeta^*) = 0$$

and

$$\lim_{\zeta \hat{\to} \beta} \langle x(\,\cdot\,, \zeta^*), x(\,\cdot\,, \zeta^*)\rangle_{\mathcal{L}(q_0)} = \lim_{\zeta \hat{\to} \beta} (\zeta - \beta)^n x(\zeta, \zeta^*) = 0.$$

That is, $x(\,\cdot\,, \zeta^*)$ converges for $\zeta \hat{\to} \beta$ weakly to the zero element. Hence we have

$$\lim_{\zeta \hat{\to} \beta} (\zeta - \beta)^n f(\zeta) = \lim_{\zeta \hat{\to} \beta} \langle f(\cdot), (\zeta^* - \beta^*)^n K_{q_0}(\,\cdot\,, \zeta)\rangle_{\mathcal{L}(q_0)} = 0. \qquad \square$$

Let the rational function $r(z)$ and the matrix function $\mathcal{M}_r(z)$ again be given as in Section 3. For simplicity we use the following notations: $\alpha_0 := \infty$, $\beta_0 := \infty$, and

$$I^+ = \left\{0 \le i \le \ell \mid \alpha_i \in \sigma_p(\widehat{A}_{q_0})\right\}, \quad J^+ = \left\{0 \le j \le k \mid \beta_j \in \sigma_p\left(A_{q_0}\right)\right\},$$

where $\widehat{A}_{q_0}$ is the self-adjoint relation in $\mathcal{L}(q_0)$ which is the isomorphic copy of the self-adjoint relation A_{-1/q_0} in $\mathcal{L}(-1/q_0)$ in the canonical model for the function $-1/q_0$ under the operator of multiplication by $q_0(\zeta)$ mapping $\mathcal{L}(-1/q_0)$ onto $\mathcal{L}(q_0)$; see Corollary 2.3. Moreover, by

$$\widehat{x}_{\alpha_i}(\zeta) := \begin{cases} \dfrac{q_0(\zeta)}{\zeta - \alpha_i} & \text{for } \; i \in I^+ \setminus \{0\}, \\ q_0(\zeta) & \text{for } \; i = 0 \in I^+ \end{cases}$$

and

$$x_{\beta_j}(\zeta) := \begin{cases} \dfrac{1}{\zeta - \beta_j} & \text{for } \; j \in J^+ \setminus \{0\}, \\ 1 & \text{for } \; j = 0 \in J^+ \end{cases}$$

we denote those eigenelements of $\widehat{A}_{q_0}$ and A_{q_0}, respectively, which correspond to the eigenvalues that are cancelled. Note that the α_i's and the β_j's are real numbers. Now we can give an explicit description of the space $\mathcal{H}_0$.

Theorem 4.2. *The space $\mathcal{H}_0$, defined in Lemma 3.1, is given by*

$$\mathcal{H}_0 = \operatorname{span}\left\{ \begin{pmatrix} \widehat{x}_{\alpha_i} \\ 0 \\ -q_0^{-1}\widehat{x}_{\alpha_i} \end{pmatrix}, \begin{pmatrix} x_{\beta_j} \\ -r^{\#}x_{\beta_j} \\ 0 \end{pmatrix} \middle|\, i \in I^+,\, j \in J^+ \right\}. \tag{4.1}$$

Proof. Note that the span on the right-hand side of (4.1) is contained in $\mathcal{H}_0$. Conversely, by definition an element $\widetilde{f} = \begin{pmatrix} f \\ g \\ h \end{pmatrix}$ with $f \in \mathcal{L}(q_0)$ and $\begin{pmatrix} g \\ h \end{pmatrix} \in \mathcal{L}(\mathcal{M}_r)$ belongs to the space $\mathcal{H}_0$ if and only if

$$f(\zeta) + \frac{g(\zeta)}{r^{\#}(\zeta)} + q_0(\zeta)h(\zeta) = 0. \tag{4.2}$$

The function $g(\zeta)/r^{\#}(\zeta)$ is a linear combination of the functions

$$(\zeta - \beta_j^*)^{-m}, \quad m = 1, \ldots, \kappa_j,\ j = 1, \ldots, k,$$

and (recall τ from (3.3)) if $\tau > 0$, of the functions ζ^{m-1}, $m = 1, \ldots, \tau$. The function $h(\zeta)$ in (4.2) is holomorphic at the points $\zeta = \beta_j$ for $j = 1, \ldots, k$ and, if $\tau > 0$, also at $\zeta = \infty$. Observe that $\lim_{z \hat{\to} \beta_j} (z - \beta_j)^n q_0(z) = 0$ for $n > 1$ and $j = 1, \ldots, k$, and, if $j \notin J^+$, also for $n = 1$. Hence multiplying both sides of the relation (4.2) by $(\zeta - \beta_j^*)^n$ for $n > 1$ and by $\zeta - \beta_j^*$ for $j \notin J^+$ and then taking the limit $\zeta \hat{\to} \beta_j^*$ we find – step by step – that $g(\zeta)/r^{\#}(\zeta)$ reduces to a sum of the form $\sum\limits_{j \in J^+} g_j x_{\beta_j}$ with $g_j \in \mathbb{C}$. Dividing both sides of (4.2) by $q_0(\zeta)$ gives

$$u(\zeta) + \frac{1}{q_0(\zeta)} \frac{g(\zeta)}{r^{\#}(\zeta)} + h(\zeta) = 0$$

with $u(\zeta) := f(\zeta)/q_0(\zeta) \in \mathcal{L}(-1/q_0)$. In the same way as above it follows that $q_0(\zeta)h(\zeta) = \sum\limits_{i \in I^+} h_i \widehat{x}_{\alpha_i}(\zeta)$ with $h_i \in \mathbb{C}$ and hence

$$f(\zeta) = -\sum_{j \in J^*} g_j x_{\beta_j}(\zeta) - \sum_{i \in I^*} h_i \widehat{x}_{\alpha_i}(\zeta),$$

which completes the proof. □

Corollary 4.3. *The dimension of $\mathcal{H}_0$ is the sum of the number of generalized poles of q that are generalized zeros of q_0 and the number of generalized zeros of q that are generalized poles of q_0.*

The following example illustrates the foregoing theorem.

Example. For $q_0 \in \mathcal{N}_0$ and $r(z) = \frac{z-\beta}{z-\overline{\alpha}}$ there appear four cases. If we identify $\mathcal{L}(\mathcal{M}_r)$ with $\mathbb{C}^2$ according to its basis given in Theorem 3.4 we obtain:

(i) If neither α is a generalized zero nor β is a generalized pole of q_0, then $\mathcal{H}_0 = \{0\}$.

(ii) If α is not a generalized zero, but β is a generalized pole of q_0, then

$$\mathcal{H}_0 = \operatorname{span}\left\{\begin{pmatrix} x_\beta \\ -1 \\ 0 \end{pmatrix}\right\}.$$

(iii) If α is a generalized zero of q_0, but β is not a generalized pole, then

$$\mathcal{H}_0 = \operatorname{span}\left\{\begin{pmatrix} \widehat{x}_\alpha \\ 0 \\ -1 \end{pmatrix}\right\}.$$

(iv) If both α is a generalized zero and β is a generalized pole of q_0, then

$$\mathcal{H}_0 = \operatorname{span}\left\{\begin{pmatrix} x_\beta \\ -1 \\ 0 \end{pmatrix}, \begin{pmatrix} \widehat{x}_\alpha \\ 0 \\ -1 \end{pmatrix}\right\}.$$

Whether a generalized zero (generalized pole) of q is also a generalized pole (generalized zero, respectively) of q_0 can also be characterized in terms of limits.

Theorem 4.4. *Let $q \in \mathcal{N}_\kappa$ be given. Then*

$$\beta \in \sigma_p(\widehat{A}_q) \cap \sigma_p(A_{q_0}) \quad \Longleftrightarrow \quad \exists\, \kappa_\beta \in \mathbb{N}: \quad \lim_{z \hat{\to} \beta} \frac{q(z)}{(z-\beta)^{2\kappa_\beta - 1}} < 0,$$

and

$$\alpha \in \sigma_p(A_q) \cap \sigma_p(\widehat{A}_{q_0}) \quad \Longleftrightarrow \quad \exists\, \nu_\alpha \in \mathbb{N}: \quad \lim_{z \hat{\to} \alpha} (z-\alpha)^{2\nu_\alpha - 1} q(z) > 0.$$

Proof. According to [L], $\beta \in \sigma_p(\widehat{A}_q)$ has degree of non-positivity κ_β if and only if

$$\lim_{z \hat{\to} \beta} \frac{q(z)}{(z-\beta)^{2\kappa_\beta - 1}} \le 0 \quad \text{and} \quad \lim_{z \hat{\to} \beta} \frac{q(z)}{(z-\beta)^{2\kappa_\beta + 1}} > 0 \text{ or } \infty.$$

From the factorization (3.2) of q we see that the first limit above is $\neq 0$ if and only if $\lim_{z \hat{\to} \beta} (z-\beta) q_0(z) \neq 0$, which happens exactly if $\beta \in \sigma_p(A_{q_0})$. The second equivalence can be proved in a similar way. □

References

[A] D. Alpay, *Reproducing kernel Kreĭn spaces of analytic functions and inverse scattering*, Doctoral dissertation, The Weizmann Institue of Science, Rehovot, Israel, 1985.

[ABDS] D. Alpay, P. Bruinsma, A. Dijksma, and H.S.V. de Snoo, *A Hilbert space associated with a Nevanlinna function*, in: M.A. Kaashoek, J.H. van Schuppen, A.C.M. Ran (editors) *Signal processing, scattering and operator theory, and numerical methods. Proceedings of the international symposium MTNS* **89**, Volume III, Birkhäuser Verlag, Basel, 1990, 115–122.

[ADRS] D. Alpay, A. Dijksma, J. Rovnyak, and H. de Snoo, *Schur functions, operator colligations, and reproducing kernel Pontryagin spaces*, Operator Theory: Adv. Appl. **96**, Birkhäuser Verlag, Basel, 1997.

[ADy] D. Alpay and H. Dym, *On applications of reproducing kernel spaces to the Schur algorithm and rational J-unitary factorization*, Operator Theory: Adv. Appl. **18**, Birkhäuser Verlag, Basel, 1986, 89–159.

[AzCD] T.Ya. Azizov, B. Ćurgus, and A. Dijksma, *Standard symmetric operators in Pontryagin spaces: A generalized von Neumann formula and minimality of boundary coefficients*, J. Functional Anal. **198** (2) (2003), 361–412.

[AzI] T.Ya. Azizov and I.S. Iokhvidov, *Foundations of the theory of linear operators in spaces with an indefinite metric*, Nauka, Moscow, 1986 (Russian); English transl. *Linear operators in spaces with an indefinite metric*, Wiley, New York, 1989.

[dB] L. de Branges, *Hilbert spaces of entire functions*, Prentice Hall, Englewood Cliffs, New Jersey, 1968.

[De] V.A. Derkach, *On generalized resolvents of Hermitian relations in Kreĭn Spaces*, J. Math. Sciences, **97** (1999), 4420–4460.

[DeH] V.A. Derkach and S. Hassi, *A reproducing kernel space model for N_κ-functions*, Proc. Amer. Math. Soc. **131** (12) (2003), 3795–3806.

[DeHS1] V. Derkach, S. Hassi, and H. de Snoo, *Operator models associated with Kac subclasses of generalized Nevanlinna functions*, Methods of Functional Analysis and Topology **5** (1) (1999), 65–87.

[DeHS2] V. Derkach, S. Hassi, and H. de Snoo, *Operator models associated with singular perturbations*, Methods of Functional Analysis and Topology **7** (3) (2001), 1–21.

[DeHS3] V. Derkach, S. Hassi, and H. de Snoo, *Rank one perturbations in a Pontryagin space with one negative square*, J. Funct. Anal. **188** (2002), 317–349.

[DLLSh] A. Dijksma, H. Langer, A. Luger, and Yu. Shondin, *A factorization result for generalized Nevanlinna functions of the class $\mathcal{N}_\kappa$*, Integral Equations Operator Theory **36** (2000), 121–125.

[DLS1] A. Dijksma, H. Langer, and H.S.V. de Snoo, *Hamiltonian systems with eigenvalue depending boundary conditions*, Operator Theory: Adv. Appl. **35** (1988), 37–83.

[DLS2] A. Dijksma, H. Langer, and H.S.V. de Snoo, *Eigenvalues and pole functions of Hamiltonian systems with eigenvalue depending boundary conditions*, Math. Nachr. **161** (1993), 107–154.

[DS] A. Dijksma and H.S.V. de Snoo, *Symmetric and self-adjoint relations in Kreĭn spaces* I, Operator Theory: Adv. Appl. **24** (1987), 145–166.

[IKL] I.S. Iokhvidov, M.G. Kreĭn, and H. Langer, *Introduction to the spectral theory of operators in spaces with indefinite metric*, Akademie-Verlag, Berlin, 1982.

[KL1] M.G. Kreĭn and H. Langer, *Über die Q-Funktion eines π-hermiteschen Operators im Raume Π_κ*, Acta Sci. Math. (Szeged) **34** (1973), 191–230.

[KL2] M.G. Kreĭn and H. Langer, *Über einige Fortsetzungsprobleme, die eng mit der Theorie hermitescher Operatoren im Raume Π_κ zusammenhängen.* I. *Einige Funktionenklassen und ihre Darstellungen*, Math. Nachr. **77** (1977), 187–236.

[KL3] M.G. Kreĭn and H. Langer, *Some propositions on analytic matrix functions related to the theory of operators on the space* Π_κ, Acta Sci. Math. (Szeged) **43** (1981), 181–205.

[L] H. Langer, *A characterization of generalized zeros of negative type of functions of the class* $\mathcal{N}_\kappa$, Operator Theory: Adv. Appl. **17**, Birkhäuser Verlag, Basel, 1986, 201–212.

[LLu] H. Langer and A. Luger, *A class of* 2×2*-matrix functions*, Glas. Mat. Ser. III **35** (55) (2000), 149–160.

[M] M.M. Malamud, *On a formula for the generalized resolvent of a nondensely defined Hermitian operator*, Ukrain. Mat. Zh. **44** (1992), 1658–1688.(Russian)

Aad Dijksma
Department of Mathematics
University of Groningen
P.O. Box 800
NL-9700 AV Groningen, The Netherlands
e-mail: a.dijksma@math.rug.nl

Heinz Langer
Institute for Analysis and Scientific Computing
Vienna University of Technology
Wiedner Hauptstrasse 8–10
A-1040 Vienna, Austria
e-mail: hlanger@mail.zserv.tuwien.ac.at

Annemarie Luger
Institute for Analysis and Scientific Computing
Vienna University of Technology
Wiedner Hauptstrasse 8–10
A-1040 Vienna, Austria
e-mail: aluger@mail.zserv.tuwien.ac.at

Yuri Shondin
Department of Theoretical Physics
Pedagogical State University
Str. Uly'anova 1, GSP 37
Nizhny Novgorod 603950, Russia
e-mail: shondin@shmath.nnov.ru

Operator Theory:
Advances and Applications, Vol. 154, 91–100

Eigenvalues and Spectral Gaps Related to Periodic Perturbations of Jacobi Matrices

Joanne Dombrowski

Abstract. This paper investigates the existence of eigenvalues in the spectral gaps of a class of Jacobi matrices resulting from periodic perturbations of Jacobi operators with smooth coefficients.

Mathematics Subject Classification (2000). 47B36, 47B15, 47B39 .

Keywords. Jacobi matrices, eigenvalues, spectral gaps.

1. Introduction

This paper considers infinite Jacobi matrices determined by a real sequence $\{a_n\}$, $a_n \neq 0$, of the following form:

$$C = \begin{bmatrix} 0 & a_1 & 0 & 0 & \dots \\ a_1 & 0 & a_2 & 0 & \dots \\ 0 & a_2 & 0 & a_3 & \dots \\ . & 0 & a_3 & 0 & \dots \\ . & . & . & . & \dots \end{bmatrix} \tag{1.1}$$

$$D_C = \left\{x = \{x_n\} \in \ell^2 : Cx \in \ell^2\right\}.$$

In general it will be assumed that $a_n = j_n + c_n$ where $\{j_n\}$ is monotone increasing with $\lim_{n\to\infty} j_n = \infty$, $\{j_n - j_{n-1}\}$ is bounded, and $c_{n+2} = c_n$. In this case, Carleman's condition $\sum\limits_{n=1}^{\infty} \frac{1}{a_n} = \infty$ implies that C is self-adjoint on the indicated maximal domain. If C is self-adjoint, then the spectral theorem asserts that it is unitarily equivalent to a multiplication operator $M_x : D \to L^2(\mu)$ defined on a dense subset D of $L^2(\mu)$ by $M_x : f(x) \to xf(x)$. If $C = \int \lambda dE_\lambda$, then the Borel measure μ is defined by $\mu(\beta) = \|E(\beta)\phi_1\|^2$, where ϕ_1 is the first standard basis vector. The vector ϕ_1 is a cyclic vector since the subdiagonal entries of the matrix are non-zero.

The given sequence $\{a_n\}$ uniquely determines a set of polynomials $\{P_n(x)\}$ defined as follows:

$$P_1(x) = 1, P_2(x) = x/a_1$$
$$P_{n+1}(x) = [xP_n(x) - a_{n-1}P_{n-1}(x)]/a_n, n \geq 2. \tag{1.2}$$

When (1.1) defines a self-adjoint operator, these polynomials form an orthonormal basis for the corresponding Hilbert space $L^2(\mu)$. They are useful in establishing the results that follow.

The results in this paper are closely related to those in [DJMP], which discusses the existence of spectral gaps under similar assumptions on the sequence $\{a_n\}$. Partial results are presented on the existence of eigenvalues in these gaps. This paper looks specifically at the eigenvalue problem, using an approach which can simplify some of the arguments in [DJMP]. Theorem 3.3 and Theorem 3.5 below extend Corollary 3.10 in [DJMP] to the case $\frac{2}{3} < p < 1$. Related ideas appear in [DP02b].

2. Preliminary results

This section considers the action of the operator C, and its spectral projections, on a particular class of vectors, which includes the eigenvectors. The next lemma generalizes some results established in [DP02a] and [DP02b].

Lemma 2.1. Let C be a self-adjoint Jacobi matrix generated by some positive sequence $\{a_n\}$. Then the spectrum of C is symmetric about the origin. If $C = \int \lambda dE_\lambda$and ψ is any vector in D_C such that $E(0,\infty)\psi = \psi(orE(-\infty,0)\psi = \psi)$ and $\psi = \sum_{n=1}^{\infty} \langle \psi, \varphi_n \rangle \varphi_n$, then

$$\sum_{n=1}^{\infty} |\langle \psi, \varphi_{2n-1} \rangle|^2 = \sum_{n=1}^{\infty} |\langle \psi, \varphi_{2n} \rangle|^2$$

Proof. Let $\{\varphi_n\}$ be the standard basis for ℓ^2. Define $U : \ell^2 \to \ell^2$ by $U\varphi_n = (-1)^{n+1}\varphi_n$. Then $UCU^{-1} = -C$ and it follows that the spectrum of C is symmetric about the origin. Viewed as an operator on $L^2(\mu)$, $UP_n(x) = (-1)^{n+1}P_n(x) = P_n(-x)$. It follows that $U\psi(x) = \psi(-x)$. If $E(0,\infty)\psi = \psi(orE(-\infty,0)\psi = \psi)$, then $\langle \psi, U\psi \rangle = 0$. Hence

$$0 = \langle \psi, C\psi \rangle = \sum_{n=1}^{\infty} |\langle \psi, \varphi_{2n-1} \rangle|^2 - \sum_{n=1}^{\infty} |\langle \psi, \varphi_{2n} \rangle|^2$$

3. Eigenvalues

The results in this section address the existence of eigenvalues. In some cases it is also possible to obtain results on absolute continuity. In the first theorem $x = y$

and so there is no gap. It will be shown that there are non-zero eigenvalues, since the spectral measure is absolutely continuous on $(-\infty, 0) \cup (0, \infty)$.

Theorem 3.1. *Let $\{j_n\}$ be a monotone increasing sequence of real numbers with difference sequence $\{d_n\}, d_n = j_n - j_{n-1}, n \geq 2$. Assume $\{d_n\}$ is bounded, $\frac{1}{2}d_3 \leq d_2$ and $\frac{1}{2}(d_{n+1} + d_{n-1}) \leq d_n, n > 2$. Let $a_{2n-1} = j_{2n-1} + x, a_{2n} = j_{2n} + x$ with x chosen so that $a_1 - \frac{1}{2}d_2 > 0$. Then C defined by (1.1) is self-adjoint and the spectral measure of C is absolutely continuous on $(-\infty, 0) \cup (0, \infty)$.*

Proof. Since $\{d_n\}$ is bounded, Carleman's condition implies that C is self-adjoint. Choose a bounded interval Δ, such that $\Delta \subset (0, \infty)$. Let $\psi = E(\Delta)\varphi_1, \psi_n = \langle \psi, \varphi_n \rangle$. Let S be the unilateral shift defined on ℓ^2 by $S\phi_n = \phi_{n+1}, n = 1, 2, \ldots$. If $J = (S - S^*)/i$ and K is defined by the commutator equation $CJ - JC = -2iK$ it follows that

$$\begin{aligned}
\langle K\psi, \psi \rangle &= a_1 |\psi_1|^2 + \sum_{n=2}^{\infty} (j_n - j_{n-1}) |\psi_n|^2 + \sum_{n=2}^{\infty} (j_n - j_{n-1}) \psi_{n-1}\psi_{n+1} \\
&\geq a_1 |\psi_1|^2 + \sum_{n=2}^{\infty} (j_n - j_{n-1}) |\psi_n|^2 \\
&\quad - \sum_{n=2}^{\infty} (j_n - j_{n-1}) \left[\frac{1}{2} |\psi_{n-1}|^2 + \frac{1}{2} |\psi_{n+1}|^2 \right] \\
&\geq a_1 |\psi_1|^2 + \sum_{n=2}^{\infty} d_n |\psi_n|^2 - \sum_{n=1}^{\infty} \frac{1}{2} d_{n+1} |\psi_n|^2 - \sum_{n=3}^{\infty} \frac{1}{2} d_{n-1} |\psi_n|^2 \\
&\geq (a_1 - \frac{1}{2}d_2) |\psi_1|^2 + (d_2 - \frac{1}{2}d_3) |\psi_2|^2 + \sum_{n=3}^{\infty} (d_n - \frac{1}{2}d_{n+1} - \frac{1}{2}d_{n-1}) |\psi_n|^2 \\
&\geq (a_1 - \frac{1}{2}d_2) |\psi_1|^2 = (a_1 - \frac{1}{2}d_2) \|E(\Delta)\varphi_1\|^4.
\end{aligned}$$

On the other hand, if λ is the midpoint of the interval Δ, then $CJ - JC = (C - \lambda I)J - J(C - \lambda I)$, and if $|\Delta|$ denotes the Lebesgue measure of the interval Δ it follows that $|\langle KE(\Delta)\phi_1, E(\Delta)\phi_1 \rangle| \leq 2\|J\| \cdot \|E(\Delta)(C - \lambda I)\phi_1\| \, \|E(\Delta)\varphi_1\| \leq \|J\| \cdot |\Delta| \cdot \|E(\Delta)\phi_1\|^2$.

The two inequalities can be combined to show that if $\mu(\Delta) = \|E(\Delta)\phi_1\|^2 \neq 0$ then $\mu(\Delta) \leq \left(\frac{\|J\|}{a_1 - \frac{1}{2}d_2} \right) |\Delta|$ so that the measure is absolutely continuous on $(-\infty, 0) \cup (0, \infty)$. □

Example 3.2. *Choose $j_n = n, x > -\frac{1}{2}$. The conditions of Theorem 3.1 are satisfied since $d_n = 1$ and $a_1 - \frac{1}{2}d_2 = 1 + x - \frac{1}{2}$.*

Theorem 3.3. *For $0 < \alpha < 1$, let $j_k = k^\alpha$. For $k \geq 1$, let $a_{2k-1} = j_{2k-1} + x, a_{2k} = j_{2k} + y$. Assume $x > y \geq -(2 - 2^\alpha)$. Then $(-|x - y|, |x - y|)$ is a gap in the essential spectrum containing no non-zero eigenvalues.*

Proof. It follows from the results in [DJMP] that $(-|x-y|, |x-y|)$ is a gap in the essential spectrum. Thus it remains to consider the issue of eigenvalues. The argument below uses the restriction of an eigenvector to the subspace spanned by the odd subscripted basis vectors to show that any non-zero eigenvalue must be outside the gap.

It follows from Lemma 2.1 above that for any vector ψ in D_C and $\lambda \neq 0$ such that $C\psi = \lambda\psi$,

$$
\begin{aligned}
\frac{1}{2}\langle C^2\psi, \psi\rangle &= \frac{1}{2}\langle C\psi, C\psi\rangle \\
&= \sum_{k=1}^{\infty} \left[((2k-1)^\alpha + x)\psi_{2k-1} + ((2k)^\alpha + y)\psi_{2k+1}\right]^2 \\
&= \sum_{k=1}^{\infty} \left[((2k-1)^\alpha + y)\psi_{2k-1} + ((2k)^\alpha + y)\psi_{2k+1}\right]^2 \\
&\quad + \sum_{k=1}^{\infty} 2(x-y)\left[((2k-1)^\alpha + y)\psi_{2k-1} + ((2k)^\alpha + y)\psi_{2k+1}\right]\psi_{2k-1} \\
&\quad + (x-y)^2 \sum_{k=1}^{\infty} \psi_{2k-1}^2 \\
&\geq \sum_{k=1}^{\infty} 2(x-y)((2k-1)^\alpha + y)\psi_{2k-1}^2 \\
&\quad - 2(x-y)\sum_{k=1}^{\infty}\left\{\frac{1}{2}((2k)^\alpha + y)\psi_{2k+1}^2 + \frac{1}{2}((2k)^\alpha + y)\psi_{2k-1}^2\right\} \\
&\quad + \sum_{k=1}^{\infty} (x-y)^2\psi_{2k-1}^2 \\
&\geq 2(x-y)\sum_{k=1}^{\infty}\left[(2k-1)^\alpha - \frac{1}{2}(2k)^\alpha + \frac{1}{2}y\right]\psi_{2k-1}^2 \\
&\quad - 2(x-y)\sum_{k=2}^{\infty}\frac{1}{2}((2k-2)^\alpha + y)\psi_{2k-1}^2 + \sum_{k=1}^{\infty}(x-y)^2\psi_{2k-1}^2 \\
&\geq 2(x-y)\sum_{k=2}^{\infty}\left[(2k-1)^\alpha - \frac{1}{2}(2k)^\alpha - \frac{1}{2}(2k-2)^\alpha\right]\psi_{2k-1}^2 \\
&\quad + \sum_{k=1}^{\infty}(x-y)^2\psi_{2k-1}^2 + 2(x-y)\left[1 - \frac{1}{2}(2)^\alpha + \frac{1}{2}y\right]\psi_1^2 \\
&\geq \sum_{k=1}^{\infty}(x-y)^2\psi_{2k-1}^2 \,.
\end{aligned}
$$

Thus it follows that $\langle C^2\psi, \psi\rangle \geq (x-y)^2\|\psi\|^2$. □

Corollary 3.4. *If $x > y > -(2-2^\alpha)$ then $\langle C^2\psi, \psi\rangle > (x-y)^2 \|\psi\|^2$ and so the endpoints of the gap $\pm|x-y|$ cannot be eigenvalues.*

Theorem 3.5. *For $0 < \alpha < 1$, let $j_k = k^\alpha$. For $k \geq 1$, let $a_{2k-1} = j_{2k-1} + x, a_{2k} = j_{2k} + y$. Assume $y > x \geq -(2^{\alpha+1} - 3^\alpha)$. Then $(-|x-y|, |x-y|)$ is a gap in the essential spectrum containing no non-zero eigenvalues.*

Proof. As in the proof above, it follows from the results in [DJMP] that $(-|x-y|, |x-y|)$ is a gap in the essential spectrum. It remains then to consider the issue of non-zero eigenvalues. The argument below uses the restriction of the eigenvector to the subspace spanned by the even subscripted basis vectors. It will again be shown that the corresponding eigenvalue must be outside the gap.

It follows from Lemma 2.1 that for any vector ψ in D_C and $\lambda \neq 0$ such that $C\psi = \lambda\psi$,

$$
\begin{aligned}
\frac{1}{2}\langle C^2\psi, \psi\rangle &= \frac{1}{2}\langle C\psi, C\psi\rangle \\
&= a_1^2 |\psi_2|^2 + \sum_{k=1}^{\infty} [a_{2k}\psi_{2k} + a_{2k+1}\psi_{2k+2}]^2 \\
&= (1+x)^2 |\psi_2|^2 + \sum_{k=1}^{\infty} [((2k)^\alpha + y)\psi_{2k} + ((2k+1)^\alpha + x)\psi_{2k+2}]^2 \\
&= (1+x)^2 |\psi_2|^2 + \sum_{k=1}^{\infty} [((2k)^\alpha + x)\psi_{2k} + ((2k+1)^\alpha + x)\psi_{2k+2}]^2 \\
&\quad + 2(y-x)\sum_{k=1}^{\infty} [((2k)^\alpha + x)\psi_{2k} + ((2k+1)^\alpha + x)\psi_{2k+2}]\psi_{2k} \\
&\quad + (y-x)^2 \sum_{k=1} |\psi_{2k}|^2 \\
&\geq (1+x)^2 |\psi_2|^2 + 2(y-x)\sum_{k=1}^{\infty} ((2k)^\alpha + x)|\psi_{2k}|^2 \\
&\quad - 2(y-x)\sum_{k=1}^{\infty} \left[\frac{1}{2}((2k+1)^\alpha + x)|\psi_{2k}|^2 + \frac{1}{2}((2k+1)^\alpha + x)|\psi_{2k+2}|^2\right] \\
&\quad + \sum_{k=1}^{\infty} (y-x)^2 |\psi_{2k}|^2 \\
&\geq (1+x)^2 |\psi_2|^2 + 2(y-x)\sum_{k=1}^{\infty} ((2k)^\alpha - \frac{1}{2}(2k+1)^\alpha + \frac{1}{2}x)|\psi_{2k}|^2 \\
&\quad - 2(y-x)\frac{1}{2}\sum_{k=2}^{\infty} ((2k-1)^\alpha + x)|\psi_{2k}|^2 + \sum_{K=1}^{\infty} (y-x)^2 |\psi_{2k}|^2
\end{aligned}
$$

$$\geq (1+x)^2 |\psi_2|^2 + 2(y-x)\sum_{k=2}^{\infty}((2k)^\alpha - \frac{1}{2}(2k+1)^\alpha - \frac{1}{2}(2k-1)^\alpha)|\psi_{2k}|^2$$

$$+2(y-x)((2)^\alpha - \frac{1}{2}(3)^\alpha + \frac{1}{2}x)|\psi_2|^2 + \sum_{k=1}^{\infty}(y-x)^2 |\psi_{2k}|^2 .$$

It follows that $\langle C^2\psi, \psi\rangle \geq (y-x)^2 \|\psi\|^2$. □

Corollary 3.6. *If $y > x > -(2^{\alpha+1} - 3^\alpha)$ then $\langle C^2\psi,\psi\rangle > (y-x)^2\|\psi\|^2$ and thus the endpoints of the gap $\pm|x-y|$ cannot be eigenvalues.*

It remains to consider the case $\alpha = 1$. The next theorem summarizes results established in [DJMP].

Theorem 3.7. *Choose $x > -1, y > -2$. Let $a_{2n-1} = (2n-1)+x, a_{2n} = (2n)+y$. Then $(-|x-y|, |x-y|)$ is a gap in the essential spectrum of C containing at most 2 eigenvalues. Furthermore, 0 is an eigenvalue if and only if $y > x$. If $y > x$ then 0 is the only eigenvalue in the spectral gap.*

With relatively minor changes in the proof, this result can be improved to allow the possibility of negative entries in (1.1) so that the sequence $\{a_n\}$ is still monotone increasing with $\lim_{n\to\infty} a_n = \infty$. An outline of the proof will be given indicating the needed changes.

Theorem 3.8. *Let $a_{2n-1} = (2n-1)+x, a_{2n} = (2n)+y$. Assume $a_n \neq 0\,\forall n$. Then $(-|x-y|, |x-y|)$ is a gap in the essential spectrum of C. Let N be the smallest positive integer so that $N + \frac{(x-1)}{2} \geq 0, N + \frac{y}{2} \geq 0$ Then the spectral gap contains at most $2N$ eigenvalues. Furthermore, 0 is an eigenvalue if and only if $y > x$.*

Proof. Carleman's condition implies that C defined by (1.1) is self-adjoint. Let A^+ denote the restriction of C^2 to the subspace spanned by the basis vectors $\{\varphi_{2k-1}\}_{k=1}^{\infty}$. Then A^+ is tridiagonal with subdiagonal sequence $a_n^+ = a_{2n-1}a_{2n}$ and diagonal sequence $b_n^+ = a_{2n-2}^2 + a_{2n-1}^2$. Obtain A^- from A^+ by negating the diagonal entries. Define the diagonal operator D so that $A^- = \tilde{A}^- + D$ where $\tilde{A}^-$ has row sums equal to 0. Then $D\varphi_n = [-(a_{2n-2}^2 + a_{2n-1}^2) + (a_{2n-3}a_{2n-2} + a_{2n-1}a_{2n})]\varphi_n$ if $a_0 = 0$.

$$\langle A^-\psi,\psi\rangle = \left\langle \tilde{A}^-\psi,\psi\right\rangle + \langle D\psi,\psi\rangle$$

$$= -\sum_{n=1}^{\infty} a_n^+ |\psi_{n+1} - \psi_n|^2 + \langle D\psi,\psi\rangle$$

$$= -\sum_{n=1}^{\infty}(2n-1+x)(2n+y)|\psi_{n+1}-\psi_n|^2 + \langle D\psi,\psi\rangle$$

$$= -4\sum_{n=1}^{\infty}(n + \frac{x-1}{2})(n+\frac{y}{2})|\psi_{n+1}-\psi_n|^2 + \langle D\psi,\psi\rangle .$$

Let N be the smallest positive integer so that $N + \frac{(x-1)}{2} \geq 0, N + \frac{y}{2} \geq 0$. Then for any $\psi = \{\psi_n\}, \psi_n = 0, n = 1, \ldots, N$,

$$\begin{aligned}\langle A^- \psi, \psi\rangle &\leq -4 \sum_{n=1}^{\infty} (n + N + \frac{x-1}{2})(n + N + \frac{y}{2}) \left|\psi_{n+1} - \psi_n\right|^2 + \langle D\psi, \psi\rangle \\ &\leq (-4)\frac{1}{4} \sum_{n=1}^{\infty} \left|\psi_n\right|^2 + \langle D\psi, \psi\rangle \\ &\leq -(x-y)^2 \sum_{n=1}^{\infty} \left|\psi_n\right|^2 .\end{aligned}$$

Hence $\langle A^+ \psi, \psi\rangle \geq (x-y)^2 \sum_{n=1}^{\infty} \left|\psi_n\right|^2$ and it follows that $(-\left|x-y\right|, \left|x-y\right|)$ is a gap in the essential spectrum of C containing at most $2N$ eigenvalues. □

The next result gives sufficient conditions for the existence of non-zero eigenvalues in the gap.

Theorem 3.9. *Let $a_{2n-1} = (2n-1) + x, a_{2n} = (2n) + y$. Assume $a_n \neq 0 \, \forall n$. If $x \geq y$ and $\left|1+x\right| \leq \left|x-y\right|$ then C defined by (1.1) has a non-zero eigenvalue in the gap $(-\left|x-y\right|, \left|x-y\right|)$.*

Proof. It was shown in [DP02b] that if $a_{n+1} - a_n \geq 0$ with $\delta = a_{2n} - a_{2n-1}, d = a_{2n+1} - a_n$, then 0 is an eigenvalue if and only if $\delta > d$. In this case, it follows that 0 is an eigenvalue if and only if $y > x$. Also, replacing a_n by $|a_n|$ leads to a unitarily equivalent operator. Note that $\lambda = |1+x|$ is a root for $P_3(\lambda)$. Let $s(\lambda) = P_n(\lambda)$. Then the modified Wronskian $W_n(s(\lambda), s(-\lambda)) = (-1)^n 2a_n P_n(\lambda) P_{n+1}(\lambda)$ has at least one node. Therefore by the results in [T96] the interval $(-\lambda, \lambda)$contains at least one eigenvalue, which in this case must be non-zero. □

Remark 3.10. *If $x > -1$, and $x \geq y$, then $|1+x| \leq |x-y|$ implies that $y \leq -1$. If $x \leq -1$ and $x \geq y$, then $|1+x| \leq |x-y|$ implies that $y \leq 2x+1$.*

Example 3.11. *Let $x = -\frac{1}{2}, y = -1$. Then the spectral gap is $(-\frac{1}{2}, \frac{1}{2})$ and it follows from the above result that there is at least one non-zero eigenvalue in the spectral gap. By symmetry there must be at least two eigenvalues in the spectral gap, and by Theorem 3.9, with $N = 2$, there are exactly two non-zero eigenvalues in this gap.*

4. Special case: $j_{2n-1} = j_{2n}$

In this section it will generally be assumed that $a_{2n-1} = n^\alpha + x, a_{2n} = n^\alpha + y, 0 < \alpha \leq 1$, with x and y chosen so that $a_n \neq 0 \, \forall n$.

Theorem 4.1. *For $x \geq y \geq 0$, let $a_{2n-1} = n + x, a_{2n} = n + y$. Then for C defined by (1.1)*

$$(-\left|\tfrac{1}{2} + x - y\right|, \left|\tfrac{1}{2} + x - y\right|)$$

is a gap in the essential spectrum no non-zero eigenvalues.

Proof. It was shown in [DJMP] that $(-\left|\frac{1}{2}+x-y\right|, \left|\frac{1}{2}+x-y\right|)$ is a gap in the essential spectrum. Thus it remains to consider the eigenvalues. The argument below considers the restriction of a possible eigenvector to the subspace spanned by the odd subscripted basis vectors. For any vector ψ in D_C and $\lambda \neq 0$ such that $C\psi = \lambda\psi$.

$$\begin{aligned}
\frac{1}{2}\left\langle C^2\psi, \psi\right\rangle &= \frac{1}{2}\left\langle C\psi, C\psi\right\rangle \\
&= \sum_{k=1}^{\infty} \left[(k+x)\psi_{2k-1} + (k+y)\psi_{2k+1}\right]^2 \\
&= \sum_{k=1}^{\infty} \left[(k+y)\psi_{2k-1} + (k+y)\psi_{2k+1} + (x-y)\psi_{2k-1}\right]^2 \\
&= \sum_{k=1}^{\infty} \left[(k+y)\psi_{2k-1} + ((k+y)\psi_{2k+1}\right]^2 \\
&\quad + 2(x-y)\sum_{k=1}^{\infty} \left[(k+y)\psi_{2k-1} + (k+y)\psi_{2k+1}\right]\psi_{2k-1} + (x-y)^2 \sum_{k=1}^{\infty} \left|\psi_{2k-1}\right|^2 \\
&\geq \sum_{k=1}^{\infty} k^2 \left|\psi_{2k-1} + \psi_{2k+1}\right|^2 \\
&\quad + 2(x-y)\sum_{k=1}^{\infty} \left[(k+y)\psi_{2k-1}^2 - \frac{1}{2}(k+y)\psi_{2k+1}^2 - \frac{1}{2}(k+y)\psi_{2k-1}^2\right] \\
&\quad + (x-y)^2 \sum_{k=1}^{\infty} \left|\psi_{2k-1}\right|^2 \\
&\geq \sum_{k=1}^{\infty} k^2 \left|\psi_{2k-1} + \psi_{2k+1}\right|^2 \\
&\quad + 2(x-y)\sum_{k=1}^{\infty} \left[\frac{1}{2}(k+y)\psi_{2k-1}^2 - \frac{1}{2}(k+y)\psi_{2k+1}^2\right] + (x-y)^2 \sum_{k=1}^{\infty} \left|\psi_{2k-1}\right|^2 \\
&\geq \sum_{k=1}^{\infty} k^2 \left|\psi_{2k-1} + \psi_{2k+1}\right|^2 + (x-y)\left[(1+y)\psi_1^2 + \sum_{k=1}^{\infty} \psi_{2k+1}^2\right] \\
&\quad + (x-y)^2 \sum_{k=1}^{\infty} \left|\psi_{2k-1}\right|^2 \\
&\geq \left(\frac{1}{2}+x-y\right)^2 \sum_{k=1}^{\infty} \left|\psi_{2k-1}\right|^2.
\end{aligned}$$

Hence $\left\langle C^2\psi, \psi\right\rangle \geq (\frac{1}{2}+x-y)^2 \sum_{k=1}^{\infty} |\psi_k|^2$. □

Corollary 4.2. *If $x > y > 0$, then the endpoints of the gap $\pm\left|\frac{1}{2}+x-y\right|$ cannot be eigenvalues.*

Proof. In this case the above proof shows that for any vector ψ in D_C and $\lambda \neq 0$ such that $C\psi = \lambda\psi$, $\langle C^2\psi, \psi\rangle > (\frac{1}{2}+x-y)^2 \sum_{k=1}^{\infty} |\psi_k|^2$. □

Remark 4.3. *It should be noted that it follows from Theorem 2.1 in [DP02] that if $x > -1, 0 \leq y-x \leq 1, y+\frac{1}{2}(y-x) > 0$ then there are no non-zero eigenvalues in the spectral gap $(-\left|\frac{1}{2}+x+y\right|, \left|\frac{1}{2}+x+y\right|)$, since the spectral measure is absolutely continuous. Furthermore, a careful look at the proof indicates that the conclusion holds if $x \neq -1$ and $0 \leq y-x \leq 1, y+\frac{1}{2}(y-x) > 0$. Thus, for example, the conclusion holds if $x = -\frac{5}{4}, y = -\frac{1}{4}$. In this case $a_1 = 1+x = -\frac{1}{4}$.*

Theorem 4.4. *For $x > y > -1$, let $a_{2n-1} = n^\alpha + x, a_{2n} = n^\alpha + y, 0 < \alpha < 1$. Then for C defined by (1.1), $(-|x-y|, |x-y|)$ is a gap in the essential spectrum, and the closed interval $[-|x-y|, |x-y|]$ contains no non-zero eigenvalues.*

Proof. It was shown in [DJMP] that $(-|x-y|, |x-y|)$ is a gap in the essential spectrum. Thus it only remains to consider the eigenvalues. The argument below uses the restriction of a possible eigenvector to the subspace spanned by the odd subscripted basis vectors. For any vector ψ in D_C and $\lambda \neq 0$ such that $C\psi = \lambda\psi$.

$$
\begin{aligned}
\frac{1}{2}\langle C^2\psi, \psi\rangle &= \frac{1}{2}\langle C\psi, C\psi\rangle \\
&= \sum_{k=1}^{\infty} \left[(k^\alpha + x)\psi_{2k-1} + (k^\alpha + y)\psi_{2k+1}\right]^2 \\
&= \sum_{k=1}^{\infty} \left[(k^\alpha + y)\psi_{2k-1} + (k^\alpha + y)\psi_{2k+1} + (x-y)\psi_{2k-1}\right]^2 \\
&= \sum_{k=1}^{\infty} \left[(k^\alpha + y)\psi_{2k-1} + (k^\alpha + y)\psi_{2k+1}\right]^2 \\
&\quad + 2(x-y)\sum_{k=1}^{\infty} \left[(k^\alpha + y)\psi_{2k-1} + (k^\alpha + y)\psi_{2k+1}\right]\psi_{2k-1} \\
&\quad + (x-y)^2 \sum_{k=1}^{\infty} |\psi_{2k-1}|^2 \\
&\geq 2(x-y)\sum_{k=1}^{\infty}\left[(k^\alpha + y)\psi_{2k-1}^2 - \frac{1}{2}(k^\alpha + y)\psi_{2k+1}^2 - \frac{1}{2}(k^\alpha + y)\psi_{2k-1}^2\right] \\
&\quad + (x-y)^2 \sum_{k=1}^{\infty} |\psi_{2k-1}|^2 \\
&\geq (x-y)\sum_{k=1}^{\infty}(k^\alpha + y)\psi_{2k-1}^2 - (x-y)\sum_{k=1}^{\infty}(k^\alpha + y)\psi_{2k+1}^2 + (x-y)^2\sum_{k=1}^{\infty}\psi_{2k-1}^2
\end{aligned}
$$

$$\geq (x-y)(1+y)\psi_1^2 + (x-y)\sum_{k=1}^{\infty}\left[((k+1)^\alpha - k^\alpha)\psi_{2k+1}^2\right] + (x-y)^2\sum_{k=1}^{\infty}\psi_{2k-1}^2$$

$$> (x-y)^2\sum_{k=1}^{\infty}\psi_{2k-1}^2.$$

Thus $\|C\psi\|^2 > (x-y)^2\,\|\psi\|^2$. □

References

[B68] J. Berezanskii, *Expansions in Eigenfunctions of Self-adjoint Operators*, Trans. Math. Monographs. **17** Amer. Math. Soc. Providence, RI 1968.

[D87] J. Dombrowski, *Cyclic operators, commutators and absolutely continuous measures*, **100** (1987), 457–463.

[DP95] J. Dombrowski and S. Pedersen, *Orthogonal polynomials, spectral measures and absolute continuity*, J. Comput. Appl. Math **65** (1995), 115–124.

[DP97] ——, *Spectral measures and Jacobi matrices related to Laguerre type systems of orthogonal polynomials*, Constr. Approx. **13** (1997), 421–433.

[DP02a] ——, *Absolute continuity for Jacobi matrices with constant row sums*, J. Math. Anal. Appl. **277** (2002), 695–713.

[DP02b] ——, *Spectral Transition parameters for a class of Jacobi matrices*, Studia Math. **152** (2002), 217–229.

[DJMP] J. Dombrowski, J. Janas. M. Moszynski and S. Pedersen, *Spectral gaps resulting from periodic perturbations of a class of Jacobi operators*, Constr. Approx. (to appear.)

[RS78] M. Reed and B. Simon, *Methods of Modern Mathematical Physics IV: Analysis of Operators*, Academic Press, New York, 1978.

[T96] G. Teschl, *Oscillation theory and renormalized oscillation theory for Jacobi operators*, Jour. Differential Equations **129** (1996), 532–558.

Joanne Dombrowski
Department of Mathematics
Wright State University
Dayton, OH 45435, USA
e-mail: joanne.dombrowski@wright.edu

Operator Theory:
Advances and Applications, Vol. 154, 101–138

WKB and Turning Point Theory for Second-order Difference Equations

Jeffrey S. Geronimo, Oscar Bruno and Walter Van Assche

Abstract. A turning point method for difference equations is developed. This method is coupled with the LG-WKB method via matching to provide approximate solutions to the initial value problem. The techniques developed are used to provide strong asymptotics for Hermite polynomials.

Mathematics Subject Classification (2000). Primary 39A10; Secondary 41A60, 65Q05.

Keywords. Turning point, difference equations, WKB, orthogonal polynomials, hermite polynomials.

1. Introduction

In this article we will be interested developing a turning point theory for difference equations of the form

$$a_{n+1}\psi_{n+1}(x) + (b_n - x)\psi_n(x) + a_n\psi_{n-1}(x) = 0, \tag{1.1}$$

with $a_n > 0$, and b_n real and matching the solutions obtained from this theory to those obtained from the LG-WKB method. Various asymptotics for the solutions of (1.1) with $\psi_{-1} = 0$, $\psi_0 = 1$ when the coefficients tend to infinity in magnitude such as nth root asymptotics, ratio asymptotics, strong asymptotics, and weak asymptotics, have been investigated by a large number of authors (Braun [B], Geronimo-Smith [GS], Maejima-Van Assche [MV], Nevai-Dehesa [ND], Schulten-Gordon [SG], Van Assche [V], Van Assche-Geronimo [VG], etc.). However with the exception of weak asymptotics these studies required x to be exterior to the region where the solutions of the above difference equation oscillate. In order to control the growth of the coefficients it was suggested in Nevai-Dehesa and further developed in Van Assche, and Van Assche et al. that the preliminary change of

J.S.G. was supported by NSF grants. J.S.G. and W.V.A. were supported by a NATO grant.

variables $x = \lambda_l y$ be performed so that $a(n,l) = \frac{a_n}{\lambda_l}$ and $b(n,l) = \frac{b_n}{\lambda_l}$ for $n \le l$ are bounded functions of n and l. This produces the difference equation

$$a(n+1,l)\tilde{\psi}(n+1,y,l) + (b(n,l) - y)\tilde{\psi}(n,y,l) + a(n,l)\tilde{\psi}(n-1,y,l) = 0. \quad (1.2)$$

To study the solutions of (1.2) in the oscillatory region and near the turning points a WKB method based upon an epsilon difference equation (see equation 2.6) was proposed by Deift and McLaughlin [DM] and further developed by Costin and Costin [CC]. While satisfactory away from the turning points, i.e., where $\frac{y-b(n,l)}{a(n,l)} \approx \pm 2$, this technique requires the matching of various solution of (1.2) in neighborhoods of the turning points, that shrink as $l^{-1/2}$ as l tends to infinity. As noted by Wang and Wong [WW] in their study of the difference equation satisfied by Bessel functions the above theory does not lead to a satisfactory uniform asymptotic expansion for Bessel functions. Other methods to study (1.2) have been proposed by Dingle and Morgan [DM] and Schulten and Gordon [SG], and we will use these methods to modify the technique of Deift and McLaughlin. We proceed as follows: in Section 2 we motivate the turning point technique to be discussed. In particular a Langer transformation for difference equations is proposed. In the next section (Section 3) this technique is put on firm ground and it is extended to the complex plane in Section 4. One of the main drawbacks of the techniques of the preceeding sections is that they cannot be used to solve the initial value problem for coefficients in (1.2) that tend to zero as l tends to infinity, for instance when the coefficients in (1.1) tend to infinity. In order to overcome this we recall, in Section 5 the WKB method proposed by Deift and McLaughlin away from the turning points and following Geronimo and Smith modify it to be applicable to the initial value problem. In Section 6 these techniques are used to obtain uniform asymptotics for Hermite polynomials. In a sequel to this article [G] the theory of external fields will be used to help obtain uniform asymptotics for other sets of orthogonal polynomials including those associated with discrete measures.

2. Motivation

a. Differential equations – the Langer transformation

Consider the differential equation,

$$\left(\frac{d^2}{dt^2} - \epsilon^{-2}k(t)^2\right)\phi(t) = 0 \quad k : \mathbf{R} \to \mathbb{C} \quad (2.1)$$

where $k^2 : \mathbf{R} \to \mathbf{R}$ is assumed to be a smooth monotonically increasing function of t in $[t_{in}, t_{fi}]$ $t_{in} = t_{initial}, t_{fi} = t_{final}$ with one simple zero, t_0, in the region. Away from t_0 the LG-WKB method ([0, p. 191]) gives two solutions

$$\phi_\pm(t) = \frac{1}{(|k^2(t)|)^{\frac{1}{4}}} e^{\pm\epsilon^{-1}\int^t k(u)du}(1 + O(\epsilon)).$$

While adequate in this region the asymptotics provided by these solutions breaks down in a neighborhood of t_0. The standard procedure introduced by Langer([L],

[O, p. 398]) to overcome this difficulty is to make the change of variables

$$\frac{2}{3}\rho^{3/2}(t) = \int_{t_0}^{t} k(u)du \tag{2.2}$$

which yields the equation,

$$\frac{d^2}{d\rho^2}\Phi = (\epsilon^{-2}\rho + \delta(\rho))\Phi,$$

where $\Phi = (\rho')^{1/2}\phi$, $\delta = \hat{k}^{-1/2}\frac{d^2}{d\rho^2}\hat{k}^{1/2}$, with $\hat{k}^2 = \frac{k^2}{\rho}$. Langer showed that under suitable conditions there are two solutions of (2.1) which provide uniform asymptotics in a neighborhood of t_0, namely

$$\phi_1 = \hat{k}^{-1/2}\mathrm{Ai}(\epsilon^{-2/3}\rho) + O(\epsilon)f(\epsilon^{-2/3}\rho),$$

and

$$\phi_2 = \hat{k}^{-1/2}\mathrm{Bi}(\epsilon^{-2/3}\rho) + O(\epsilon)f(\epsilon^{-2/3}\rho)^{-1},$$

where the Airy functions $\mathrm{Ai}(\epsilon^{-2/3}\rho)$ and $\mathrm{Bi}(\epsilon^{-2/3}\rho)$ satisfy the differential equation,

$$\frac{d^2}{d\rho^2}\chi = \epsilon^{-2}\rho\chi, \tag{2.3}$$

and f is a function that decays at the same rate as Ai for large values of its argument.

b. A model equation

In order to find an analog of the Langer transformation for difference equations as well as the corresponding solutions we begin by considering the model equation

$$\psi(t+\epsilon) + \psi(t-\epsilon) - 2\cosh k(t)\psi = 0.$$

Following Schulten and Gordon we search for approximate solutions by making the ansatz $\psi(t) = g(t)\mathrm{Ai}(\epsilon^{-2/3}\rho(t))$ where the functions g and ρ will be chosen so that ψ satisfies the difference equation up to order ϵ^2. Roughly speaking this will allow us to show that there are actual solutions to the above difference equation that are within ϵ of these approximate solutions after ϵ^{-1} steps. Using the asymptotic formula for Ai (Olver[O] p. 392),

$$\mathrm{Ai}(\epsilon^{-2/3}\rho) = \frac{e^{-2/3\epsilon^{-1}\rho^{3/2}}}{2\sqrt{\pi}\epsilon^{-1/6}\rho^{1/4}}\left(1 - \frac{\mu_1\epsilon}{\frac{2}{3}\rho^{3/2}}\right) + O(\epsilon^2),$$

we find

$$\psi\,(t\pm\epsilon) = (g(t) \pm \epsilon g'(t))$$
$$\frac{e^{-2/3\epsilon^{-1}\rho^{3/2}(t\pm\epsilon)}}{2\sqrt{\pi}\epsilon^{-1/6}\rho(t\pm\epsilon)^{1/4}}\left(1 - \frac{\mu_1\epsilon}{\frac{2}{3}\rho(t\pm\epsilon)^{3/2}}\right) + O(\epsilon^2).$$

With $f = \frac{2}{3}\rho^{3/2}$ write,

$$e^{\epsilon^{-1}f(t)}e^{-\epsilon^{-1}f(t\pm\epsilon)} = e^{\pm f'(t)}\left(1 - \frac{\epsilon f''(t)}{2}\right) + O(\epsilon^2).$$

Thus,

$$e^{\epsilon^{-1}f}\psi(t\pm\epsilon) = (g(t)\pm\epsilon g'(t))\left(1 - \frac{\epsilon f''(t)}{2}\right)\left(1 \mp \frac{\epsilon f'(t)}{6f(t)}\right)$$
$$\left(1 - \frac{\epsilon\mu_1}{f(t)}\right)\frac{e^{\pm f'(t)}}{2\sqrt{\pi}\epsilon^{-1/6}\rho^{1/4}(t)} + O(\epsilon^2),$$

which implies

$$\psi(t+\epsilon) + \psi(t-\epsilon) - 2\cosh k(t)\psi(t)$$
$$= (e^{-f'(t)} + e^{f'(t)} - 2\cosh k(t))\psi(t) + \epsilon\left(\frac{f'(t)}{3f(t)} - \frac{\cosh f'(t)}{\sinh f'(t)}f''(t) - \frac{2g'(t)}{g(t)}\right)$$
$$\times g(t)\sinh f'(t)\frac{e^{-\epsilon^{-1}f(t)}}{2\sqrt{\pi}\epsilon^{-1/6}\rho(t)^{1/4}}\left(1 - \frac{\epsilon\mu_1}{f(t)}\right) + O(\epsilon^2).$$

In order to have only the $O(\epsilon^2)$ term remain choose f and g so that

$$f'(t) = k(t),$$

and

$$\frac{g'(t)}{g(t)} = -\frac{1}{2}\frac{\cosh k(t)k(t)'}{\sinh k(t)} + \frac{\rho'(t)}{4\rho(t)}.$$

Thus

$$\frac{2}{3}\rho^{3/2} = \int k(t)dt,$$

and

$$g(t) = \left(\frac{\rho(t)}{\sinh^2 k(t)}\right)^{1/4}.$$

With this choice of ρ and g, ψ satisfies the second order difference equation up to order ϵ^2.

c. The Langer transformation for difference equations

For difference equations of the form

$$a_{n+1}\psi_{n+1}(x) + (b_n - x)\psi_n(x) + a_n\psi_{n-1}(x) = 0, \tag{2.4}$$

with $a_n > 0$, and b_n real, we assume that a_n and b_n are discretizations of the at least three times differentiable functions $a(u)$ and $b(u)$ respectively. With $u = \frac{t}{\epsilon}$ set

$$a(t,\epsilon) = \frac{a(\frac{t}{\epsilon})}{\lambda_\epsilon} \quad \text{and} \quad b(t,\epsilon) = \frac{b(\frac{t}{\epsilon})}{\lambda_\epsilon}, \tag{2.5}$$

then (2.4) becomes the *epsilon* difference equation

$$a(t_{n+1},\epsilon)\tilde{\psi}(t_{n+1},y,\epsilon) + (b(t_n,\epsilon) - y)\tilde{\psi}(t_n,y,\epsilon) + a(t_n,\epsilon)\tilde{\psi}(t_{n-1},y,\epsilon) = 0, \tag{2.6}$$

where $t_n = n\epsilon$, $x = \lambda_\epsilon y$, $\tilde{\psi}(t_n, y, \epsilon) = \psi_n(\lambda_\epsilon y)$, and λ_ϵ is chosen so that $a(t,\epsilon)$ and $b(t,\epsilon)$ are bounded functions of ϵ. In order to find the modifications to the g and k above write $a(t+\epsilon,\epsilon) = a(t+\epsilon/2,\epsilon) + \frac{\epsilon a'(t+\epsilon/2,\epsilon)}{2} + O(\epsilon^2)$ and $a(t,\epsilon) = a(t+\epsilon/2,\epsilon) - \frac{\epsilon a'(t+\epsilon/2,\epsilon)}{2} + O(\epsilon^2)$ and set

$$\cosh k(t,y,\epsilon) = \frac{y - b(t,\epsilon)}{2a\left(t+\frac{\epsilon}{2},\epsilon\right)} \qquad \text{for } t \geq t_0, \tag{2.7}$$
$$\cos k(t,y,\epsilon) = \frac{y - b(t,\epsilon)}{2a\left(t+\frac{\epsilon}{2},\epsilon\right)} \qquad \text{for } t \leq t_0.$$

Here t_0 is such that $\frac{y-b(t_0,\epsilon)}{2a(t_0+\frac{\epsilon}{2})} = 1$. We now define the Langer transformation for the above difference equation as

$$\frac{2}{3}\rho^{3/2}(t,y,\epsilon) = \int_{t_0}^{t} \cosh^{-1}\left(\frac{y-b(u,\epsilon)}{2a\left(u+\frac{\epsilon}{2},\epsilon\right)}\right)du \qquad \text{for } t \geq t_0, \tag{2.8}$$
$$\frac{2}{3}(-\rho)^{3/2}(t,y,\epsilon) = \int_{t}^{t_0} \cos^{-1}\frac{y-b(u,\epsilon)}{2a\left(u+\frac{\epsilon}{2},\epsilon\right)}du \qquad \text{for } t \leq t_0.$$

The function g can be written as

$$g(t,y,\epsilon) = \begin{cases} \left(\frac{\rho(t,y,\epsilon)}{a^2\left(t+\frac{\epsilon}{2},\epsilon\right)\sinh^2 k(t,y,\epsilon)}\right)^{1/4} & t \geq t_0, \\ \left(\frac{\rho(t,y,\epsilon)}{a^2\left(t+\frac{\epsilon}{2},\epsilon\right)\sin^2 k(t,y,\epsilon)}\right)^{1/4} & t < t_0, \end{cases} \tag{2.9}$$

and the corresponding approximate solutions are given by

$$\psi^1(t,y,\epsilon) = g(t,y,\epsilon)\mathrm{Ai}(\epsilon^{-\frac{2}{3}}\rho(t,y,\epsilon)), \tag{2.10}$$

and

$$\psi^2(t,y,\epsilon) = g(t,y,\epsilon)\mathrm{Bi}(\epsilon^{-\frac{2}{3}}\rho(t,y,\epsilon)). \tag{2.11}$$

From (2.7) we can rewrite k as

$$k(t,y,\epsilon) = \ln\left(\frac{y-b(t,\epsilon)}{2a(t+\frac{\epsilon}{2},\epsilon)} + \sqrt{\left(\frac{y-b(t,\epsilon)}{2a(t+\frac{\epsilon}{2},\epsilon)}\right)^2 - 1}\right), \tag{2.12}$$

where the branch of the square root is chosen so that $\ln(z+\sqrt{z^2-1}) \sim \ln z$ for z large. Then (2.8) can be written in the compact form

$$\frac{2}{3}\rho^{\frac{3}{2}}(t,y,\epsilon) = \int_{t_0}^{t} k(u,y,\epsilon)du. \tag{2.13}$$

We will suppose that for each fixed $y \in [y_1, y_2]$, there is an $\epsilon_0(y) > 0$ so that for each $\epsilon \in [0, \epsilon_0(y)]$, $k^2(t,y,\epsilon)$ is a monotonically increasing function of t with one zero, t_0, in the interval $[t_{in}, t_{fi}]$.

Also

i. $\frac{\partial^i}{\partial t^i} k^2(t, y, \epsilon) \in C^0([t_{in}, t_{fi}] \times [0, \epsilon_0])$ $i = 0, \ldots, 3$,

ii. $\frac{\partial^i}{\partial t^i} \frac{k^2(t,y,\epsilon)}{t - t_0(y,\epsilon)} \in C^0([t_{in}, t_{fi}] \times [0, \epsilon_0])$, $i = 0, 1, 2$ with $\frac{k^2(t,y,\epsilon)}{t-t_0(y,\epsilon)}$ strictly positive in $[t_{in}, t_{fi}] \times [0, \epsilon_0]$ and,

iii. $|\frac{\sinh^2(k(t,y,\epsilon))}{t-t_0(y,\epsilon)}| > 0$ for $(t, \epsilon) \in [t_{in}, t_{fi}] \times [0, \epsilon_0]$.

In order to obtain solutions that are continuous in y we will suppose,

ia. $\frac{\partial^i}{\partial t^i} k^2(t, y, \epsilon) \in C^0([t_{in}, t_{fi}] \times [y_1, y_2] \times [0, \epsilon_0])$, $i = 0, \ldots, 3$

iia. $\frac{\partial^i}{\partial t^i} \frac{k^2(t,y,\epsilon)}{t-t_0(y,\epsilon)} \in C^0([t_{in}, t_{fi}] \times [y_1, y_2] \times [0, \epsilon_0])$, $i = 0, 1, 2$ with $\frac{k^2(t,y,\epsilon)}{t-t_0(y,\epsilon)}$ strictly positive in $[t_{in}, t_{fi}] \times [y_1, y_2] \times [0, \epsilon_0]$, and

iiia. $|\frac{\sinh^2(k(t,y,\epsilon))}{t-t_0(y,\epsilon)}| > 0$ for $(y, t, \epsilon) \in [t_{in}, t_{fi}] \times [y_1, y_2] \times [0, \epsilon_0]$.

Finally we will suppose that

$$\frac{\partial^i}{\partial t^i} a(t, \epsilon) \in C^0([t_{in}, t_{fi}] \times [0, \epsilon_0]), \ i = 0, \ldots, 3. \tag{2.14}$$

The turning points are located at $(\frac{y-b(t,\epsilon)}{2a(t+\frac{\epsilon}{2},\epsilon)})^2 = 1$ which will be either at $y = 2a(t+\frac{\epsilon}{2}, \epsilon) + b(t, \epsilon) = \gamma_+(t, \epsilon)$ or $y = b(t, \epsilon) - 2a(t+\frac{\epsilon}{2}, \epsilon) = \gamma_-(t, \epsilon)$. Conditions ii and iii impose that for each fixed y there is a unique simple zero of k^2, $t_0 \in [t_{in}, t_{fi}]$ such that $y = \gamma_+(t_0, \epsilon)$ and the second turning point is not encountered.

If

$$A(\epsilon) = \inf_{t \in [t_{in}, t_{fi}]} \gamma_-(t, \epsilon), \tag{2.15}$$

and

$$B(\epsilon) = \sup_{t \in [t_{in}, t_{fi}]} \gamma_+(t, \epsilon), \tag{2.16}$$

then we see from above that for each fixed ϵ in order for y to be a turning point it must be in the interval $[A(\epsilon), B(\epsilon)]$.

The next section is devoted to making the arguments leading to (2.10) and (2.11) rigorous.

3. Error analysis

We begin by considering the Langer transformation (2.13). The lemma below which follows from a lemma of Olver ([O] p. 399) shows that smoothness in k is transferred to smoothness in ρ.

Lemma 3.1. *Suppose that for* $(y, \epsilon) \in [y_1, y_2] \times [0, \epsilon_0]$, $k^2(t, y, \epsilon)$ *is montonically increasing for* t *in* $[t_{in}, t_{fi}]$ *and satisfies* i *and* ii. *Then equation* (2.13) *gives a one to one map between* t *and* ρ *such that* ρ *satisfies* i *and* ii. *If it is assumed that* k^2 *satisfies* ia *and* iia *then so does* ρ.

We now examine the solution to the Airy differential equation. From (2.3) it can be seen using Taylor series that $\chi(\epsilon^{-2/3}\rho(t+\epsilon))$ and $\chi(\epsilon^{-2/3}\rho(t-\epsilon))$ can be written in terms of $\chi(\epsilon^{-2/3}\rho(t))$ and $\chi'(\epsilon^{-2/3}\rho(t))$. With

$$\tilde{\rho}_\pm = \pm\frac{\rho(t\pm\epsilon)-\rho(t)}{\epsilon}, \tag{3.1}$$

the methods of Wang and Wong [WW, Lemma 1] can be slightly modified to obtain

Lemma 3.2. *Suppose χ is any solution of (2.3) with ρ and $\tilde{\rho}$ continuous then*

$$\begin{aligned}\chi(\epsilon^{-2/3}\rho(t\pm\epsilon)) &= \chi(\epsilon^{-2/3}\rho(t)\pm\epsilon^{1/3}\tilde{\rho}_\pm) \\ &= \chi(\epsilon^{-2/3}\rho(t))X_1(\rho,\tilde{\rho}_\pm,\pm\epsilon)\pm\epsilon^{1/3}\chi'(\epsilon^{-2/3}\rho(t))X_2(\rho,\tilde{\rho}_\pm,\pm\epsilon),\end{aligned} \tag{3.2}$$

with

$$X_i(\rho,\tilde{\rho}_\pm,\pm\epsilon) = \sum_{n=0}^{\infty}(\pm\epsilon)^n X_{i,n}(\rho,\tilde{\rho}_\pm)\ i=1,2, \tag{3.3}$$

where for $\rho>0$,

$$X_{1,0}(\rho,\tilde{\rho}) = \cosh\sqrt{\rho}\tilde{\rho}\ ,\ X_{2,0}(\rho,\tilde{\rho}) = \frac{\sinh\sqrt{\rho}\tilde{\rho}}{\sqrt{\rho}}, \tag{3.4}$$

and

$$X_{i,k}(\rho,\tilde{\rho}) = \frac{1}{\sqrt{\rho}}\int_0^{\tilde{\rho}} sX_{i,k-1}(\rho,s)\sinh\sqrt{\rho}(\tilde{\rho}-s)ds. \tag{3.5}$$

Above if $\rho<0$ then $\sqrt{\rho}$ is replaced by $i\sqrt{-\rho}$.

Also using induction one finds (see Wang and Wong [WW])

$$|X_{1,k}(\rho,\tilde{\rho})| \le 3^k\left(\frac{1}{3}\right)_k\frac{|\tilde{\rho}|^{3k}}{(3k)!}e^{|\Re(\sqrt{\rho}\tilde{\rho})|}, \tag{3.6}$$

and

$$|X_{2,k}(\rho,\tilde{\rho})| \le 3^k\left(\frac{2}{3}\right)_k\frac{|\tilde{\rho}|^{3k+1}}{(3k+1)!}e^{|\Re(\sqrt{\rho}\tilde{\rho})|}. \tag{3.7}$$

By taking the partial of (3.2) with respect to $\tilde{\rho}_\pm$ it is not difficult to see that

$$\begin{aligned}\chi'(\epsilon^{-2/3}\rho(t)\pm\epsilon^{1/3}\tilde{\rho}_\pm) &= \pm\epsilon^{-1/3}\chi(\epsilon^{-2/3}\rho(t))X_3(\rho,\rho_\pm,\pm\epsilon) \\ &\quad+\chi'(\epsilon^{-2/3}\rho(t))X_4(\rho,\rho_\pm,\pm\epsilon),\end{aligned} \tag{3.8}$$

where

$$X_3 = \frac{\partial}{\partial\tilde{\rho}}X_1, \tag{3.9}$$

and

$$X_4 = \frac{\partial}{\partial\tilde{\rho}}X_2. \tag{3.10}$$

Equations (3.4) and (3.5) imply that

$$X_{1,1}(\rho,\tilde{\rho}_\pm) = \frac{\tilde{\rho}_\pm^2}{4\sqrt{\rho}}\sinh\sqrt{\rho}\tilde{\rho}_\pm - \frac{\tilde{\rho}_\pm}{4\rho}\cosh\sqrt{\rho}\tilde{\rho}_\pm + \frac{1}{4\rho^{3/2}}\sinh\sqrt{\rho}\tilde{\rho}_\pm,$$

and

$$X_{2,1}(\rho,\tilde{\rho}_\pm) = \frac{\tilde{\rho}_\pm^2}{4\sqrt{\rho}} \cosh\sqrt{\rho}\tilde{\rho}_\pm - \frac{\tilde{\rho}_\pm}{4\rho^{3/2}} \sinh\sqrt{\rho}\tilde{\rho}_\pm.$$

If $\tilde{\rho}_\pm$ can be expanded as

$$\tilde{\rho}_\pm = \rho(t)' \pm \epsilon \frac{\rho(t)''}{2} + \hat{r}_\pm, \tag{3.11}$$

with

$$\hat{r}_\pm = \pm\frac{1}{2\epsilon}\int_t^{t\pm\epsilon} \rho^{(3)}(s)(t\pm\epsilon - s)^2 ds, \tag{3.12}$$

then

$$\begin{aligned} X_{1,0} - \cosh\sqrt{\rho}\rho' \mp \epsilon \frac{\sqrt{\rho}\rho''}{2} \sinh\sqrt{\rho}\rho' \\ = \hat{r}_\pm \sinh\sqrt{\rho}\rho' + \frac{\rho}{2}\int_{\rho'}^{\tilde{\rho}_\pm} \cosh\sqrt{\rho}s(\tilde{\rho}_\pm - s)ds, \end{aligned} \tag{3.13}$$

$$\begin{aligned} X_{1,1} - \frac{(\rho')^2}{4\sqrt{\rho}} \sinh\sqrt{\rho}\rho' + \frac{\rho'}{4\rho} \cosh\sqrt{\rho}\rho' - \frac{1}{4\rho^{3/2}} \sinh\sqrt{\rho}\rho' \\ = \int_{\rho'}^{\tilde{\rho}_\pm} \left(\frac{s}{4\sqrt{\rho}} \sinh\sqrt{\rho}s + \frac{s^2}{4} \cosh\sqrt{\rho}s\right) ds, \end{aligned} \tag{3.14}$$

$$\begin{aligned} X_{2,0} - \frac{\sinh\sqrt{\rho}\rho'}{\sqrt{\rho}} \mp \epsilon\frac{\rho''}{2} \cosh\sqrt{\rho}\rho' = \hat{r}_\pm \cosh\sqrt{\rho}\rho' \\ + \frac{\sqrt{\rho}}{2}\int_{\rho'}^{\tilde{\rho}_\pm} \sinh\sqrt{\rho}s(\tilde{\rho}_\pm - s)ds, \end{aligned} \tag{3.15}$$

and

$$\begin{aligned} X_{2,1} - \frac{(\rho')^2}{4\rho} \cosh\sqrt{\rho}\rho' + \frac{\rho'}{4\rho^{3/2}} \sinh\sqrt{\rho}\rho' \\ = \int_{\rho'}^{\tilde{\rho}_\pm} \left(\left(\frac{s^2}{4} - \frac{1}{4\rho}\right) \sinh\sqrt{\rho}s + \frac{s}{4\sqrt{\rho}} \cosh\sqrt{\rho}s\right)ds. \end{aligned} \tag{3.16}$$

If $\rho < 0$ then replace $\sqrt{\rho}$ by $i\sqrt{-\rho}$ in the above formulas.

The theorems below makes rigorous the argument of Shulten and Gorden and show that the Langer transformation plays a role in difference equations similar to that in differential equations.

Set

$$W = \{(t,\epsilon), t \in [t_{in}+\epsilon, t_{fi}-\epsilon] \subset [t_{in}, t_{fi}], \epsilon \in (0,\epsilon_0]\}$$

and

$$\overline{W} = clos(W).$$

Theorem 3.3. *Suppose that for each fixed $y \in [y_1, y_2]$, k^2 is monotonically increasing in t with a single zero, $t_0 \in [t_{in}, t_{fi}]$, and satisfies* i, ii, *and* iii. *Also suppose $a(t,\epsilon)$ is strictly positive and satisfies* (2.14). *Let $\psi^i(t,y,\epsilon)$ $i = 1,2$ be given by equations* (2.10) *and* (2.11) *respectively. Then for all $(t,\epsilon) \in W$,*

$$\begin{aligned} &a(t+\epsilon,\epsilon)\psi^i(t+\epsilon,y,\epsilon) + a(t,\epsilon)\psi^i(t-\epsilon,y,\epsilon) \\ &\quad - 2a(t+\epsilon/2,\epsilon)\cosh k(t,y,\epsilon)\psi^i(t,y,\epsilon) \\ &\quad = \beta^{(i)}(t,y,\epsilon) \qquad i = 1,2 \end{aligned} \tag{3.17}$$

where for each $y \in [y_1, y_2]$, $\psi^i \in C^0([t_{in}, t_{fi}] \times (0, \epsilon_0])$, and $\beta^{(i)} \in C^0(W)$. Furthermore,

$$|\beta^{(1)}(t,y,\epsilon)| \le c(y)\epsilon^2 \sup_{u\in(t-\epsilon,t+\epsilon)} [|Ai(\epsilon^{-2/3}\rho(u,y,\epsilon))| + \epsilon^{1/3}|Ai'(\epsilon^{-2/3}\rho(u,y,\epsilon))|],$$

while

$$|\beta^{(2)}(t,y,\epsilon)| \le c(y)\epsilon^2 \sup_{u\in(t-\epsilon,t+\epsilon)} [|Bi(\epsilon^{-2/3}\rho(u,y,\epsilon))| + \epsilon^{1/3}|Bi'(\epsilon^{-2/3}\rho(u,y,\epsilon))|].$$

If it is assumed that ia–iiia. *hold then $\psi^i \in C^0([t_{in}, t_{fi}] \times [y_1, y_2] \times (0, \epsilon_0])$, and $\beta^i \in C^0(W \times [y_1, y_2])$.*

Proof. We will only consider the case $i = 1$ as the $i = 2$ case is proved in a similar manner. From the Lemma 3.1 it follows that $\frac{k^2(y,t)}{\rho(t)}$ satisfies i. and is nonzero for $y \in [y_1, y_2]$ and $(t,\epsilon) \in [t_{in}, t_{fi}] \times [0, \epsilon_0]$. From the hypotheses on k we see also that $\left(\frac{k}{\sinh k}\right)^2$ satisfies i in the same region. This implies $g(t)$ satisfies i and is nonzero. Lemma 3.2 says that

$$\mathrm{Ai}(\epsilon^{-2/3}\rho(t\pm\epsilon)) = \mathrm{Ai}(\epsilon^{-2/3}\rho(t))X_1(\rho,\tilde{\rho}_\pm,\pm\epsilon) \pm \epsilon^{1/3}\mathrm{Ai}'(\epsilon^{-2/3}\rho(t))X_2(\rho,\tilde{\rho}_\pm,\pm\epsilon) \tag{3.18}$$

where $\tilde{\rho}_\pm$ has the expansions given in (3.11). Equations (3.13) through (3.16), conditions i and ii and equations (3.6) and (3.7) imply that

$$|X_{1,0} - \cosh\sqrt{\rho}\rho' \mp \epsilon \sinh\sqrt{\rho}\rho' \frac{\sqrt{\rho}\rho''}{2}| < c(y)\epsilon^2, \tag{3.19}$$

$$|X_{1,1} - \frac{(\rho')^2}{4\sqrt{\rho}}\sinh\sqrt{\rho}\rho' + \frac{\rho'}{4\rho}\cosh\sqrt{\rho}\rho' - \frac{1}{4\rho^{3/2}}\sinh\sqrt{\rho}\rho'| < c(y)\epsilon, \tag{3.20}$$

$$|X_{2,0} - \frac{\sinh\sqrt{\rho}\rho'}{\sqrt{\rho}} \mp \epsilon\cosh\sqrt{\rho}\rho'\frac{\rho''}{2}| < c(y)\epsilon^2, \tag{3.21}$$

$$|X_{2,1} - \frac{(\rho')^2}{4\rho}\cosh\sqrt{\rho}\rho' - \frac{\rho''}{2}\cosh\sqrt{\rho}\rho'' + \frac{\rho'}{4\rho^{3/2}}\sinh\sqrt{\rho}\rho'| < c(y)\epsilon, \tag{3.22}$$

and

$$|X_{1,k}| < 3^k \left(\frac{1}{3}\right)_k \frac{c(y)^k}{(3k)!}, \quad |X_{2,k}| < 3^k \left(\frac{2}{3}\right)_k \frac{c(y)^k}{(3k)!},$$

where $c(y)$ is independent of $(t,\epsilon) \in \overline{W}$. Thus,

$$\begin{aligned}\mathrm{Ai}(\epsilon^{-2/3}\rho(t\pm\epsilon)) = \Bigg(&\cosh\sqrt{\rho}\rho' \\ &\pm\epsilon\left[\frac{(\rho')^2}{4\sqrt{\rho}}\sinh\sqrt{\rho}\rho' + \frac{\sqrt{\rho}\rho''}{2}\sinh\sqrt{\rho}\rho' - \frac{\rho'}{4\rho}\cosh\sqrt{\rho}\rho' + \frac{1}{4\rho^{3/2}}\sinh\sqrt{\rho}\rho'\right]\Bigg)\mathrm{Ai} \\ &\pm\epsilon^{1/3}\left(\frac{\sinh\sqrt{\rho}\rho'}{\sqrt{\rho}} \pm\epsilon\left[\frac{(\rho')^2}{4\rho}\cosh\sqrt{\rho}\rho' + \frac{\rho''}{2}\cosh\sqrt{\rho}\rho' - \frac{\rho'}{4\rho^{3/2}}\sinh\sqrt{\rho}\rho'\right]\right)\mathrm{Ai}' \\ &+ e(t,y,\epsilon).\end{aligned} \tag{3.23}$$

From the above estimates and the convergence of the series (3.3) we obtain the bound

$$|e(t,y,\epsilon)| \le \epsilon^2 c_1(y) \sup_{u\in(t-\epsilon,t+\epsilon)} [|Ai(\epsilon^{-2/3}\rho(u,y,\epsilon))| + \epsilon^{1/3}|Ai'(\epsilon^{-2/3}\rho(u,y,\epsilon))|].$$

Suppose for now that $\rho \neq 0$. If $g(t\pm\epsilon)$ and $a(t\pm\frac{\epsilon}{2})$ are expanded out to second order in ϵ we find using the equation, $\frac{g'}{g} = \frac{1}{4}\frac{\rho'}{\rho} - \frac{1}{2}\frac{a'(t+\epsilon 2)}{a(t+\frac{\epsilon}{2})} - \frac{\cosh(\rho^{1/2}\rho')}{2\sinh(\rho^{1/2}\rho')}(\frac{\rho^{-1/2}\rho'^2}{2} + \rho^{1/2}\rho'')$ that

$$\begin{aligned}&\tilde{a}(t+\epsilon)(g\mathrm{Ai})(t+\epsilon) + \tilde{a}(t)(g\mathrm{Ai})(t-\epsilon) - (y-b(t,\epsilon))(g\mathrm{Ai})(t) \\ &= a\left(t+\frac{\epsilon}{2}\right)g(t)\Bigg[\mathrm{Ai}(\epsilon^{-2/3}\rho(t+\epsilon)) + \mathrm{Ai}(\epsilon^{-2/3}\rho(t-\epsilon)) \\ &\quad - 2\cosh(\rho^{1/2}\rho')\mathrm{Ai}(\epsilon^{-2/3}\rho(t)) \\ &\quad + \epsilon\left(\frac{1}{4}\frac{\rho'}{\rho} - \frac{\cosh(\rho^{1/2}\rho')}{2\sinh(\rho^{1/2}\rho')}\left(\frac{\rho^{-1/2}\rho'^2}{2} + \rho^{1/2}\rho''\right)\right) \\ &\quad \times \left(\mathrm{Ai}(\epsilon^{-2/3}\rho(t+\epsilon)) - \mathrm{Ai}(\epsilon^{-2/3}\rho(t-\epsilon))\right)\Bigg] \\ &+ e_1(t,y,\epsilon),\end{aligned} \tag{3.24}$$

where

$$\begin{aligned}|e_1(t,y,\epsilon)| &\le 2\epsilon^2 \sup_{u\in(t-\epsilon,t+\epsilon)} [|a'g'| + |ag''| + |ga''| + \epsilon(|a'g''| + |a''g'|) \\ &\qquad + \epsilon^2|a''g''|]|\mathrm{Ai}(\epsilon^{-2/3}\rho(u,y,\epsilon))| \\ &\le \epsilon^2 c_2(y) \sup_{u\in(t-\epsilon,t+\epsilon)} |\mathrm{Ai}(\epsilon^{-2/3}\rho(u,y,\epsilon))|.\end{aligned}$$

Here,

$$c_2(y) = \sup_{(t,\epsilon)\in\overline{W}} [|a'g'| + |ag''| + |ga''| + \epsilon(|a'g''| + |a''g'|) + \epsilon^2|a''g''|].$$

This as well as (3.23) show that there is cancellation on the right-hand side of (3.24) out to order ϵ^2 so that

$$a(t+\epsilon)(g\mathrm{Ai})(t+\epsilon)+a(t)(g\mathrm{Ai})(t-\epsilon)-2a\,(t+\epsilon 2)\cosh(\rho^{1/2}\rho')(g\mathrm{Ai})(t) \\ =\beta^{(1)}(t,y,\epsilon), \tag{3.25}$$

where

$$|\beta^{(1)}(t,y,\epsilon)|\le\epsilon^2c(y)\sup_{u\in(t-\epsilon,t+\epsilon)}[|\mathrm{Ai}(\epsilon^{-2/3}\rho(u,y,\epsilon))|+\epsilon^{1/3}|\mathrm{Ai}'(\epsilon^{-2/3}\rho(u,y,\epsilon))|].$$

Since

$$\left(a\left(t+\frac{\epsilon}{2}\right)^{1/2}g(t)\right)' \\ =a\left(t+\frac{\epsilon}{2}\right)g(t)\left(\frac{1}{4}\frac{\rho'}{\rho}-\frac{\cosh(\rho^{1/2}\rho')}{2\sinh(\rho^{1/2}\rho')}\right)\left(\rho^{-1/2}\rho'^2 2+\rho^{1/2}\rho''\right),$$

the restriction $\rho\neq 0$ can now be removed. An analogous bound for $\beta^{(2)}(t,y,\epsilon)$ holds with Ai replaced by Bi. The continuity properties follow from conditions i–iii, (3.2) and the fact that the Airy functions are continuous (in fact entire) functions of their arguments. The uniformity of the error when ia–iiia are imposed follow by taking the supremum over y in the constants given and the continuity properties follow as above. □

To show that the above functions stay close to real solutions of the difference equation we introduce some auxiliary functions.

Let $u^{(j)}(x)=g(x)\tilde{u}^{(j)}(x)$ with $\tilde{u}^{(i)}(x)=\left(\frac{-x}{3}\right)^{1/2}e^{(-1)^{j+1}i\pi/6}H^{(j)}_{1/3}(\frac{2}{3}(-x)^{3/2})$. where $H^{(1)}_{1/3}, H^{(2)}_{1/3}$ are the first and second kind Hankel functions of order one third. Here we define $\ln z$ so that it is positive for large enough positive z. From the series representation for $H^{(i)}_{1/3}$ we see that $\tilde{u}^{(i)}(x)$ is continuous and infinitely differentiable for x real and each satisfies the Airy differential equation. The following give the relations between $\tilde{u}^{(j)}$, Ai and Bi [O, pp. 250 and 392]. For $x>0$,

$$\mathrm{Ai}(x)=-\frac{1}{2}\tilde{u}^{(1)}(x)=\frac{1}{2}\left(\frac{x}{3}\right)^{1/2}e^{2i\pi/3}H^{(1)}_{1/3}\left(\frac{2}{3}x^{3/2}e^{i\pi/2}\right),$$

and

$$\mathrm{Bi}(x)=(\frac{x}{3})^{1/2}\mathrm{Re}\left(e^{-i\pi/6}H^{(2)}_{1/3}\left(\frac{2}{3}x^{3/2}e^{i\pi/2}\right)\right),$$

while for $x<0$

$$\mathrm{Ai}(x)=\mathrm{Re}(\tilde{u}^{(1)}(x)),$$

and

$$\mathrm{Bi}(x)=-\mathrm{Im}(\tilde{u}^{(1)}(x)).$$

For $x < 0, \overline{\tilde{u}^{(2)}(x)} = \tilde{u}^{(1)}(x)$ (Olver [O] p. 238). Note that for large negative x we find that

$$\tilde{u}^{(j)}(x) = \frac{1}{\sqrt{\pi}}(-x)^{-1/4} e^{(\pm 1)^{(j+1)}(i\frac{2}{3}(-x)^{3/2} - \frac{i\pi}{4})}\left(1 + \frac{1}{|x|^{2/3}}\right), \tag{3.26}$$

and

$$\tilde{u}^{(j)}(x)' = \frac{1}{\sqrt{\pi}}(-x)^{1/4} e^{(\pm 1)^{(j+1)}(i\frac{2}{3}(-x)^{3/2} - \frac{i\pi}{4})}\left(1 + \frac{1}{|x|^{2/3}}\right). \tag{3.27}$$

An important property of $\tilde{u}^{(j)}$, is that they do not vanish for all x real [W]. Some useful auxiliary functions we will use below are the modulus function, M, and weight function, E, associated with the Airy functions. If $c_0 < 0, c_0 \sim -.36$ is the largest point on the x axis where $Ai(x) = Bi(x)$ we find from Olver [O, p. 395] that

$$M(x) = (2Ai(x)\mathrm{Bi}(x))^{1/2}, \quad E(x) = \left(\frac{\mathrm{Bi}(x)}{\mathrm{Ai}(x)}\right)^{1/2}, \qquad x \geq c_0, \tag{3.28}$$

and

$$M(x) = (\mathrm{Ai}(x)^2 + \mathrm{Bi}(x)^2)^{1/2}, \quad E(x) = 1, \quad x \leq c_0.$$

The Airy functions Ai and Bi can be expressed in terms of these functions as

$$E(x)\mathrm{Ai}(x) = M(x)\sin\theta(x), \qquad E^{-1}(x)\mathrm{Bi}(x) = M(x)\sin\theta(x), \tag{3.29}$$

where θ is called the phase function. The modulus and weight functions have the following asymptotic properties

$$E(x) \sim \sqrt{2}e^{\frac{2}{3}x^{\frac{3}{2}}},$$

for large positive x while, $M(x) \sim \sqrt{\pi}|x|^{-1/4}$, for large x. The weight function is a nondecreasing function of its argument while the modulus function is increasing for $x \leq c_0$ ([O, Lemma 5.1]). Since M is continuous there is a positive constant c_3 such that for all x,

$$M(x) \leq c_3. \tag{3.30}$$

We also have [O, p. 395] that

$$\pi|x|^{1/2}M(x)^2 \leq \lambda_1 < 2. \tag{3.31}$$

Formulas similar to (3.29) also hold for the derivatives of the Airy functions. In this case the functions M and θ are replaced by N and ω respectively. For large x, $N(x) \sim \sqrt{\pi}|x|^{1/4}$ so from the continuity of $N(x)$ we find that

$$N(x) \leq c_4(1 + |x|^{1/4}) \tag{3.32}$$

Furthermore it also follows from the above asymptotic formulas that there is a constant, say c_5, such that

$$\left|\frac{E^2(x)u^{(1)}(x)}{\tilde{u}^{(2)}(x)}\right|, \quad \left|\frac{E^{-2}(x)\tilde{u}^{(2)}(x)}{\tilde{u}^{(1)}(x)}\right| \leq c_5. \tag{3.33}$$

That $\tilde{u}^{(j)}$ solve the same differential equations as the Airy functions implies using the above arguments,

Lemma 3.4. *Suppose that for each fixed* $y \in [y_1, y_2]$, k^2 *is monotonically increasing in* t *with a single zero ,* $t_0 \in [t_{in}, t_{fi}]$, *and satisfies* i, ii, *and* iii. *Also suppose* $a(t,\epsilon)$ *is strictly positive and satisfies* (2.14). *Then for all* $(t,\epsilon) \in W$,

$$\begin{aligned} &a(t+\epsilon,\epsilon)u^{(j)}(t+\epsilon,y,\epsilon) + a(t,\epsilon)u^{(j)}(t-\epsilon,y,\epsilon) \\ &\quad - 2a(t+\epsilon/2)\cosh k(t,y,\epsilon)u^{(j)}(t,y,\epsilon) = \beta_1^{(j)}(t,y,\epsilon), \quad j=1,2 \end{aligned} \tag{3.34}$$

where for each $y \in [y_1, y_2]$, $u^{(j)} \in C^0([t_{in}, t_{fi}] \times (0, \epsilon_0])$, *and* $\beta_1^{(j)} \in C^0(W)$. *Furthermore,*

$$\begin{aligned} |\beta_1^{(j)}(t,y,\epsilon)| \le c(y)\epsilon^2 &\sup_{u\in(t-\epsilon,t+\epsilon)} \Big(M(\epsilon^{-2/3}\rho(u,y,\epsilon)) \\ &+ \epsilon^{1/3}N(\epsilon^{-2/3}\rho(u,y,\epsilon))\Big)E^{(-1)^j}(\epsilon^{-2/3}\rho(u,y,\epsilon)). \end{aligned}$$

If it is assumed that ia.–iiia. *hold then* $u^{(j)} \in C^0([t_{in}, t_{fi}] \times [y_1, y_2] \times (0, \epsilon_0])$, *and* $\beta^{(j)} \in C^0(W \times [y_1, y_2])$.

Since $\tilde{u}^{(j)}$ satisfy the Airy equation it is not difficult to compute their Wronskian

$$W(\tilde{u}^{(1)}, \tilde{u}^{(2)}) = \tilde{u}^{(1)}(x)'\tilde{u}^{(2)}(x) - \tilde{u}^{(2)}(x)'\tilde{u}^{(1)}(x) = \frac{2i}{\pi}.$$

With this we can obtain,

Lemma 3.5. *Suppose that for each fixed* $y \in [y_1, y_2]$, k^2 *is monotonically increasing in* t *with a single zero ,* $t_0 \in [t_{in}, t_{fi}]$, *and satisfies* i, ii, *and* iii. *Also suppose* $a(t,\epsilon)$ *is strictly positive and satisfies* (2.14). *Then for all* $(t,\epsilon) \in W$,

$$\begin{aligned} &u^{(1)}(t+\epsilon,y,\epsilon)u^{(2)}(t,y,\epsilon) - u^{(1)}(t,y,\epsilon)u^{(2)}(t+\epsilon,y,\epsilon) \\ &\quad = \epsilon^{1/3}\frac{2i}{\pi}g(t+\epsilon,y,\epsilon)g(t,y,\epsilon)X_2(\rho,\tilde{\rho}_+,\epsilon). \end{aligned} \tag{3.35}$$

For ϵ_0 *sufficiently small there is a constant* $c_6(y)$ *such that*

$$|\frac{2}{\pi}g(t+1,y,\epsilon)g(t,y,\epsilon)X_2(\rho,\tilde{\rho}_+,\epsilon)|^{-1} < c_6(y) \tag{3.36}$$

If it is assumed that ia.-iiia. *hold then* c_6 *can be choosen uniform in* y.

Proof. Equation (3.35) follows from the Wronskian identity above and (3.2) with χ replaced by $u^{(i)}$, $i = 1,2$. Conditions ii and iii imply that for ϵ_0 sufficiently small $X_2(\rho,\tilde{\rho}_+)$ ((3.4)) is nonzero. The convergence of the series for X_2 which follows from (3.7) shows that X_2 is nonzero for ϵ_0 sufficiently small which leads to (3.36). □

In the rest of the paper we will assume that ϵ_0 is chosen sufficiently small so that (3.36) holds.

We now apply the above results to show how close $\psi^j(t_n, y, \epsilon)$ stays to a solution of the difference equation

$$a_1((n+1)\epsilon, \epsilon)f(n+1) + a_1(n\epsilon, \epsilon)f(n-1) - (y - b_1(n\epsilon, \epsilon))f(n) = 0\,. \tag{3.37}$$

With $t_n = \epsilon n$, $\sigma_i^j = \frac{f_i(n) - \psi^j(n)}{u^{(j)}(n)}$ $i = 1 \ldots 2$, $\hat{\psi}^j(n) = \frac{\psi^j(n)}{u^{(j)}(n)}$, $\hat{\beta}^{(j)} = \frac{\beta^{(j)}(n)}{u^{(j)}(n)}$, and

$$\Delta(w(n)) \equiv w(n) - w_1(n),$$

we find using (3.17) and (3.37),

$$u^{(j)}(n-1)\left(\sigma_1^j(n) - \sigma_1^j(n-1)\right) - u^{(j)}(n+1)\frac{a((n+1)\epsilon, \epsilon)}{a(n\epsilon, \epsilon)}\left(\sigma_1^j(n+1) - \sigma_1^j(n)\right) \tag{3.38}$$
$$= h_1^j(n) + q_1^j(n)$$

where

$$h_1^j(n) = \left(\Delta\left(\frac{y - b(n\epsilon, \epsilon)}{a(n\epsilon, \epsilon)}\right) + \frac{\hat{\beta}_1^{(j)}(n\epsilon)}{a(n\epsilon, \epsilon)}\right) u^{(j)}(n)\sigma^j(n)$$
$$- \Delta\left(\frac{a((n+1)\epsilon, \epsilon)}{a(n\epsilon, \epsilon)}\right) u^{(j)}(n+1)\sigma^j(n+1),$$

and

$$q_1^j(n) = \Delta\left(\frac{y - b(n\epsilon, \epsilon)}{a(n\epsilon, \epsilon)}\right) u^{(j)}(n)\hat{\psi}^j(n)$$
$$- \Delta\left(\frac{a((n+1)\epsilon, \epsilon)}{a(n\epsilon, \epsilon)}\right) u^{(j)}(n+1)\hat{\psi}^j(n+1) + \hat{\beta}^{(j)}(n\epsilon)\frac{u^{(j)}(n)}{a(n\epsilon, \epsilon)}\,.$$

Selecting a solution of f_1 of (3.37) such that $\sigma_1^j(n_2) = 0 = \sigma_1^j(n_2 - 1)$ yields

$$\sigma_1^j(n) = \sum_{i=n+1}^{n_2-1} G_1^j(n,i)\frac{h^j(i) + q^j(i)}{u^{(j)}(i-1)}$$

where

$$G_1^j(n,i) = -\sum_{k=n}^{i-1} \frac{a(i\epsilon, \epsilon)u^{(j)}(i-1)u^{(j)}(i)}{a((k+1)\epsilon, \epsilon)u^{(j)}(k)u^{(j)}(k+1)}\,. \tag{3.39}$$

The above formula for σ can be recast as

$$\sigma_1^j(n) = \sum_{i=n+1}^{n_2-1} \tilde{G}_1^j(n,i)\frac{q_1^j(i)}{u^{(j)}(i-1)} + \sum_{i=n+1}^{n_2} K_1^j(n,i)\sigma_1^j(i), \tag{3.40}$$

where

$$K_1^j(n,i) = \tilde{G}_1^j(n,i)\left(\Delta\left(\frac{y-b(i\epsilon,\epsilon)}{a(i\epsilon,\epsilon)}\right) + \frac{\hat{\beta}_1^{(j)}(i\epsilon)}{a(i\epsilon,\epsilon)}\right)\frac{u^{(j)}(i)}{u^{(j)}(i-1)}$$
$$+ \tilde{G}_1^j(n,i-1)\Delta\left(\frac{a((i)\epsilon,\epsilon)}{a((i-1)\epsilon,\epsilon)}\right)\frac{u^{(j)}(i)}{u^{(j)}(i-2)}.$$

Here $\tilde{G}_1^j(n,i) = G_1^j(n,i)$ for $i \le n_2 - 1$ and zero otherwise. Note that $G_1^j(n,i) = 0$ for $i \le n$.

Applying a similar argument to that above on the equation

$$u^{(j)}(n+1)(\sigma_2^j(n+1) - \sigma_2^j(n)) - \frac{a(n\epsilon,\epsilon)}{a((n+1)\epsilon,\epsilon)}u^{(j)}(n-1)\left(\sigma_2^j(n) - \sigma_2^j(n-1)\right) \tag{3.41}$$
$$= h_2^j(n) + q_2^j(n),$$

where

$$h_2^j(n) = -\left(\Delta\left(\frac{y-b(n\epsilon,\epsilon)}{a((n+1)\epsilon,\epsilon)}\right) + \frac{\hat{\beta}_1^{(j)}(n\epsilon)}{a((n+1)\epsilon,\epsilon)}\right)u^{(j)}(n)\sigma^j(n)$$
$$+\Delta\left(\frac{(a(n\epsilon,\epsilon)}{a((n+1)\epsilon,\epsilon)}\right)u^{(j)}(n-1)\sigma^j(n-1),$$

and

$$q_2^j(n) = -\Delta\left(\frac{y-b(n\epsilon,\epsilon)}{a_1((n+1)\epsilon,\epsilon)}\right)u^{(j)}(n)\hat{\psi}^j(n)$$
$$+\Delta\left(\frac{a(n\epsilon,\epsilon)}{a((n+1)\epsilon,\epsilon)}\right)u^{(j)}(n-1)\hat{\psi}^j(n-1) - \frac{\hat{\beta}^{(j)}(n\epsilon)u^{(j)}(n)}{a((n+1)\epsilon,\epsilon)},$$

and selecting a solution f_2 of (3.37) such that $\sigma_2^j(n_1) = 0 = \sigma_2^j(n_1+1)$ gives

$$\sigma_2^j(n) = \sum_{i=n_1+1}^{n-1} G_2^j(n,i)\frac{q_2^j(i)}{u^{(j)}(i+1)} + \sum_{i=n_1}^{n-1} K_2^j(n,i)\sigma_2^j(i). \tag{3.42}$$

Here $\tilde{G}_2^j(n,i) = G_2^j(n,i)$ for $i \ge n_1 + 1$ and zero otherwise,

$$G_2^j(n,i) = \sum_{k=i}^{n-1}\frac{a((i+1)\epsilon,\epsilon)u^{(j)}(i+1)u^{(j)}(i)}{a((k+1)\epsilon,\epsilon)u^{(j)}(k)u^{(j)}(k+1)}, \tag{3.43}$$

and

$$K_2^j(n,i) = -\tilde{G}_2^j(n,i)\left(\Delta\left(\frac{y-b(i\epsilon,\epsilon)}{a((i+1)\epsilon,\epsilon)}\right) + \frac{\hat{\beta}_1^{(j)}(i\epsilon)}{a((i+1)\epsilon,\epsilon)}\right)\frac{u^{(j)}(i)}{u^{(j)}(i+1)}$$
$$+ \tilde{G}_2^j(n,i+1)\Delta\left(\frac{a((i+1)\epsilon,\epsilon)}{a(i+2)\epsilon,\epsilon)}\right)\frac{u^{(j)}(i)}{u^{(j)}(i+2)}.$$

The lemma below gives a bound on the above Green's functions.

Lemma 3.6. *Set* $z(i) = \epsilon^{-2/3}\rho(i\epsilon, y, \epsilon)$ *and suppose that the hypotheses of Lemma 3.5 hold. Then for fixed* y *such that* $[t_n, t_i] \subset [t_{in} + \epsilon, t_{fi} - \epsilon]$, $\epsilon \in (0, \epsilon_0]$ *the following inequalities hold*

$$|G_j^j(n,i)| \le G(i) = c(y)\epsilon^{-1/3} M(i)M(i-1), \qquad j = 1,2 \tag{3.44}$$

If ia–iiia hold then the above constant can be chosen uniform in y.

Proof. Set

$$c_7(y) = \sup_{(t,\epsilon)\in\overline{W}} |g(t+\epsilon, y, \epsilon)|,$$

$$c_8(y) = \sup_{(t,\epsilon)\in\overline{W}} E^{-1}(x(t))E(x(t+\epsilon)), \quad x = \epsilon^{-2/3}\rho(t, y, \epsilon).$$

Since g is strictly bounded (from ii and iii) we find that c_7 is finite. The continuity and asymptotic properties of E show that c_8 is also finite. If conditions ia–iiia hold then the above constants can be made uniform in y. The hypothesis that k^2 is a monotonically increasing function of t implies the $e^{-\frac{2}{3\epsilon}\rho^{\frac{3}{2}}(t)}$ will be exponentially decreasing for $t > t_0$ and of magnitude one for $t < t_0$. Thus it is appropriate to set $j = 1$ when considering the solution to (3.38) and $j = 2$ when considering the solutions (3.41).

We now take up the case $i = 1$ and we will temporarily suppress the dependence on all variables except k. Observe that the denominators in each term of G_1^1 can be recast as

$$\begin{aligned}&\frac{1}{a(k+1)u^{(1)}(k)u^{(1)}(k+1)} \\ &= \frac{1}{a(k+1)}\frac{1}{u^{(2)}(k+1)u^{(1)}(k) - u^{(2)}(k)u^{(1)}(k+1)}\left(\frac{u^{(2)}(k+1)}{u^{(1)}(k+1)} - \frac{u^{(2)}(k)}{u^{(1)}(k)}\right).\end{aligned} \tag{3.45}$$

Since

$$\begin{aligned}&\frac{1}{a(k+1)}\frac{1}{u^{(2)}(k+1)u^{(1)}(k) - u^{(2)}(k)u^{(1)}(k+1)} \\ &- \frac{1}{a(k+2)}\frac{1}{u^{(2)}(k+2)u^{(1)}(k+1) - u^{(2)}(k+1)u^{(1)}(k+2)} \\ &= -u^{(2)}(k+1)\frac{a(k+2)u^{(1)}(k+2) + a(k+1)u^{(1)}(k))}{D_k} \\ &+ u^{(1)}(k+1)\frac{a(k+2)u^{(2)}(k+2) + a(k+1)u^{(2)}(k))}{D_k}),\end{aligned} \tag{3.46}$$

where D_k is the common denominator of the left-hand side of the above equation, equation (3.34) can be use to recast (3.46) as

$$= \frac{\beta_1^{(2)}(k+1)u^{(1)}(k+1) - \beta_1^{(1)}(k+1)u^{(2)}(k+1)}{D_k}.$$

With the above equations, summation by parts yields

$$\sum_{k=n}^{i-1} \frac{1}{a(k+1)u^{(1)}(k)u^{(1)}(k+1)} \tag{3.47}$$

$$= \sum_{k=n}^{i-2} \left(\frac{u^{(2)}(k+1)}{u^{(1)}(k+1)} - \frac{u^{(2)}(n)}{u^{(1)}(n)} \right) \frac{\beta_1^{(2)}(k+1)u^{(1)}(k+1) - \beta_1^{(1)}(k+1)u^{(2)}(k+1)}{D_k}$$

$$+ \left(\frac{u^{(2)}(i)}{u^{(1)}(i)} - \frac{u^{(2)}(n)}{u^{(1)}(n)} \right) \frac{1}{a(i)} \frac{1}{u^{(2)}(i)u^{(1)}(i-1) - u^{(2)}(i-1)u^{(1)}(i)} .$$

From equation (3.33) we find using the monotonicity of E that

$$\left|\frac{u^{(2)}(m)}{u^{(1)}(m)}\right| \leq c_5 E^2(i),$$

for $m \leq i$. Also the definitions of β and β_1, equations (3.32) and (3.30) yield,

$$|\beta_1^{(2)}(k+1)u^{(1)}(k+1)| < c(y)\epsilon^2 c_3(c_3 + 2\epsilon^{1/6}c_4(y))c_7(y),$$

and

$$|\beta_1^{(1)}(k+1)u^{(2)}(k+1)| < c(y)\epsilon^2 c_3(c_3 + 2\epsilon^{1/6}c_4(y))c_7(y)c_8(y).$$

Finally from Lemma 3.5 we find that if ϵ_0 is chosen sufficiently small so that equation (3.36) holds for all $(t,\epsilon) \in W$, then

$$\left|\frac{1}{a(k+1)(u^{(2)}(k+1)u^{(1)}(k) - u^{(2)}(k)u^{(1)}(k+1))}\right| \leq \epsilon^{-1/3} c_6(y) \sup \frac{1}{a(t,\epsilon)}.$$

Since there are at most ϵ^{-1} terms in the sum on the right-hand side of (3.47) we find using the above bounds that it is bounded by a constant times $\epsilon^{1/3}E^2(i)$. The above inequalities also show that the boundary term in (3.47) is bounded by a constant times $\epsilon^{-1/3}E^2(i)$. From the relations between $u^{(1)}$ and the Airy functions we find the inequality

$$|a(i)u^{(1)}(i-1)u^{(1)}(i)| \leq c_7(y)^2 \sup a(t,\epsilon) M(i-1)M(i)E^{-1}(i-1)E^{-1}(i),$$

which used with c_8 yields

$$|G_1^1(n,i)| \leq \sum_{k=n}^{i-1} \left|\frac{a(i)u^{(1)}(i-1)u^{(1)}(i)}{a(k+1)u^{(1)}(k)u^{(1)}(k+1)}\right| \leq c(y)\epsilon^{-1/3} M(i)M(i-1),$$

which gives the result for $i = 1$. The result for $i = 2$ follows in an analogous manner with the roles of $u^{(1)}$ and $u^{(2)}$ interchanged. □

With the above lemma we can now show

Lemma 3.7. *Suppose that the hypotheses of Lemma* 3.5 *hold and* $[t_{N_1-1}, t_{N_2+1}] \subset [t_{in}, t_{fi}]$, $\epsilon \in (0, \epsilon_0]$. *Then,*

$$\sum_{j=N_1}^{N_2} G(i) < c(y)\epsilon^{-1}.$$

If ia–iiia *hold then the above constant can be chosen uniform in* y.

Proof. From Lemma 3.6 we find

$$\sum_{i=N_1}^{N_2} G(i) \le \tilde{c}(y)\epsilon^{-1/3} \sum_{i=N_1}^{N_2} M(i-1)M(i).$$

The sum on the right-hand side of the above equation can be bounded by

$$\sum_{i=N_1}^{N_2} M(i-1)M(i) \le c_3^2 + \sum_{(i:\rho(i-1)>0)} M(i-1)M(i) + \sum_{(i:\rho(i)<0)} M(i-1)M(i).$$

If (3.31) is now used we find

$$\sum_{(i:\rho(i-1)>0)} M(i-1)M(i) \le \frac{\lambda_1^2}{c_9(y)}\epsilon^{-2/3} \int_{t_0}^{t_{N_2}} \frac{dt}{(t-t_0)^{1/2}} \le c(y)\epsilon^{-2/3},$$

where

$$c_9(y) = \inf_{t,\epsilon} \left|\frac{\rho(t,y,\epsilon)}{(t-t_0(y,\epsilon))}\right|^{\frac{1}{2}}.$$

Condition ii assures that $c_9(y) > 0$. Applying the same reasoning to the remaining sum yields the result. □

With the above we will now show that there is an actual solution of the difference equation (3.37) that is close to the approximate solution. First set

$$c_{10}(y) = \max\left(\sup_{(t,\epsilon)\in\overline{W}} \left|\frac{u^{(j)}(t+1,y,\epsilon)}{u^{(j)}(t,y,\epsilon)}\right|, \sup_{(t,\epsilon)\in\overline{W}} \left|\frac{u^{(j)}(t,y,\epsilon)}{u^{(j)}(t+1,y,\epsilon)}\right|\right) \qquad j=1,2$$

Theorem 3.8. *Suppose that for each fixed* $y \in (y_1, y_2)$, k^2 *is monotonically increasing in* t *with a single zero* $t_0 \in [t_{in}, t_{fi}]$ *and satisfies* i, ii, *and* iii,. *Also suppose* $a(t,\epsilon)$ *is strictly positive and satisfies* (2.14), *and that* $a_1(t,\epsilon)$ *and* $b_1(t,\epsilon) \in C^0([t_{in}, t_{fi}]) \times C^0([0,\epsilon_0])$, *are uniformly bounded on* $[t_{in}, t_{fi}] \times [0, \epsilon_0]$, *and* $a_1(t,\epsilon)$ *is strictly positive. Let* $[t_{N_1-1}, t_{N_2+1}] \subset [t_{in}, t_{fi}]$, $\epsilon \in (0, \epsilon_0]$, *and set*

$$K(i) = c_{10}(y)G(i)\left(\left|\Delta\left(\frac{a(i\epsilon)}{a((i-1)\epsilon)}\right)\right| + \left|\Delta\left(\frac{y-b(i\epsilon)}{a(i\epsilon)}\right)\right|\right), \qquad j=1,2.$$

Then there exists solutions f_1 and f_2 of (3.37) such that

$$|\sigma_1^1(n\epsilon)| = |\frac{f_1(n\epsilon) - \psi^1(n\epsilon)}{u^{(1)}(n\epsilon)}| \qquad (3.48)$$
$$\leq c(y)\sum_{N_1}^{N_2}(|K(i) + G(i)|\hat{\beta}^{(1)}(i)|)e^{c(y)\sum_{N_1}^{N_2}(K(i)+G(i)|\hat{\beta}_1^{(1)}(i))|)},$$

and

$$|\sigma_2^2(n\epsilon)| = |\frac{f_2(n\epsilon) - \psi^2(n\epsilon)}{u^{(2)}(n\epsilon)}| \qquad (3.49)$$
$$\leq c(y)\sum_{N_1}^{N_2}(K(i) + G(i)|\hat{\beta}^{(2)}(i)|)e^{c(y)\sum_{N_1}^{N_2}(K(i)+G(i)|\hat{\beta}_1^{(2)}(i))|)}.$$

Thus, if

$$\sum K(i) = o(1), j = 1, 2, \qquad (3.50)$$

then

$$\frac{f_j(n, y, \epsilon) - \psi^j(n, y, \epsilon)}{u^{(j)}(n, y, \epsilon)} = o(1), \qquad j = 1, 2, \qquad (3.51)$$

uniformly in $\overline{W}$. If

$$\sup_{(t,\epsilon)\in\overline{W}} |a(t, \epsilon) - a_1(t, \epsilon)| = 0(\epsilon^2) = \sup_{t,\epsilon} |b(t, \epsilon) - b_1(t, \epsilon)|, \qquad (3.52)$$

then the $o(1)$ in equation (3.51) is in fact $0(\epsilon)$. If it is assumed that ia.–iiia. *hold then the above convergence is uniform on $\overline{W} \times [y_1, y_2]$.*

Proof. The interval $[t_{N_1}, t_{N_2}] \subset [t_{in}, t_{fi}]$, $\epsilon \in (0, \epsilon_0]$ so that Theorem 3.3 and Lemma 3.4 can be used. Take the magnitude of both sides of equation (3.40) with $j = 1$. Replace $|\tilde{G}_1^1(n, i)|$ by $G(i)$ and note that $|\hat{\psi}^j|$ is uniformly bounded as are $|a(n\epsilon, \epsilon)|$, $|a(n\epsilon, \epsilon)|^{-1}$, $|a_1(n\epsilon, \epsilon)|$, and $|a_1(n\epsilon, \epsilon)|^{-1}$. Inequality (3.48) now follows from the Picard iteration. The inequality (3.49) follows from equation (3.42) (with $j = 2$) in a similar manner. Suppose now that (3.50) holds. From Theorem 3.3 and Lemma 3.4 we see that $|\hat{\beta}^{(1)}(i)| = c(y)\epsilon^2 = |\hat{\beta}_1^{(1)}(i)|$ for all $i \in [N_1 + 1, N_2 - 1]$. Thus

$$\sum_{n_1}^{n_2} G(i)|\hat{\beta}^{(1)}(i)|) \leq c(y)\epsilon,$$

where Lemma 3.7 has been used. Since the same inequality holds with $\hat{\beta}^{(1)}$ replaced by $\hat{\beta}_1^{(1)}$ we see that there is a solution f_1 satisfying (3.51). If (3.52) holds then $\sum K^1(i) \leq c\epsilon^2 \sum G(i) = c(y)\epsilon$ so that (3.51) is true for $j = 1$ with $o(1)$ replaced by $0(\epsilon)$. That the convergence is uniform on $\overline{W}$ or uniform on $\overline{W} \times [y_1, y_2]$ follows from the fact that the constants in the error terms may be chosen uniform on $\overline{W}$ or $\overline{W} \times [y_1, y_2]$ respectively. Similar reasoning can be used for f_2. □

4. Extension to complex values

The above considerations can be extended to complex values of the parameter y. We first extend Theorem 3.3. We remind the reader that $(\rho^{\frac{1}{2}}\rho')(t,y,\epsilon) = -k(t,y,\epsilon)$, so g can be written as,

$$g(t,y,\epsilon) = \left(\frac{\rho(t,y,\epsilon)}{a^2\left(t+\frac{\epsilon}{2},\epsilon\right)\sinh^2(\rho^{1/2}\rho')(t,y,\epsilon)}\right)^{1/4},$$

where the above differentiation is with respect to t. Let Ω be a region in the complex plane and $H(\Omega)$ the set of functions holomorphic in Ω. We will assume

ib. g and $\rho \in C^\infty([t_{in},t_{fi}] \times [0,\epsilon_0]) \times H(\Omega)$, and $\frac{\partial^i}{\partial t^i}g$ and $\frac{\partial^i}{\partial t^i}\rho \in C^0([t_{in},t_{fi}] \times [0,\epsilon_0] \times \Omega)$, $i = 0,\dots,3$,

iib. there is a point $t_0(y,\epsilon)$ such that $\frac{\rho}{t-t_0} \in C^2([t_{in},t_{fi}]) \times C^0([0,\epsilon_0] \times \Omega)$ and $|\frac{\rho}{t-t_0(y,\epsilon)}|$ is uniformly bounded away from zero for $(t,y,\epsilon) \in [t_{in},t_{fi}] \times \Omega \times [0,\epsilon_0]$,

iiib. for fixed $(y,\epsilon) \in \Omega \times [0,\epsilon_0]$, $\mathrm{Re}\rho(t,y,\epsilon)^{3/2}$ is a nonincreasing function of t.

Remark. The choice of iiib. is dictated by the examples discussed below.

Theorem 4.1. *Let Ω be a region in the complex plane. Suppose* ib *holds and $a \in C^\infty([t_{in},t_{fi}] \times [0,\epsilon_0])$ is strictly positive for $(t,\epsilon) \in [t_{in},t_{fi}] \times [0,\epsilon_0]$. Let f be any entire function solution of the Airy differential equation and $\psi(t,y,\epsilon) = g(\rho(t,y,\epsilon))f(\epsilon^{-2/3}\rho(t,y,\epsilon))$. Then $\psi \in C^\infty([t_{in},t_{fi}] \times (0,\epsilon_0]) \times H(\Omega)$ and for all $(t,\epsilon) \in W$ satisfies*

$$\begin{aligned} &a(t+\epsilon,\epsilon)\psi(t+\epsilon,y,\epsilon) + a(t,\epsilon)\psi(t-\epsilon,y,\epsilon) \\ &\quad - 2a(t+\epsilon/2,\epsilon)\cosh(\rho^{-1/2}\rho'(t,y,\epsilon))\psi(t,y,\epsilon) = \beta^{(f)}(t,y,\epsilon) \end{aligned} \tag{4.1}$$

where $\beta^{(f)} \in C(W) \times H(\Omega)$, and $\beta^{(f)} \in C^\infty(W \times \Omega)$. Furthermore,

$$|\beta^{(f)}(t,y,\epsilon)| \le d(y)\epsilon^2 \sup_{u\in(t-\epsilon,t+\epsilon)} [|f(\epsilon^{-2/3}\rho(u,y,\epsilon))| + \epsilon^{1/3}|f'(\epsilon^{-2/3}\rho(u,y,\epsilon))|].$$

In the above formulas the constant $d(y)$ may be chosen uniform on compact subsets of Ω.

Proof. Let K be a compact subset of Ω and set $d_i(K) = \sup_K c_i$, $i = 1,2$, and $d = \sup_K c$. The proof follows as in Theorem 3.3 replacing c_i and c by d_i and d. □

In the extension of the Airy differential equation to the complex plane it is customary to introduce the solutions $\mathrm{Ai}_0 = \mathrm{Ai}$, and

$$\mathrm{Ai}_{\pm1}(z) = \mathrm{Ai}(ze^{\mp 2i\pi/3}), \tag{4.2}$$

and the regions $S_0 = \{z : |\arg z| \le \frac{\pi}{3}\}$ and

$$S_{\pm1} = e^{\pm 2\pi/3}S_0.$$

Since the above function and Bi all satisfy the same differential equation there is a relation among them given by ([O, p. 414]),

$$\mathrm{Ai}_{\pm} = \frac{1}{2}e^{\mp\frac{i\pi}{3}}(\mathrm{Ai}(z) \mp i\mathrm{Bi}(z)). \tag{4.3}$$

Thus it follows from the asymptotic expansions of Ai and Bi [O, p. 413] that Ai_j is recessive in S_j and dominant in S_{j+1}, and S_{j-1} where the suffix j is enumerated mod 3. Furthermore since the zeros of Ai are all real and negative [O, p. 418], Ai_1 is nonzero in $S_0 \cup S_1$. Two other solutions of the Airy equation that will be useful are $\tilde{w}^{(i)}$ which for complex values of z are defined by

$$\tilde{w}^{(j)}(y) = \frac{1}{2}(\frac{y}{3})^{\frac{1}{2}}e^{(-1)^{j+1}i\frac{\pi}{6}}H^{(j)}_{1/3}(\frac{2}{3}y^{\frac{2}{3}}e^{\frac{i\pi}{2}}), \qquad j = 1,2 \tag{4.4}$$

where we take the branch of the square root so that $\mathrm{Re}(z^{3/2}) \geq 0$ for $z \in S_0$ and $\mathrm{Re}(z^{3/2}) \leq 0$ for $z \in S_1$ which is the principal branch of $z^{3/2}$. It follows from the asymptotic expansions for Hankel functions ([O, p. 238]) that $\tilde{w}^{(1)}$ is recessive in S_0 while $\tilde{w}^{(2)}$ dominant in S_0. The equations ([O, p. 239])

$$H^1_\nu(ze^{m\pi i}) = -[\sin\{(m-1)\nu\pi\}H^1_\nu(z) + e^{-\nu\pi i}\sin\{m\nu\pi\}H^2_\nu(z)]/\sin\nu\pi, \tag{4.5}$$

and

$$H^2_\nu(ze^{m\pi i}) = -[e^{\nu\pi i}\sin\{m\nu\pi\}H^1_\nu(z) + \sin\{(m+1)\nu\pi\}H^2_\nu(z)]/\sin\nu\pi, \tag{4.6}$$

with $\nu = 1/3$ and $m = 1$ show that $\tilde{w}^{(1)}$ is dominant in S_1 as is $\tilde{w}^{(2)}$. Since $K_{1/3}(z)$ is nonzero for $|\arg(z)| \leq \pi$ (Watson [W], Olver [O, p. 254]) and $\frac{\pi i}{2}e^{\frac{\pi i}{6}}H^{(1)}_{1/3}(ze^{\frac{i\pi}{2}}) = K_{1/3}(z)$, we find that $\tilde{w}^1(z)$ is nonzero for $|\arg(z)| \leq \frac{\pi}{3}$. Lommel's method [O, p. 414] applied to $\tilde{w}^{(1)}$ and $h = \frac{1}{2}(\frac{y}{3})^{\frac{1}{2}}e^{-i\frac{\pi}{6}}H^{(2)}_{1/3}(\frac{2}{3}y^{\frac{2}{3}}e^{-\frac{i\pi}{2}})$ can now be used to show that $\tilde{w}^{(1)}$ has no zeros in $S_1 \cup S_0 \cup S_{-1}$. Another application of Lommel's method with $\tilde{w}^{(2)}$ in place of $\tilde{w}^{(1)}$ above and a corresponding change to h shows that $\tilde{w}^{(2)}$ is also nonzero in $S_1 \cup S_0 \cup S_{-1}$.

Set $S = S_0 \cup S_1$, and let Ω a region in the complex y plane. We will restrict Ω so that

$$\rho([t_1, t_2], \Omega, [0, \epsilon_0])^{3/2} \subset S_0 \cup S_1. \tag{4.7}$$

Extensions to $S_0 \cup S_{-1}$ can be accomplished using the symmetry properties of the solutions. Because $\tilde{w}^{(2)}$ is dominant in $S_1 \cup S_0$ it is not a numerically satisfactory auxillary function in this region thus we set

$$\tilde{u}^{(1)} = \tilde{w}^{(1)},$$

$$\tilde{u}^{(2)} = \mathrm{Ai}_1,$$

$$\hat{u}^{(1)}(z) = e^{2/3z^{3/2}}\tilde{u}^{(1)}(z),\ \hat{u}^{(2)}(z) = e^{-2/3z^{3/2}}\tilde{u}^{(2)}(z), \widehat{\mathrm{Ai}}_0(z) = e^{2/3z^{3/2}}\mathrm{Ai}_0(z),$$

and

$$\widehat{\mathrm{Ai}}_1(z) = e^{-2/3z^{3/2}}\mathrm{Ai}_1(z).$$

It is not difficult to see from the continuity and asymptotic expansions for the above functions ([O, p. 392, 413]) that the following constants are finite,

$$d_3 = \max\left(\sup_S |\hat{u}^{(1)}(z)|, \sup_S |\widehat{\mathrm{Ai}}_1(z)|, \sup_S |\widehat{\mathrm{Ai}}_0(z)|\right),$$

$$\nu_1 = \max\left(\sup_S |z^{1/2}\hat{u}^{(1)}(z)|,\ \sup_S |z^{1/2}\hat{u}^{(2)}(z)|\right),$$

$$d_4 = \max\left(\sup_S \frac{|e^{2/3z^{3/2}}\tilde{u}^{(1)}(z)'|}{1+|z|^{1/4}},\ \sup_S \frac{|e^{-2/3z^{3/2}}\tilde{u}^{(2)}(z)'|}{1+|z|^{1/4}}\right),$$

and

$$d_5 = \max\left(\sup_S |e^{-4/3z^{3/2}}\frac{\tilde{u}^{(2)}(z)}{\tilde{u}^{(1)}(z)}|,\ \sup_S |e^{4/3z^{3/2}}\frac{\tilde{u}^{(1)}(z)}{\tilde{u}^{(2)}(z)}|\right).$$

Set

$$\psi^1 = g\mathrm{Ai}_0, \tag{4.8}$$

and

$$\psi^2 = g\mathrm{Ai}_1. \tag{4.9}$$

With $k(t,y,\epsilon) = \rho(t,y,\epsilon)^{1/2}\rho'(t,y,\epsilon)$, we now obtain an analog of Lemma 3.6. Because of iib. the appropriate functions to use are G_1^2 and G_2^1 in 3.39 and 3.43 respectively, also the role of the weight function will be played by

$$\tilde{E}(t) = |e^{2/3\rho(t,y,\epsilon)^{3/2}}|. \tag{4.10}$$

Lemma 4.2. *Suppose that* (4.7) *holds,* ib–iiib *are satisfied, and*

$$a(t,\epsilon) \in C^\infty([t_{in}, t_{fi}] \times [0,\epsilon_0])$$

is strictly positive. Let K be a compact subset of Ω. Then there is a constant $d(K)$ such that for $[n\epsilon, i\epsilon] \subset [t_{in}+\epsilon, t_{fi}-\epsilon], \epsilon \in (0,\epsilon_0]$ the following inequality holds

$$|G_2^1(n,i)| \le \tilde{G}^1(i) = d(K)\epsilon^{-1/3}|\hat{u}^{(1)}(i)\hat{u}^{(1)}(i-1)|, \tag{4.11}$$

and

$$|G_1^2(n,i)| \le \tilde{G}^2(i) = d(K)\epsilon^{-1/3}|\hat{u}^{(2)}(i)\hat{u}^{(2)}(i-1)|. \tag{4.12}$$

Proof. The Wronskian of A_1 and $\tilde{u}^{(1)}$ is

$$W(\mathrm{A}_1, \tilde{u}^{(1)}) = -\frac{e^{-\frac{2\pi i}{3}}}{2\sqrt{3}\pi},$$

which is less in magnitude than $2/\pi$.

Thus let $d_i(K) = \sup_K c_i(y)$ $i = 6, 7$ where c_6 is given in (3.36) and c_7 in Lemma 3.6. Let $d_8(K) = \sup_{(t,y,\epsilon)\in\overline{W}\times K} \tilde{E}(x(t,Y,\epsilon))E^{-1}(x(t+\epsilon,y,\epsilon))$, $x(t,y,\epsilon) = \epsilon^{-2/3}\rho(t,y,\epsilon)$. Conditions ib and iib insure that these constants are finite. We begin with G_1^2 and observe that for y fixed the steps in Lemma 3.6 can be followed to equation (3.47) with the roles of $u^{(1)}$ and $u^{(2)}$ interchanged. From iiib we find the bound

$$|\frac{u^{(1)}(m)}{u^{(2)}(m)}| \leq d_5(K)\tilde{E}^{-1}(i)^2,$$

for $m \leq i$. The rest of the steps to (4.11) now follow as in Lemma 3.6. A similar analysis gives (4.12). □

With the above lemma we can now show

Lemma 4.3. *Suppose the hypotheses of Lemma 4.2. Let K be a compact subset of Ω. Then there is a constant $d(K)$ such that such that for $[t_{N_1-1}, t_{N_2+1}] \subset [t_{in}, t_{fi}], \epsilon \in (0, \epsilon_0]$,*

$$\sum_{j=N_1}^{N_2} \tilde{G}^j(i) < d(K)\epsilon^{-1}, \qquad j = 1, 2.$$

Proof. Set $d_9 = \inf_{[t_{in},t_{fi}]\times\Omega\times[0,\epsilon_0]} |\frac{\rho}{t-t_0}|^{1/2}$ which by iiib is nonzero. From Lemma 4.2

$$\sum_{i=N_1}^{N_2} \tilde{G}^1(i) \leq \tilde{d}(K)\epsilon^{-1/3} \sum_{i=N_1}^{N_2} |\hat{u}^{(1)}(i-1)\hat{u}^{(1)}(i)|,$$

which can be bounded by

$$\sum_{i=N_1}^{N_2} |\hat{u}^{(1)}(i-1)\hat{u}^{(1)}(i)| \leq d_3^2 + \sum_{(i:\Re(\rho(i)^{3/2})<0)} |\hat{u}^{(1)}(i-1)\hat{u}^{(1)}(i)| + \sum_{(i:\Re(\rho(i-1)^{3/2})>0)} |\hat{u}^{(1)}(i-1)\hat{u}^{(1)}(i)|.$$

If ν_1 is now used we find

$$\sum_{(i:\Re(\rho(i-1)^{3/2})>0)} |\hat{u}^{(1)}(i-1)\hat{u}^{(1)}(i)| \leq \frac{\nu_1^2}{d_9(K)}\epsilon^{-2/3}\int_{t_{N_1}}^{t_{N_2}} \frac{dt}{|t-t_0|^{1/2}} \leq d(K)\epsilon^{-2/3}.$$

A similar argument bounds the remaining sum which gives the result for $j = 1$. The result for $j = 2$ follows in a similar manner. □

With the above lemmas we can now prove the main result of this section

Theorem 4.4. *Suppose that* (4.7) *holds,* ib–iiib *are satisfied, and*

$$a(t,\epsilon) \in C^\infty([t_{in}, t_{fi}] \times [0, \epsilon_0])$$

is strictly positive. Furthermore suppose that $a_1(t,\epsilon)$ and $b_1(t,\epsilon)$ are continuous on $[t_{in}, t_{fi}] \times [0, \epsilon_0]$ and $a_1(t,\epsilon)$ is strictly positive. Let $[(N_1+1)\epsilon, (N_2-1)\epsilon] \subset [t_{in}, t_{fi}]$, $\epsilon \in (0, \epsilon_0]$, and set

$$K^j(i) = d_{15}(K)G^j(i)\left(|\Delta\left(\frac{a(i\epsilon)}{a((i-1)\epsilon)}\right)| + |\Delta\left(\frac{y-b(i\epsilon)}{a(i\epsilon)}\right)|\right), \qquad j = 1, 2.$$

Then there exist solutions f_1 and f_2 of (3.37) such that

$$|\sigma_2^1(n\epsilon)| = |\frac{f_1(n\epsilon) - \psi^1(n\epsilon)}{u^{(1)}(n\epsilon)}| \tag{4.13}$$
$$\leq d(K) \sum_{n_1}^{n_2} (|K^1(i) + G^1(i)|\hat{\beta}^{(1)}(i)|)e^{d(K)\sum_{n_1}^{n_2}(K^1(i)+G^1(i)|\hat{\beta}_1^{(1)}(i))|)},$$

and

$$|\sigma_1^2(n\epsilon)| = |\frac{f_2(n\epsilon) - \psi^2(n\epsilon)}{u^{(2)}(n\epsilon)}| \tag{4.14}$$
$$\leq d(K) \sum_{n_1}^{n_2} (K^2(i) + G^2(i)|\hat{\beta}^{(2)}(i)|)e^{d(K)\sum_{n_1}^{n_2}(K^2(i)+G^2(i)|\hat{\beta}_1^{(2)}(i))|)}.$$

Here ψ^j, $j = 1, 2$ are defined by (4.8) and (4.9) respectively. Thus if

$$\sum K^j(i) = o(1), \qquad j = 1, 2, \tag{4.15}$$

then

$$f^j(n) = \psi^j(y, n) + o(1). \tag{4.16}$$

If

$$\sup_{t,\epsilon} |a(t, \epsilon) - a_1(t, \epsilon)| = 0(\epsilon^2) = \sup_{t,\epsilon} |b(t, \epsilon) - b_1(t, \epsilon)|, \tag{4.17}$$

with the sup *be taken over $[t_{in}, t_{fi}] \times [0, \epsilon_0]$ then the $o(1)$ in equation (4.16) is in fact* $0(\epsilon)$.

Proof. The proof follows like the one given in Theorem 3.8 with $d_{10} = \sup_{y\in K} c_{10}$. □

In the next section we consider the initial value problem.

5. Initial value problem singular case

If t_{in} can be taken to be equal to zero then the above results can be used to solve the initial value problem and thus give uniform asymptotics for orthogonal polynomials. Unfortunately in the case when $a(n)$ and $|b(n)|$ tend to infinity, in general $k(t, y, \epsilon)^2$ will not satisfy condition i and $a(t, \epsilon)$ will not satisfy (2.14) for t in a neighborhood of zero. This is because if $a(t, \epsilon)$ and $b(t, \epsilon)$ are to be bounded then λ_ϵ must increase to infinity (see Section 6 for an example). Another case where condition i or (2.14) can be violated is in the case of varying recurrence coefficients [DM], [KV]. Finally it may not by evident where ρ is a single-valued analytic function so a LG-WKB approach is useful. We begin by switching to the equation satisfied by the polynomials $\tilde{p}_n = 2^n p_n/k_n$,

$$\tilde{p}_{n+1}(x) + 2(b_n - x)\tilde{p}_n(x) + 4a_n^2 p_{n-1}(x) = 0.$$

If the scaling indicated in the introduction is performed we arrive at the difference equation

$$\tilde{p}_{n+1}(y) + 2(b(n\epsilon,\epsilon) - y)\tilde{p}_n(y) + 4a^2(n\epsilon,\epsilon)\tilde{p}_{n-1}(y) = 0. \tag{5.1}$$

We now consider the ϵ-difference equation

$$\psi(t+\epsilon,y,\epsilon) + 2(b(t,\epsilon) - y)\psi(t,y,\epsilon) + 4a^2(t,\epsilon)\psi(t-\epsilon,y,\epsilon) = 0. \tag{5.2}$$

Following Deift and McLaughlin [DM, Appendix I] we look for solutions of (5.2) of the form

$$f(t,y,\epsilon) = e^{\frac{1}{\epsilon}s_0(t,y,\epsilon)+s_1(t,y,\epsilon)}(1+O(\epsilon^2)). \tag{5.3}$$

Substitute the expressions for f in (5.2) then expand $f(t\pm\epsilon,\epsilon)$ (here again we suppress the dependence upon y) in powers of ϵ. The coefficient of ϵ^0 is

$$e^{s_0(t,\epsilon)'} + 4a(t,\epsilon)^2 e^{-s_0(t,\epsilon)'} = 2(y - b(t,\epsilon)), \tag{5.4}$$

while the coefficient of ϵ^1 is

$$s_1(t,\epsilon)' = -\frac{e^{s_0(t,\epsilon)'} + 4a^2(t,\epsilon)e^{-s_0(t,\epsilon)'}}{2(e^{s_0(t,\epsilon)'} - 4a^2(t,\epsilon)e^{-s_0(t,\epsilon)'})}\, s_0(t,\epsilon)''. \tag{5.5}$$

With the use of

$$(e^{s_0'} - a^2 e^{-s_0'})' = (e^{s_0'} + 4a^2 e^{-s_0'})s_0'' - 8aa'e^{-s_0'},$$

and the derivative of (5.4) we find

$$\begin{aligned} s_1(t,\epsilon)' = & -\frac{(e^{s_0(t,\epsilon)'} - 4a(t,\epsilon)^2 e^{-s_0(t,\epsilon)'})'}{2(e^{s_0(t,\epsilon)'} - 4a^2(t,\epsilon)e^{-s_0(t,\epsilon)'})} \\ & + \frac{b(t,\epsilon)'}{e^{s_0(t,\epsilon)'} - 4a^2(t,\epsilon)e^{-s_0(t,\epsilon)'}} + \frac{s_0(t,\epsilon)''}{2}. \end{aligned}$$

Solving the relevant equations yields

$$s_0^{\pm}(t,\epsilon) = \int^t \ln\left(y - b(u,\epsilon) \pm \sqrt{(y-b(u,\epsilon))^2 - 4a(u,\epsilon)^2}\right) du, \tag{5.6}$$

and

$$\begin{aligned} s_1^{\pm}(t,\epsilon) = & -\frac{1}{4}\ln 2^2((y-b(t,\epsilon))^2 - 4a(t,\epsilon)^2) \\ & + \frac{1}{2}\ln(y - b(t,\epsilon) \pm \sqrt{(y-b(t,\epsilon))^2 - 4a(t,\epsilon)^2}) \\ & \pm \frac{1}{2}\int^t \frac{b'(u,\epsilon)du}{\sqrt{(y-b(u,\epsilon))^2 - 4a(u,\epsilon)^2}}. \end{aligned} \tag{5.7}$$

This gives us two approximate solutions to the difference equation given by

$$f^{\pm}(t_n,y,\epsilon) = e^{\frac{1}{\epsilon}s_0^{\pm}(t_n,y,\epsilon)+s_1^{\pm}(t_n,y,\epsilon)},$$

with $s_0^{\pm}$ and $s_1^{\pm}$ given above and $t_n = n\epsilon$.

Set

$$V_{t_i,t_j}(g) = \int_{t_i}^{t_j} |g(t,\epsilon)|dt, \tag{5.8}$$

$$\begin{aligned} V_1(t,\epsilon) = {} & V_{t,t+\epsilon}(b''(\cdot)(t+\epsilon-\cdot)) + V_{t,t+\epsilon}(a^2(\cdot)''(t+\epsilon-\cdot)) \\ & + V_{t,t+\epsilon}(b'(\cdot)^2(t+\epsilon-\cdot)) \\ & + V_{t,t+\epsilon}((4a'a(\cdot))^2(t+\epsilon-\cdot))), \end{aligned} \tag{5.9}$$

$$\begin{aligned} V_2(t,\epsilon) = {} & (V_{t-\epsilon,t}(b''(\cdot)(\cdot-(t-\epsilon)) + V_{t-\epsilon,t}(|a^2(\cdot)''|(\cdot-(t-\epsilon))) \\ & + V_{t-\epsilon,t}(b'(\cdot)^2(\cdot-(t-\epsilon))) + V_{t-\epsilon,t}((4a'a(\cdot))^2(\cdot-(t-\epsilon)))), \end{aligned} \tag{5.10}$$

and

$$V_3(t,\epsilon) = V_{t,t+\epsilon}(aa')^2 + V_{t,t+\epsilon}(b')^2. \tag{5.11}$$

We will impose the following assumptions on $a(t,\epsilon)$ and $b(t,\epsilon)$

ic. $a(t,\epsilon),\ b(t,\epsilon) \in C^0([0,t_{fi}]\times[0,\epsilon_0])$,
iic. $\frac{\partial^i}{\partial t^i}a(t,\epsilon),\ \frac{\partial^i}{\partial t^i}b(t,\epsilon) \in C^0((0,t_{fi}]\times[0,\epsilon_0])\ i=1,2$,
iiic. $\sup\int_0^{t_{fi}}|b'(t,\epsilon)|<\infty, \sup\int_0^{t_{fi}}|a'a(t,\epsilon)|<\infty$
vic. $a(0,0)=0$ but $a(t,\epsilon)>0, (t,\epsilon)\in(0,t_f]\times[0,\epsilon_0]$,
vc. $V_1(t,\epsilon)\in C^0([0,t_{fi}-\epsilon]\times[0,\epsilon_0])$, $V_2(t,\epsilon)\in C^0([\epsilon,t_{fi}]\times[0,\epsilon_0])$, and $V_3(t,\epsilon)\in C^0([0,t_{fi}]\times[0,\epsilon_0])$

Let

$$A(\epsilon) = \inf_{[0,t_{fi}]}(b(t,\epsilon)-2a(t,\epsilon)),\ B(\epsilon) = \sup_{[0,t_{fi}]}(b(t,\epsilon)+2a(t,\epsilon)),$$

$$I(\epsilon) = [A(\epsilon),B(\epsilon)],$$

$$A_{a,b}(\epsilon) = \inf_{[t_{in},t_{fi}]}(b(t,\epsilon)-2a(t,\epsilon)),\ B_{a,b}(\epsilon) = \sup_{[t_{in},t_{fi}]}(b(t,\epsilon)+2a(t,\epsilon)),$$

and

$$I_{a,b}(\epsilon) = [A_{a,b}(\epsilon),B_{a,b}(\epsilon)].$$

We now examine the analytic properties of the above approximate solution. Set

$$f_1(t,y,\epsilon) = \sqrt{(y-b(t,\epsilon))^2-4a(t,\epsilon)^2}$$

and,

$$f_2^{\pm}(t,y,\epsilon) = y-b(t,\epsilon)\pm\sqrt{(y-b(t,\epsilon))^2-4a(t,\epsilon)^2}.$$

Lemma 5.1. *Suppose conditions* ic. *holds. Then* f_1, *and* f_2^+ *are nonvanishing in* $[0,t_{fi}]\times C\backslash I(\epsilon)$, *and* f_1, f_2^+, *and* $\ln f_2^+ \in C^0([0,t_{fi}])\times H(C\backslash I(\epsilon))$. *Likewise* f_2^- *is nonvanishing in* $[t_{in},t_{fi}]\times C$ *and* $\ln f_2^- \in C^0([t_{in},t_{fi}])\times H(C\backslash I_{a,b}(\epsilon))$ *for* $t_{in}>0$. *For any compact set* K *such that* $I(0)\subset \mathrm{int}(K)$ *there is an* ϵ_K *such that* f_1 *and* f_2^+ *are non-zero in* $[0,t_{fi}]\times[0,\epsilon_K]\times C\backslash K$, *and* f_1, f_2^+, *and* $\ln f_2^+\in C^0([0,t_{fi}]\times[0,\epsilon_K])\times H(C\backslash K)$. *If* $\tilde K$ *is a compact set in* $C\setminus K$ *then* f_1, f_2^+ *and* $\ln f_2^+ \in C^0([0,t_{fi}]\times[0,\epsilon_K]\times\tilde K)$. *Finally if* $t_{in}>0$ *and* K *a compact set such that* $I_{a,b}(0)\subset$ int(K), *then there is an* ϵ_K *such that* $f_2^{\pm}$ *and* $\ln f_2^{\pm}\in C^2([t_{in},t_{fi}]\times[0,\epsilon_K])\times H(C\backslash$

K), f_2^- is nonzero in $[t_{in}, t_{fi}] \times [0, \epsilon_K] \times C \setminus K$, and $f_2^\pm$, $\ln f_2^\pm \in C^2([t_{in}, t_{fi}] \times [0, \epsilon_K] \times \tilde{K})$ where $\tilde{K}$ is a compact subset of $C \backslash K$.

Proof. From the definition of $I(\epsilon)$ we see that $(y - b(t,\epsilon))^2 - 4a(t,\epsilon)^2$ is nonzero for all $(t,y) \in [0, t_{fi}] \times C\backslash I(\epsilon)$ hence a branch of the square root function can be choosen so that $f_1(t,y,\epsilon) \in H(C \setminus I(\epsilon))$ for each $t \in [0, t_{fi}]$ and such that for large y, $f_2^+ \sim 2y$. That $f_1(t,y,\epsilon) \in C^0([0,t_{fi}]) \times H(C\backslash I(\epsilon))$ follows from the continuity properties of a, b, and the square root function. The above argument also shows that $f_2^+ \in C^0([0,t_{fi}]) \times H(C\backslash I(\epsilon))$. Since $x + \sqrt{x^2-1}$ maps $C\backslash[-1,1]$ to the exterior of the unit circle we see that f_2^+ is nonvanishing in $[0,t_{fi}] \times C\backslash I(\epsilon)$. This implies that $\ln f_2^+ \in C^0([0,t_{fi}]) \times H(C\backslash I(\epsilon))$. For $t > t_{in}$ iic and the differentiability of ln and square root functions show that $\frac{\partial^i}{\partial t^i} f_1$, $\frac{\partial^i}{\partial t^i} f_2^\pm$ and $\frac{\partial^i}{\partial t^i} \ln f_2^+ \in C^0([t_{in}, t_{fi}]) \times H(C \setminus I(\epsilon))$ $i = 1 \ldots 2$. Since $a(t,\epsilon)$ is strictly greater than zero in $[t_{in}, t_{fi}]$ we find

$$f_2^- = \frac{4a(\epsilon,t)^2}{f_2^+},$$

which does not vanish for each

$$(t,y) \in [t_{in}, t_{fi}] \times C\backslash I(\epsilon)\ , \quad t_{in} > 0.$$

The properties of $\ln f_2^-$ now follow. Let K be a compact set such that $I(0) \subset$ int(K). The continuity of $a(t,\epsilon)$ and $b(t,\epsilon)$ imply that there exists an $\epsilon_K \leq \epsilon_0$ such that $I(\epsilon) \subset \text{int}(K)$ for all $\epsilon \leq \epsilon_K$. This plus the continuity properties of the ln and square root functions give that f_1, f_2^+, and $\ln f_2^+ \in C^0([0,t_{fi}] \times [0,\epsilon_K]) \times H(C\backslash K)$ and f_1, f_2^+ and $\ln f_2^+ \in C^0([0,t_{fi}] \times [0,\epsilon_K] \times \tilde{K})$ for any compact set $\tilde{K} \subset C\backslash K$. For $t \in [t_{in}, t_{fi}]$ the properties of $f_2^\pm$ and $\ln f_2^\pm$ follow in a similar manner. □

With this we now discuss the approximate solutions and their relation to the initial value problem. Define

$$H^\pm_{s_0,s_1}(t) = \epsilon(s_0(t)''/2 \pm s_1'(t)),$$

$$J^\pm_{s_0,s_1}(t) = \frac{1}{2\epsilon}\int_t^{t\pm\epsilon} s_0'''(u)(t \pm \epsilon - u)^2 du + \int_t^{t\pm\epsilon} s_1''(u)(t\pm\epsilon - u)du,$$

and

$$R^\pm_{s_0,s_1}(t) = (H^\pm_{s_0,s_1}(t) + J^\pm_{s_0,s_1}(t))^2 \int_0^1 e^{(H^\pm_{s_0,s_1}(t)+J^\pm_{s_0,s_1}(t))u}(1-u)du.$$

With the above we have

Lemma 5.2. *Suppose* ic–ivc *hold and set*

$$f^+(t,y,\epsilon) = \frac{(y-b(t,\epsilon)+\sqrt{(y-b(t,\epsilon))^2-4a(t,\epsilon)^2})^{1/2}}{((y-b(t,\epsilon))^2-4a(t,\epsilon)^2)^{1/4}} \times \exp\Big(\int_0^t \frac{b'(u,\epsilon)du}{2\sqrt{(y-b(u,\epsilon))^2-4a(u,\epsilon)^2}}\Big) \times \exp\Big(1/\epsilon\int_0^t \ln(y-b(u,\epsilon)+\sqrt{(y-b(u,\epsilon))^2-4a(u,\epsilon)^2})du\Big), \tag{5.12}$$

and

$$f^-(t,y,\epsilon) = \frac{(y-b(t,\epsilon)-\sqrt{(y-b(t,\epsilon))^2-4a(t,\epsilon)^2})^{1/2}}{((y-b(t,\epsilon))^2-4a(t,\epsilon)^2)^{1/4}} \times \exp\Big(-\int_t^{t_{fi}} \frac{b'(u,\epsilon)du}{2\sqrt{(y-b(u,\epsilon))^2-4a(u,\epsilon)^2}}\Big) \times \exp\Big(1/\epsilon\int_t^{t_{fi}} \ln(y-b(u,\epsilon)-\sqrt{(y-b(u,\epsilon))^2-4a(u,\epsilon)^2})du\Big). \tag{5.13}$$

Then for $n\epsilon \in [0, t_{fi}]$, $0 < \epsilon \le \epsilon_0$, $f^+ \in C^0([0,t_{fi}]) \times H(C\backslash I(\epsilon))$, *is nonzero and satisfies the difference equation*

$$f^+((n+1)\epsilon) + 2(b(n\epsilon,\epsilon)-y)f^+(n\epsilon) + 4a^2(n\epsilon,\epsilon)f^+((n-1)\epsilon = \eta^+(n\epsilon)f^+(n\epsilon), \quad \epsilon \le n\epsilon < t_{fi}-\epsilon \tag{5.14}$$

where

$$\eta^+(t) = e^{s_0^+(t)'}(J^+_{s_0^+,s_1^+}(t) + 4a(t,\epsilon)^2R^+_{s_0^+,s_1^+}(t)) + e^{-s_0^+(t)'}(J^-_{s_0^+,s_1^+}(t) + 4a(t,\epsilon)^2R^-_{s_0^+,s_1^+}(t)). \tag{5.15}$$

Likewise for $n\epsilon \in [t_{in}, t_{fi}]$, $t_{in} > 0$ *and* $0 < \epsilon \le \epsilon_0$, $f^- \in C[t_{in}, t_{fi}] \times H(C\backslash I_{a,b}(\epsilon))$ *is nonzero and satisfies the above difference equation for* $t_{in}+\epsilon < n\epsilon < t_{fi}-\epsilon$ *with* η^+ *replaced by*

$$\eta^-(t) = e^{s_0^-(t)'}(J^+_{s_0^-,s_1^-}(t) + R^+_{s_0^-,s_1^-}(t)) + e^{-s_0^-(t)'}(J^-_{s_0^-,s_1^-}(t) + R^-_{s_0^-,s_1^-}(t)). \tag{5.16}$$

Let K *be compact subset of* C *such that* $I(0) \subset \mathrm{int}(K)$ *and* $\tilde{K}$ *be a compact subset of* $C\backslash K$. *Then there is an* ϵ_K *such that* $f^+ \in C^0([0,t_{fi}] \times (0,\epsilon_K]) \times H(C\backslash K)$, $f^+ \in C^0([0,t_{fi}] \times (0,\epsilon_K] \times \tilde{K})$, $\eta^+ \in C^0([\epsilon, t_{fi}-\epsilon] \times [0,\epsilon_K] \times \tilde{K})$, *and* $\eta^+ \in C^0([\epsilon, t_{fi}-\epsilon] \times [0,\epsilon_K]) \times H(C \setminus K)$. *If* $t_{in} > 0$, K *and* $\tilde{K}$ *are compact sets such that* $I_{a,b}(0) \subset \mathrm{int}(K)$, *and* $\tilde{K} \subset C\backslash K$, *then there exists an* ϵ_K *such that* $f^- \in C^0([t_{in}, t_{fi}] \times (0,\epsilon_K]) \times H(C\backslash K)$, *and* $f^- \in C^0([t_{in}, t_{fi}] \times (0,\epsilon_K] \times \tilde{K})$ *where* $t_{in} > 0$. *Also* $\eta^- \in C^0([t_{in}, t_{fi}] \times [0,\epsilon_K]) \times H(C\backslash K)$ *and* $\eta^- \in C^0([t_{in}, t_{fi}] \times [0,\epsilon_K] \times \tilde{K})$.

Proof. It follows from Morera's Theorem, Lemma 5.1 and the integrability of b' that the integrals in (5.12) yield functions analytic in $C\backslash I(\epsilon)$ for each $t \in [0, t_{fi}]$. The fact that $f^+ \in C^0[0, t_{fi}] \times H(C\backslash I(\epsilon))$ for $\epsilon > 0$ now follows from Lemma 5.1 and the integrability of b'. The nonvanishing of f_1 and f_+^2 in $[0, t_{fi}] \times C\backslash I(\epsilon)$ shows that f^+ is nonvanishing in this set also. Let K be a compact set in C such that $I(0) \subset \operatorname{int}(K)$ then Lemma 5.1 and condition ic imply that there is an ϵ_K so that the integral

$$I_1(t, y, \epsilon) = \int_0^t \ln\left(y - b(u, \epsilon) + \sqrt{(y - b(u,t))^2 - 4a^2(u,t)}\right) du,$$

is $C^0([0, t_{fi}] \times [0, \epsilon_K]) \times H(C\backslash K)$. Also for any compact set $\tilde{K} \in C\backslash K$, $I_1 \in C^0([0, t_{fi}] \times [0, \epsilon_K] \times \tilde{K})$. From the integral representation,

$$\frac{1}{\sqrt{(y-b)^2 - 4a^2}} = \frac{1}{\pi}\int_{b-2a}^{b+2a} \frac{dx}{\sqrt{(x-b)^2 - 4a^2}}, \tag{5.17}$$

we find that $\frac{1}{f_1}$ is uniformly bounded for $(t, \epsilon, y) \in [0, t_{fi}] \times [0, \epsilon_K] \times C\backslash K$. The continuity conditions ic and iic, and the uniform integrability of b' imply that

$$I_2(t, y, \epsilon) = \int_0^{t_{fi}} \frac{b'(u, \epsilon)}{\sqrt{(y - b(u,\epsilon))^2 - 4a(u,\epsilon)^2}},$$

is in $C^0([0, t_{fi}] \times [0, \epsilon_K]) \times H(C\backslash K)$ and $I_2 \in C^0([0, t_{fi}] \times [0, \epsilon_K] \times \tilde{K})$ for $\tilde{K}$ as above. This plus Lemma 5.1 and the fact that $\frac{1}{\epsilon}$ is continuous for $\epsilon > 0$ gives the continuity properties of f^+. A similar argument gives the continuity properties of f^-.

The equations for $\eta^\pm$ are obtained by substituting (5.12) or (5.13) into (5.14) expanding to first order using Taylor's Theorem with the remainder and equations (5.4) and (5.5). From the definition of H and J we find

$$H^\pm_{s_0^\pm, s_1^\pm}(t) + J^\pm_{s_0^\pm, s_1^\pm}(t) = \frac{1}{\epsilon}\int_t^{t\pm\epsilon} s_0^\pm(u)''(t \pm \epsilon - u)du + \int_t^{t\pm\epsilon} s_1^\pm(u)'du. \tag{5.18}$$

Also

$$s_0^\pm(t)'' = \mp\frac{b'(t,\epsilon)}{f_1} \mp \frac{4a'(t,\epsilon)a(t,\epsilon)}{f_2^\pm(t) f_1}. \tag{5.19}$$

Lemma 5.1, iic, vc, and (5.18) show that $R^+_{s_0^+, s_1^+} \in C^0([0, t_{fi} - \epsilon] \times [0, \epsilon_0]) \times H(C\backslash K)$, and $R^+_{s_0^+, s_1^+} \in C^0([0, t_{fi} - \epsilon] \times [0, \epsilon_0] \times \tilde{K})$. In a similar manner it follows that $R^-_{s_0^+, s_1^+} \in C^0([\epsilon, t_{fi}] \times [0, \epsilon_0]) \times H(C\backslash K)$, and $R^-_{s_0^+, s_1^+} \in C^0([\epsilon, t_{fi}] \times [0, \epsilon_0] \times \tilde{K})$. By differentiating $s_0^+(t)''$ with respect to t and likewise $s_1^+(t)'$ it follows from vc. after a tedious computation that $J^+_{s_0^+, s_1^+} \in C^0([\epsilon, t_{fi} - \epsilon] \times [0, \epsilon_0]) \times H(C\backslash K)$ and $J^+_{s_0^+, s_1^+} \in C^0([\epsilon, t_{fi} - \epsilon] \times [0, \epsilon_0] \times \tilde{K})$ while $J^-_{s_0^+, s_1^+} \in C^0([\epsilon, t_{fi}] \times [0, \epsilon_0]) \times H(C\backslash K)$ and $J^-_{s_0^+, s_1^+} \in C^0([\epsilon, t_{fi}] \times [0, \epsilon_0] \times \tilde{K})$. Thus η^+ has the continuity properties claimed in the Lemma. A similar analysis follows for η^-. □

The following Lemma will be useful in controlling the error between the approximate solutions and bona fide solutions of the initial value problem.

Lemma 5.3. *Suppose* ic–vc *hold and let* $\tilde{K}$ *be a compact subset of* $C\backslash I(0)$. *Then there exists an* $\epsilon_{\tilde{K}}$ *such that for all* $\epsilon \in [0, \epsilon_{\tilde{K}}]$,

$$\sum_{i=1}^{n-1} |R^{+}_{s_0^+, s_1^+}(i\epsilon)| \leq \sum_{i=1}^{n-1} V_{t_i,t_{i+1}}(s_0^{+\prime\prime}) + V_{t_i,t_{i+1}}(s_1^{+\prime}))^2 e^{d(\tilde{K})} \tag{5.20}$$

$$\leq d(\tilde{K}) \sum_{i=1}^{n} V_3(t_i, \epsilon),$$

$$\sum_{i=1}^{n-1} |R^{-}_{s_0^+, s_1^+}(i\epsilon)| \leq \sum_{i=1}^{n-1} (V_{t_{i-1},t_i}(s_0^{+\prime\prime}) + V_{t_{i-1},t_i}(s_1^{+\prime}))^2 e^{d(\tilde{K})} \tag{5.21}$$

$$\leq d(\tilde{K}) \sum_{i=0}^{n-1} V_3(t_i, \epsilon)$$

$$\sum_{i=1}^{n-1} |J^{+}(s_0^+, s_1^+)(t_i)| \leq \sum_{i=1}^{n-1} (\frac{1}{2} V_{t_i,t_{i+1}}(s_0^{+\prime\prime\prime}(t_{i+1} - \cdot)) + V_{t_i,t_{i+1}}(s_1^{+\prime\prime}(t_{i+1} - \cdot)) \tag{5.22}$$

$$\leq d(\tilde{K}) \sum_{i=1}^{n-1} V_1(t_i, \epsilon),$$

and

$$\sum_{i=1}^{n-1} |J^{-}(s_0^+, s_1^+)(t_i)| \leq \sum_{i=0}^{n-2} \frac{1}{2} V_{t_i,t_{i+1}}(s_0^{+\prime\prime\prime}(\cdot - t_i)) + V_{t_i,t_{i+\epsilon}}(s_1^{+\prime\prime}(\cdot - t_i)) \tag{5.23}$$

$$\leq d(\tilde{K}) \sum_{i=1}^{n-2} V_2(t_i, \epsilon).$$

Here $d(\tilde{K})$ *is a constant depending only upon* $\tilde{K}$ *and* $t_i = i\epsilon$.

Proof. From Lemma 5.1 there exists a constant M such that

$$|\frac{1}{f_1}|, |\frac{1}{f_2^+}| < M$$

for all $(t, \epsilon, y) \in [0, t_{fi}] \times [0, \epsilon_{\tilde{K}}] \times \tilde{K}$. Thus from Lemma 5.1, equation (5.19) and equation (5.5) we find, $|s_0^+(t)''| \leq \frac{|b'(t,\epsilon)|}{M} + \frac{4|a'(t,\epsilon)a(t,\epsilon)|}{M^2}$ and $|s_1^+(t)'| < \tilde{d}(K)\left(\frac{|b'(t,\epsilon)|}{M} + \frac{4|a'(t,\epsilon)a(t,\epsilon)|}{M^2}\right)$, where $d(\tilde{K})$ depends only upon $\tilde{K}$. Thus

$$|H^{+}_{s_0^+, s_1^+}(t) + J^{+}_{s_0^+, s_1^+}(t)| \leq d(K) \int_t^{t+\epsilon} |b'(u, \epsilon)| + |a'(u, \epsilon)a(u, \epsilon)| du,$$

and

$$|H^-_{s_0^+,s_1^+}(t) + J^-_{s_0^+,s_1^+}(t)| \leq d(K)\int_{t-\epsilon}^{t} |b'(u,\epsilon)| + |a'(u,\epsilon)a(u,\epsilon)|du.$$

Therefore,

$$|R^+_{s_0^+,s_1^+}(t_i)| \leq \Big(\frac{1}{\epsilon}V_{t_i,t_{i+1}}(s_0^{+''}(t_{i+1}-\cdot)) + V_{t_i,t_{i+1}}(s_1^{+'})\Big)^2 e^{2d(K)M_1}, \qquad t_i \in [0, t_f-\epsilon] \tag{5.24}$$

and

$$|R^-_{s_0^+,s_1^+}(t_i)| \leq (\frac{1}{\epsilon}V_{t_{i-1},t_i}(s_0^{+''}(\cdot - t_{i-1})) + V_{t_{i-1},t_i}(s_1^{+'}))^2 e^{2d(K)M_1}, \quad t_i \in [\epsilon, t_f]. \tag{5.25}$$

Summing (5.24) from 1 to $n-1$ then using standard inequalities gives the first part of (5.20). Substituting the above inequalities for $|s_0^+(t)''|$ and $|s_1^+(t)'|$ in (5.24) and using standard inequalities gives the second part of (5.20). Equation (5.21) is obtained in the same way. Lemma 5.1 and elementary calculus gives the bound

$$|s_0^+(t)'''| \leq d(K)(|b(t)''| + 4(|a(t)a''(t)| + |a'(t)^2|) + (|b'(t)| + 4|a(t)a(t)'|)^2),$$

and likewise for $|s_1^+(t)|$. These bounds give the second inequalities in (5.22) and (5.23). □

Lemma 5.4. *Suppose that* $\frac{\partial^i}{\partial t^i}a(t,\epsilon), \frac{\partial^i}{\partial t^i}b(t,\epsilon) \in C^0([t_{in}, t_{fi}] \times [0,\epsilon_0]), i = 0\ldots 2$ *then* $\sum_{i=[\frac{t_{in}}{\epsilon}]}^{[\frac{t_{fi}}{\epsilon}]} |\eta^\pm(t_i)| < c\epsilon$.

Proof. The boundedness of a, b and the first and second partial time derivatives imply that

$$\sum_{i=[\frac{t_{in}}{\epsilon}]}^{[\frac{t_{fi}}{\epsilon}]} |\eta^\pm(t_i)| < \sum_{i=[\frac{t_{in}}{\epsilon}]}^{[\frac{t_{fi}}{\epsilon}]} c\epsilon^2 < \tilde{c}\epsilon.$$

□

We now obtain error bounds on how close f^+ stays to a solution of the initial value problem. Similar results were obtained (without error bounds) by [GS] in the case when the coefficients in the recurrence formula were in certain classes of regularly varying or slowly varying functions. For the case when $a(t,\epsilon)$ are strictly bounded away from zero and $a(t,\epsilon), b(t,\epsilon) \in C^\infty$ similar results were obtained by Costin and Costin [CC] and Deift and McLaughlin [DM].

Theorem 5.5. *Suppose* ic–vc. *hold, and suppose that* $a_1(t,\epsilon)$ *and* $b_1(t,\epsilon)$ *are in* $C[0,t_{fi}] \times C[0,\epsilon_0])$ *and* $a_1(t,\epsilon) > 0$, *for* $(t,\epsilon) \in (0,t_f] \times [0,\epsilon_0]$. *Let* K *be a compact set in* $C\backslash I(0)$ *and* $\tilde{p}_n(y,\epsilon)$ *be a solution of*

$$\tilde{p}_{n+1}(y,\epsilon) + 2(b_1(n\epsilon,\epsilon) - y)\tilde{p}_n(y,\epsilon) + 4a_1(n\epsilon,\epsilon)^2\tilde{p}_{n-1}(y,\epsilon) = 0, \tag{5.26}$$
$$\tilde{p}_0(y,\epsilon) = 1, \ \tilde{p}_1(y,\epsilon) = 2(y - b_1(0,\epsilon)).$$

Then there exists an ϵ_K *such that for* $(n\epsilon, \epsilon) \in [0, t_f] \times (0, \epsilon_K]$

$$\tilde{p}_n(y, \epsilon) = f^+(n\epsilon, y, \epsilon)/f^+(0, y, \epsilon)(1 + \phi(n\epsilon, y, \epsilon)), \tag{5.27}$$

where

$$|\phi(n\epsilon, y, \epsilon)| < d(K)\Big(\Delta(b(0, \epsilon)) + 4a^2(0, \epsilon) + e^{d(K)(V_{0,\epsilon}(b') + V_{0,\epsilon}(aa'))} - 1 + L(1, n-1)\Big), \tag{5.28}$$

where

$$S(n-1) = \left(\sum_{i=1}^{n-1} V_1(t_i, \epsilon) + V_2(t_i, \epsilon) + \Delta(b(i\epsilon)) + \Delta(a(i\epsilon)^2) + \sum_{i=0}^{n} V_3(t_i, \epsilon)\right).$$

Proof. The proof is very close to that given in the earlier sections and we will only indicate the differences. Substituting (5.27) into (5.26) and following the steps that led to (3.41) it is not difficult to see that ϕ satisfies the equation,

$$\begin{aligned} \phi(n+1) - \phi(n) - 4a(n\epsilon)^2 \frac{f^+(n-1)}{f^+(n+1)} (\phi(n) - \phi(n-1)) \\ = (\Delta(b(n\epsilon)) - \eta^+(n)) \frac{f^+(n)}{f^+(n+1)} (1 + \phi(n)) \\ + 4\Delta(a(n\epsilon)^2\Big) \frac{f^+(n-1)}{f^+(n+1)} (1 + \phi(n-1)), \qquad 1 \le n \end{aligned} \tag{5.29}$$

where we have suppressed the dependence on all variables but n. Solve the above equation with the initial conditions $\phi(0) = 0, \phi(1) = f^+(0)\tilde{p}_1(y, \epsilon)/f^+(1) - 1$ to obtain,

$$\begin{aligned} \phi(n) = \left(\frac{f^+(0)p(1)}{f^+(1)} - 1\right) G(n, 0) + 4\Delta\Big(a(\epsilon)^2\Big) \frac{f^+(0)}{f^+(2)} G(n, 1) \\ + \sum_{i=1}^{n-1} G(n, i)(\Delta(b(i\epsilon)) - \eta^+(i)) \frac{f^+(i)}{f^+(i+1)} (1 + \phi(i)) \\ \sum_{i=1}^{n-2} G(n, i+1) 4\Delta(a((i+1)\epsilon)^2) \frac{f^+(i)}{f^+(i+2)} (1 + \phi(i)) \end{aligned} \tag{5.30}$$

where

$$G(n, i) = \sum_{j=i}^{n-1} \frac{f^+(i+1)f^+(i)}{f^+(j+1)f^+(j)} \prod_{k=i+1}^{j} 4a(k\epsilon, \epsilon)^2.$$

With

$$\frac{f^+(i+1)f^+(i)}{f^+(j+1)f^+(j)} = \prod_{i+2}^{j+1} \frac{f^+(k-2)}{f^+(k)},$$

so that

$$\frac{f^+(i+1)f^+(i)}{f^+(j+1)f^+(j)}\prod_{k=i+1}^{j}4a(k\epsilon,\epsilon)^2=\frac{f^+(j-1)}{f^+(j+1)}4a((i+1)\epsilon,\epsilon)^2\prod_{k=i+2}^{j}\frac{4a(k\epsilon,\epsilon)^2}{f^+(k)^2}e^{g(k\epsilon,\epsilon)},$$

where

$$g(k\epsilon,\epsilon)=\frac{1}{2\epsilon}\int_{k\epsilon}^{(k-2)\epsilon}\frac{\partial}{\partial u}\log f_+^2(u,\epsilon)((k-2)\epsilon-u)du+s_1^+((k-2)\epsilon,\epsilon)-s_1^+(k\epsilon,\epsilon).$$

From Lemma 5.1, the above hypothesis, and the mapping properties of $x+\sqrt{x^2-1}$ mentioned above we find that there is an ϵ_K, such that for all $(t,\epsilon,y)\in[0,t_{fi}]\times[0,\epsilon_K]\times K, \ln|f_2^+/2a(t,\epsilon)|>\ln r$ with $r>1$. Also from the nonvanishing of f_2^+ we find

$$|\frac{f^+(k)}{f^+(k+1)}|<e^{\frac{1}{\epsilon}\int_{t_k}^{t_{k+1}}|\ln(f_2^+(u))|du+d(K)(V_{t_k,t_{k+1}}(b')+V_{t_k,t_{k+1}}(aa')}\leq d(\tilde{K}).$$

Thus,

$$|G(n,i)|\leq d(K)e^{d(K)(V_{0,t_{fi}}(b')+V_{0,t_{fi}}(aa')}\frac{1-r^{-2(n-i)}}{1-r^{-1}}$$

From (5.4) the inequality

$$|\frac{p(1)f^+(0)}{f^+(1)}-1|<d(K)(\Delta(b(0,\epsilon))+4a^2(0,\epsilon)+e^{d(K)(V_{0,\epsilon}(b')+V_{0,\epsilon}(aa'))}-1)$$

can be obtained, and (5.15) and Lemma 5.3 can be used to obtain

$$\sum_{i=1}^{n-1}|e^{-s_0^+(i\epsilon)'}\eta^+(i\epsilon)|\leq d(K)\left(\sum_{i=1}^{n-1}V_1(t_i,\epsilon)+V_2(t_i,\epsilon)+V_3(t_{i-1},\epsilon)\right),\quad n>1.$$

Thus the Picard iteration applied to (5.30) gives

$$\begin{aligned}|\phi(n)|\leq d(K)(\Delta(b(0,\epsilon))+4a^2(0,\epsilon)+e^{d(K)(V_{0,\epsilon}(b')+V_{0,\epsilon}(aa'))}-1\\ +S(n-1))e^{d(K)S(n-1)},\end{aligned}\tag{5.31}$$

where $S(n-1)$ is given above. This leads to the required bound for $|p(n)|$. □

A formula that will be useful for the matching of various solutions is

$$\begin{aligned}&\int_0^{t_n}\ln y-b(u)+\sqrt{(y-b(u))^2-4a^2(u+\frac{\epsilon}{2})}du\\ &=\int_0^{t_n}\ln y-b(u)+\sqrt{(y-b(u))^2-4a^2(u)}du\\ &-\frac{\epsilon}{2}\int_0^{t_n}\frac{4a(u)a'(u)}{f(u)\sqrt{(y-b(u))^2-4a(u)^2}}+\phi^1,\end{aligned}\tag{5.32}$$

where ϕ^1 has properties similar to ϕ above.

6. Applications

The previous results can be combined to obtain uniform asymptotics for solutions the difference equation of the form (2.6) including the initial value problem. This is accomplished by matching solutions in various overlapping regions. Here we will study the case of Hermite polynomials whose asymptotics can be found in [S] and more recently [DKMVZ]. More general problems will be considered at a later time [G].

The orthonormal Hermite polynomials $\{\hat{H}_n(x)\}$ (Olver [O, p. 403], Szegő [S]) have recurrence coefficients $b_n = 0$, $n \geq 0$, $a_n = \sqrt{n/2}$, $n \geq 1$ and we choose $\epsilon = 1/l$, $t_n = n/l$ and $\hat{\lambda}_l = \lambda_\epsilon = \sqrt{2l+1}$ which yields $a\left(t, \frac{1}{l}\right) = \frac{1}{2}\sqrt{\frac{2lt}{2l+1}}$. Sterling's formula implies

$$\frac{e^{l\int_0^{\bar{l}} \ln 2a(u+\frac{1}{2l},\frac{1}{l})du}}{\prod_1^n \frac{2a(i)}{\lambda_{1/l}}} = \frac{e^{1/4}}{\pi^{1/4}}(1+O(\frac{1}{l})),$$

so that for $y \in C\backslash\left[-\sqrt{\frac{2lt_{if}+1}{2l+1}}, \sqrt{\frac{2lt_{fi}+1}{2l+1}}\right]$, $t_{fi} \geq 1$ equation (5.27) and (5.32) give

$$p_n(y, \frac{1}{l})$$
$$= \frac{e^{1/4}}{\pi^{1/4}}\left(\frac{y^2}{y^2 - 4a^2(t_n + \frac{1}{2l}, \frac{1}{l})}\right)^{1/4} e^{l\int_0^{t_n} \ln\left(\frac{\hat{\lambda}_l y}{\sqrt{2lu+1}} + \sqrt{\left(\frac{\hat{\lambda}_l y}{\sqrt{2lu+1}}\right)^2 - 1}\right)du}\left(1+O\left(\frac{1}{l}\right)\right),$$

where $p_n(y, \frac{1}{l}) = \hat{H}_n(\hat{\lambda}_l y)$. Since

$$l\int_{-\frac{1}{2l}}^{0} \ln\left(\frac{\hat{\lambda}_l y}{\sqrt{2lu+1}} + \sqrt{\left(\frac{\hat{\lambda}_l y}{\sqrt{2lu+1}}\right)^2 - 1}\right)du$$
$$= \frac{1}{2}\ln\left(\hat{\lambda}_l y + \sqrt{(\hat{\lambda}_l y)^2 - 1}\right) + \frac{1}{2}(\hat{\lambda}_l y)^2 - \frac{1}{2}\hat{\lambda}_l y\sqrt{(\hat{\lambda}_l y)^2 - 1},$$

we find

$$p_n(y, \frac{1}{l}) \tag{6.1}$$
$$= \frac{1}{(\hat{\lambda}_l 2)^{1/2}\pi^{1/4}}\left(\frac{1}{y^2 - 4a^2(t_n + \frac{1}{2l}, \frac{1}{l})}\right)^{1/2} e^{l\int_{-\frac{1}{2l}}^{t_n} \ln\left(\frac{\hat{\lambda}_l y}{\sqrt{2lu+1}} + \sqrt{\left(\frac{\hat{\lambda}_l y}{\sqrt{2lu+1}}\right)^2 - 1}\right)du}$$
$$\times\left(1+O\left(\frac{\ln l}{l}\right)\right).$$

Thus for $t_n = 1$, $p_l(y, \frac{1}{l}) = \hat{H}_l(\hat{\lambda}_l y)$ and for $y \in C\backslash[-1, 1]$,

$$\frac{\hat{H}_l(\hat{\lambda}_l y)}{e^{\hat{\lambda}_l^2 y^2/2}} = \frac{\left(y + \sqrt{y^2 - 1}\right)^{\hat{\lambda}_l^2/2}}{\sqrt{2\hat{\lambda}_l}\pi^{1/4}(y^2 - 1)^{1/4}} e^{-\hat{\lambda}_l^2 y\frac{\sqrt{y^2-1}}{2}}\left(1+O\left(\frac{1}{l}\right)\right).$$

The convergence in the above formula is uniform on compact subsets of $C\backslash[-1,1]$. We now analyze the behavior of the Hermite polynomials in the oscillatory region. With $t_0 = \frac{(\hat{\lambda}_l y)^2-1}{2l}$, the Langer transformation (2.2) gives for $y > \sqrt{2lt_{fi}+1}$

$$\frac{2}{3}\rho^{3/2}(t,y,l) = \int_t^{t_0} \ln\left(\frac{\hat{\lambda}_l y}{\sqrt{2lu+1}} + \sqrt{\left(\frac{\hat{\lambda}_l y}{\sqrt{2lu+1}}\right)^2 - 1}\right) du. \tag{6.2}$$

The change of variables $v = \sqrt{2lu+1}$ in the above integral gives

$$\begin{aligned}\frac{2}{3}\rho^{3/2}(t,y,l) &= \frac{1}{l}\int_0^{\hat{\lambda}_l y} \ln\left(\frac{\hat{\lambda}_l y}{v} + \sqrt{\left(\frac{\hat{\lambda}_l y}{v}\right)^2 - 1}\right) v\,dv \\ &\quad - \frac{1}{l}\int_0^{\sqrt{2lt+1}} \ln\left(\frac{\hat{\lambda}_l y}{v} + \sqrt{\left(\frac{\hat{\lambda}_l y}{v}\right)^2 - 1}\right) v\,dv \\ &= \frac{(\hat{\lambda}_l y)^2}{2l} - \frac{1}{l}\int_0^{\sqrt{2lt+1}} \ln\left(\hat{\lambda}_l y + \sqrt{(\hat{\lambda}_l y)^2 - v^2}\right) dv.\end{aligned}$$

The last integral in the above formula gives an analytic extension to $\rho^{3/2}$ which shows that $\mathrm{Re}\rho^{3/2}$ is a decreasing function of t and $\lim_{\mathrm{Im}y\to 0,\mathrm{Im}y>0}\rho^{3/2}$ is a continuous function which for y real and positive has only one zero at t_0. Also for large enough y, $\rho^{3/2}$ is nonzero. Thus Theorems 3.8 and 4.4 can be used. Let $S = \{y \in C : \Re(y) > 0, 0 \le y \le \delta\}$ where δ is chosen sufficiently small so that $\frac{\rho^{\frac{3}{2}}}{(t-t_0(y,\epsilon))^{\frac{3}{2}}}$ is nonzero in S. For fixed $y \in S$ there exists an interval $[t_a, t_b] \subset (0, t_{fi})$ $t_{fi} > 1$ and an $\tilde{\epsilon}(y)$ such that for all $\epsilon < \tilde{\epsilon}(y)$, $t_0(y,\epsilon) \notin [t_a, t_b]$. From equation (6.1) we find

$$\frac{p_n(y,\frac{1}{l})}{e^{(\hat{\lambda}_l y)^2/2}} = \frac{1}{\sqrt{2\hat{\lambda}_l}\pi^{1/4}}\left(\frac{1}{y^2 - 4a^2(t_n + \frac{1}{2l})}\right)^{1/4} e^{-\frac{2}{3}l\rho^{3/2}(t_n,y,\frac{1}{l})}(1 + O(\frac{1}{l})). \tag{6.3}$$

Since the function on the left-hand side of the above equation satisfies the same difference equation as $p_n(y,\frac{1}{l})$ Theorem 4.4 implies that it can be written as $c^1\psi^1 + c^2\psi^2$ with c^1 and c^2 independent of t_n. Equation 4.3, the Wronskian of Ai and Bi,

$$W[\mathrm{Ai},\mathrm{Bi}] = \frac{1}{\pi}$$

and equation 3.4 gives

$$\psi_1(t_{n+1})\psi_2(t_n) - \psi_1(t_n)\psi_2(t_{n+1}) = -g(t_n)^2\frac{ie^{-\frac{i\pi}{3}}}{2\pi l^{\frac{1}{3}}}\frac{\sinh(\sqrt{\rho(t_n)}\rho(t_n)')}{\sqrt{\rho(t_n)}}.$$

The asymptotic expansions of the Airy functions,

$$\mathrm{Ai}(l^{\frac{2}{3}}\rho(t,y,\frac{1}{l})) = \frac{1}{2\pi^{\frac{1}{2}}l^{\frac{1}{6}}\rho(t_n,y,\frac{1}{l})^{\frac{1}{4}}}e^{-\frac{2}{3}l\rho^{3/2}(t_n,y,\frac{1}{l})}(1+O(\frac{1}{l})),$$

$$\mathrm{Bi}(l^{\frac{2}{3}}\rho(t,y,\frac{1}{l})) = \frac{1}{\pi^{\frac{1}{2}}l^{\frac{1}{6}}\rho(t_n,y,\frac{1}{l})^{\frac{1}{4}}}e^{\frac{2}{3}l\rho^{3/2}(t_n,y,\frac{1}{l})}(1+O(\frac{1}{l})),$$

coupled with the previous equation yields

$$c^1 = \frac{\pi^{1/4}\sqrt{2\pi}\,l^{\frac{1}{6}}}{\hat{\lambda}_l^{1/2}}(1+O(\frac{1}{l})),$$

and

$$|c^2| < c(y)\frac{\pi^{1/4}\sqrt{2}}{\hat{\lambda}_l^{1/12}}e^{-\frac{2}{3}\mathbf{Re}(\rho^{3/2}(t_n,y,\epsilon)+\rho^{3/2}(t_n+\epsilon,y,\epsilon))}O(\frac{1}{l}).$$

Thus for $t_a \leq t_n$,

$$|c^2\psi^2| < c(y)e^{-\frac{2}{3}\mathbf{Re}(\rho^{3/2}(t_a+\epsilon,y,\epsilon))}O(\frac{1}{l}),$$

where the fact that $\mathrm{Re}(\rho^{3/2})$ is a decreasing function of t has been used. This leads to

$$\frac{p_n(y,\frac{1}{l})}{e^{(\hat{\lambda}_l y)^2/2}} = \frac{\pi^{1/4}\sqrt{2\pi}\,l^{\frac{1}{6}}}{\hat{\lambda}^{1/2}}\,g\left(t_n,y,\frac{1}{l}\right)\mathrm{Ai}\left(\hat{\lambda}_l^{2/3}\rho\left(t_n,y,\frac{1}{l}\right)\right)\left(1+O\left(\frac{1}{l}\right)\right),$$

The above formula is valid for $t_b < t_n < t_{fi}$ with $t_{fi} > 1$ so that for $t_n = 1$,

$$\frac{\hat{H}_l(\hat{\lambda}_l y)}{e^{(2l+1)y^2/2}} = \frac{\sqrt{2}\pi^{1/4}}{(2l+1)^{1/12}}\left(\frac{\hat{\rho}(y)}{(y^2-1)^2}\right)^{1/4}\mathrm{Ai}((2l+1)^{1/3}\hat{\rho}(y))\left(1+O\left(\frac{1}{l}\right)\right),$$

where $\hat{\rho} = \frac{l^{2/3}}{\hat{\lambda}_l^{4/3}}\rho$. For $1 < y$ we find

$$l^{2/3}\rho\left(1,y,\frac{1}{l}\right) = (2l+1)^{2/3}\left(\frac{3}{4}y\sqrt{y^2-1} - \frac{3}{4}\cosh^{-1}(y)\right)^{2/3},$$

while for $0 < y < 1$ we find

$$l^{2/3}\rho\left(1,y,\frac{1}{l}\right) = -(2l+1)^{2/3}\left(\frac{3}{4}\cos^{-1}(y) - \frac{3}{4}y\sqrt{1-y^2}\right)^{2/3}.$$

The above formula is similar to the one found in Olver [O, p. 403]. Examination of the errors show that these formulas convergence uniformly on compact subsets of S. Extensions to other regions of the complex plane may be accomplished by using the symmetries $H_n(-z) = (-1)^n H_n(z)$ and $H_n(\bar{z}) = \overline{H_n(z)}$.

Acknowledgment

J.S.G. would like to thank K. McLaughlin, and P. Miller for discussion pertaining to the matching problem S. Garoufalidis for comments regarding the introduction, A. Martinez Finkelstein for pointing out reference [WW], and F. Olver for discussion pertaining to Airy and Hankel functions in the complex plane.

References

[B] P.A. Braun, WKB method for three-term recurrence relations and quasienergies of an anharmonic oscillator, *Theoret. and Math. Phys.* **37** (1978) 1070–1081.

[CC] O. Costin, R. Costin, Rigorous WKB method for finite-order linear recurrence relations with smooth coefficients, *SIAM J. Math. Anal.* **27** (1996) 110–134.

[DM] P. Deift, K. T.-R. McLaughlin, A Continuum Limit of the Toda Lattice, *Memoirs of AMS* **#624** (1998) Providence, RI.

[DMo] R.B. Dingle, G.J. Morgan, WKB methods for difference equations I, *Appl. Sci. Res.* **18** (1967) 221–237.

[DKMVZ] P. Deift, T, Kriecherbauer, K. T.-R. McLaughlin, S. Venakides, X. Zhou, Strong Asymptotics of Orthogonal Polynomials with respect to exponential weights, *Comm. Pure. Appl. Math.***52** (1999) 1491–1552.

[G] J.S. Geronimo, WKB and Turning point Theory for second order difference equations: External Fields and Strong Asymptotics for Orthogonal Polynomials, In prep.

[GS] J.S. Geronimo, D. Smith, WKB (Liouville-Green) Analysis of Second Order Difference Equations and Applications, *JAT* **69** (1992) 269–301.

[L] R.E. Langer On the asymptotic solutions of ordinary differential equations, with an application to the Bessel function of large order, *Trans. Amer. Math. Soc.* **33** (1931) 23-64.

[MV] M. Maejina, W. Van Assche, Probabilistic proofs of asymptotic formulas for some classical polynomials, *Math. Proc. Cambridge Philos. Soc.* **97** (1985) 499–510.

[ND] P. Nevai, J.S. Dehisa, On asymptotic properties of zeros of orthogonal polynomials, *SIAM J. Math. Anal.* **10** (1979) 1184–1192.

[O] F.W.J. Olver, *Asymptotics and Special Functions* (1974) Academic Press London.

[SG] K. Schulten, R.G. Gordon, Semiclassical analysis of 3γ and 6γ coefficients for quantum mechanical of angular momentum, *JMP* **16** (1975) 1971–1988.

[S] G. Szegő, *Orthogonal Polynomials*, AMS Colloq. Pub. Vol. 23, 4th ed, Providence, RI (1978).

[V] W. Van Assche, Asymptotics for Orthogonal Polynomial, *Lectures Notes in Mathematics* **1265** Springer-Verlag (1967) London.

[VG] W. Van Assche, J.S. Geronimo, Asymptotics for orthogonal polynomials with regularly varying recurrence coefficients, *Rocky Mtn. J. Math.* **19** (1989) 39–49.

[WW] Z. Wang, R. Wong, Uniform asymptotic expansion of $J_v(va)$ via a difference equation, *Numer. Math.* **91** (2002) 147–193.

[W] G.N. Watson, *A treatise on the theory of Bessel functions* (2nd ed), Cambridge Univ. Press (1944) London.

Jeffrey S. Geronimo
School of Mathematics
Georgia Institute of Technology
Atlanta, GA 30332-0160, USA
e-mail: geronimo@math.gatech.edu

Oscar Bruno
Applied & Computational Mathematics
California Institute of Technology
Pasadena, CA 91125, USA

Walter Van Assche
Department Wiskunde
Celestijnenlaan 200B
B-3030 Leuven, Belgium

Operator Theory:
Advances and Applications, Vol. 154, 139–149

Bounds for the Points of Spectral Concentration of One-dimensional Schrödinger Operators

Daphne J. Gilbert, Bernard J. Harris and Suzanne M. Riehl

Abstract. We investigate the phenomenon of spectral concentration for one-dimensional Schrödinger operators with decaying potentials on the half-line. For suitable classes of short range and long range potentials, we outline systematic procedures which enable numerical estimates of upper bounds for points of spectral concentration to be obtained. Our approach involves use of the Riccati equation to construct appropriate convergent series for a generalised Dirichlet m-function, from which the existence and properties of derivatives of the corresponding spectral functions can be established. An incidental outcome in the case of long range potentials is that upper bounds for embedded singular spectrum can also be obtained.

Mathematics Subject Classification (2000). 34L05, 47E05.

Keywords. Spectral concentration, Sturm-Liouville operators.

1. Introduction

In the context of this paper, we use the term spectral concentration to refer to a significant localised intensification of the spectrum occuring within an interval of purely absolutely continuous spectrum. This terminology and usage in connection with the one-dimensional Schrödinger operator goes back to Titchmarsh, who provided a detailed mathematical analysis in a number of well-known cases [22]. Titchmarsh's investigations were themselves developments of earlier work by Schrödinger, Oppenheimer and others on such problems as ionisation of the hydrogen atom in a weak electric field (the Stark effect), and the behavior of the hydrogen atom in a uniform magnetic field (the Zeeman effect) [16], [19]. A common feature of these problems is that a discrete spectrum is replaced by a continuous spectrum after the introduction of a 'perturbing' field while, in the vicinity of the eigenvalues of the original problem, there are poles of the meromorphic continuation of the perturbed Green's function located just below the real axis, on the so-called unphysical sheet. These poles, also known as resonances or pseudo-eigenvalues, are of significant interest to physicists in connection with impedance theory, resonance scattering and spectral stability, and are correlated with the existence of scattering states which remain localised for a long time [5], [12].

In Titchmarsh's examples, the influence of the eigenvalues of the unperturbed problem is also reflected in the occurence nearby of sharp peaks in the perturbed spectral density function, and away from these peaks the perturbed spectral density is small. Thus close to the eigenvalues of the original operator, there are so-called points of spectral concentration of the perturbed operator, so that in some sense the spectrum of the perturbed operator is 'close' to that of the original [11]. Moreover, the fact that both resonance poles and points of spectral concentration are located in the vicinity of the eigenvalues of the unperturbed problems suggests that a correlation between spectral concentration and resonances may hold more widely, and indeed there is considerable evidence to support this idea, whether or no the phenomena arise in connection with the perturbation of a discrete spectrum. For example, in those cases for which the Kodaira formula for the spectral density holds ([13], p.940), it is evident that if a resonance pole in the meromorphic continuation of the Green's function into the unphysical sheet lies sufficiently close to the real axis, then it is liable to increase the spectral density on part of the real axis closest to the pole. These related phenomena have given rise to a considerable literature over the years and useful references may be found in [12], [17].

Independently of the connection with resonance poles, the study of spectral concentration in its own right introduces a valuable additional dimension into analysis of the absolutely continuous part of the spectrum. In recent years, there has been a significant focus on the development of analytical and numerical methods for investgating the existence and location of this phenomenon (see, e.g., [3], [6]). The present paper provides an overview of some recent contributions of the authors in this direction.

2. Mathematical background

We consider the one-dimensional Schrödinger operator H_α associated with the system

$$Ly(x) := -y''(x) + q(x)y(x) = zy(x), \qquad x \in [0, \infty), \;\; z \in \mathbf{C},$$
$$y(0)\cos\alpha + y'(0)\sin\alpha = 0, \qquad \alpha \in [0, \pi),$$

where the potential $q(x)$ is real-valued, locally square integrable and decays to zero as $x \to \infty$. In this situation the self-adjoint operator H_α acting on $\mathcal{H} = L_2[0, \infty)$ is defined by

$$H_\alpha f = Lf, \; f \in D(H_\alpha),$$

where

$$D(H_\alpha) = \{f \in \mathcal{H} : Lf \in \mathcal{H}; f, f' \text{ locally a.c.}; \cos\alpha f(0) + \sin\alpha f'(0) = 0\}.$$

It is well known that L is in Weyl's limit point case at infinity, so that for each $z \in \mathbf{C}\backslash\mathbf{R}$ there exists precisely one linearly independent solution of $Lu = zu$ in $L_2[0,\infty)$, and that the essential spectrum of H_α fills the semi-axis $[0,\infty)$.

Associated with H_α is a non-decreasing spectral function, $\rho_\alpha(\lambda)$, and we define the spectrum, $\sigma(H_\alpha)$, to be the complement in $\mathbf{R}$ of the set of points in a neighbourhood of which $\rho_\alpha(\lambda)$ is constant (which is consistent with the more usual definition in terms of the resolvent operator). Subject to the normalisation $\rho_\alpha(0) = 0$, the spectral function can be decomposed uniquely into absolutely continuous, singular continuous and pure point parts, in terms of which the corresponding components of the spectrum, $\sigma_{a.c.}(H_\alpha)$, $\sigma_{s.c.}(H_\alpha)$ and $\sigma_{p.p.}(H_\alpha)$, are defined in a similar way.

The spectrum of H_α may also be studied through properties of the related Titchmarsh-Weyl m-function, $m_\alpha(z)$, which is defined and analytic on $\mathbf{C}\backslash\mathbf{R}$ and is a Herglotz function in the upper half-plane. Formulae connecting $\rho_\alpha(\lambda)$ and $m_\alpha(z)$ include

$$m_\alpha(z_2) - m_\alpha(z_1) = \int_{-\infty}^{\infty} \left(\frac{1}{\lambda - z_2} - \frac{1}{\lambda - z_1} \right) d\rho_\alpha(\lambda),$$

which holds for $z_1, z_2 \in \mathbf{C}^+$, and

$$\rho'_\alpha(\lambda) = \lim_{\epsilon \to 0} \frac{1}{\pi} \Im m_\alpha(\lambda + i\epsilon), \tag{2.1}$$

which holds for all $\lambda \in \mathbf{R}$ for which the respective limits exist. The m-function is closely related to the Green's function for H_α and hence reflects the analyticity properties of the resolvent operator. It is also related to solutions of the differential equation in various ways. For example, for $z \in \mathbf{C}^+$, let $\psi(x,z)$ denote the so-called Weyl solution of $Lu = zu$, which is in $L_2[0,\infty)$ and satisfies $\psi(x,z) \sim \exp ikx$ as $x \to \infty$, where $k^2 = z$ and the principal branch is chosen. Then the logarithmic derivative of $\psi(x,z)$ with respect to x, evaluated at $x = 0$, is equal to the value of the Dirichlet m-function, $m_0(z)$, at z, i.e.,

$$\frac{\psi'(0,z)}{\psi(0,z)} = m_0(z)$$

We remark that for $z \in \mathbf{C}^+$, $\psi(x,z)$ cannot vanish at $x = 0$, since to do so would imply that H_0 had a non-real eigenvalue at z, and thus contradict the self-adjointness of H_0. In a similar way, $\psi(x_0,z) \neq 0$ for any $x_0 > 0$, $z \in \mathbf{C}^+$. We may therefore set

$$m(x,z) := \frac{\psi'(x,z)}{\psi(x,z)} \tag{2.2}$$

for $x \geq 0, z \in \mathbf{C}^+$, and it is straightforward to check that for each $x_0 \geq 0$, $m(x_0,z)$ is the Dirichlet m-function associated with the system: $Lu = zu$, $x \geq x_0$, $u(x_0,z) = 0$. We will refer to $m(x,z)$ as the generalised Dirichlet m-function, and note that in terms of our earlier notation, $m(0,z) \equiv m_0(z)$. Since $\psi(x,z)$ is a solution of $Lu = zu$, it follows from (2.2) that $m(x,z)$ satisfies the Riccati equation

$$\frac{\partial}{\partial x} m(x,z) = -z + q(x) - (m(x,z))^2 \tag{2.3}$$

for $x \in [0,\infty)$, $z \in \mathbf{C}^+$.

In the present context, we formally define spectral concentration as follows (cf. [3]).

Definition 2.1. The point $\lambda_c \in \mathbf{R}$ is said to be a *point of spectral concentration of* H_α *if*

(i) $\rho'_\alpha(\lambda)$ *exists finitely and is continuous in a neighbourhood of* λ_c*, and*
(ii) $\rho'_\alpha(\lambda)$ *has a local maximum at* λ_c.

We note that since $\rho_\alpha(\lambda)$ is non-decreasing, the definition implies that $\rho'_\alpha(\lambda)$ exists and satisfies $0 \le \rho'_\alpha(\lambda) < \rho'_\alpha(\lambda_c) < \infty$ in a deleted neighbourhood of λ_c, from which it follows by the local continuity of $\rho'_\alpha(\lambda)$ that $\rho'_\alpha(\lambda) > 0$ in a neighbourhood of λ_c. Thus the definition effectively restricts attention to points of spectral concentration which occur in subintervals of the essential spectrum in which the spectral function is purely absolutely continuous and strictly increasing. A further consequence of the definition is that if $\rho''_\alpha(\lambda)$ exists and has one sign for $\lambda > M$, then $\rho'_\alpha(\lambda)$ exists, is absolutely continuous, but has no local maxima in (M,∞), so that M is an upper bound for points of spectral concentration of H_α.

3. Short range potentials

In the case where $q \in L_1[0,\infty)$, it was shown by Titchmarsh [21] that for $\lambda > 0$,

$$m_\alpha(\lambda) := \lim_{\epsilon \downarrow 0} m_\alpha(\lambda + i\epsilon)$$

exists and satisfies

$$m_\alpha(\lambda) = \frac{\mu_1(\lambda) + i\nu_1(\lambda)}{\mu(\lambda) + i\nu(\lambda)},$$

where $\mu_1(\lambda), \nu_1(\lambda), \mu(\lambda), \nu(\lambda)$ are continuous functions of λ, and $\mu(\lambda), \nu(\lambda)$ do not vanish simultaneously. It follows from the properties of these functions and (2.1) above that

$$\rho'_\alpha(\lambda) = \frac{1}{\pi}\Im m_\alpha(\lambda) = \frac{1}{\pi\sqrt{\lambda}(\mu^2(\lambda) + \nu^2(\lambda))},$$

so that $\rho'_\alpha(\lambda)$ is continuous with $0 < \rho'_\alpha(\lambda) < \infty$ for $\lambda > 0$, and hence $\sigma(H_\alpha)$ is purely absolutely continuous on $(0,\infty)$ [9]. It is not hard to show that the generalised Dirichlet m-function and corresponding Dirichlet spectral functions on $L_2[x,\infty)$ have similar properties, so that $m(x,z)$ may be continuously extended on to the non-negative real axis, $z = \lambda \in \mathbf{R}^+$, for $x \ge 0$. We then have (cf. (2.2)) that $m(x,\lambda)$ is well defined, continuous and non-real for $x \ge 0, \lambda > 0$, and satisfies

$$m(x,\lambda) = \frac{\psi'(x,\lambda)}{\psi(x,\lambda)},$$

where $\psi(x,\lambda) \notin L_2[0,\infty)$ is the pointwise limit as $z \downarrow \lambda$ of the Weyl solution, and is itself a solution of $Lu = \lambda u$. It follows that the Riccati equation (2.3) is also satisfied for $z = \lambda > 0$, so that

$$\frac{\partial}{\partial x} m(x,\lambda) = -\lambda + q(x) - (m(x,\lambda))^2 , \tag{3.1}$$

where $m(x,\lambda)$ is the finite non-real limit as $z \downarrow \lambda$ of the generalised Dirichlet m-function, $m(x,z)$.

We now show that we can investigate the behavior of the spectral density, $\rho_0'(\lambda)$, using the Riccati equation. For $x \geq 0, \lambda > 0$, we have

$$m(x,\lambda) := \Re m(x,\lambda) + i\Im m(x,\lambda),$$

from which by (2.1),

$$m(0,\lambda) := m_0(\lambda) = \Re m_0(\lambda) + i\pi\rho_0'(\lambda). \tag{3.2}$$

Hence, in principle, $\rho_0'(\lambda)$ can be obtained for $\lambda > 0$ by finding the appropriate solution of the Riccati equation and evaluating at $x = 0$. If in addition it can be shown that $m(x,\lambda)$ is differentiable with respect to λ for sufficiently large λ, we can also seek conditions under which $\rho_0''(\lambda)$ exists and satisfies

$$\rho_0''(\lambda) = \frac{1}{\pi}\left[\Im \frac{\partial}{\partial \lambda} m(x,\lambda)\right]_{x=0} \tag{3.3}$$

Equations (3.1) and (3.3) will form the basis for our investigation into the existence of upper bounds for points of spectral concentration of H_0.

It is rarely possible to solve the Riccati equation explicitly, so we proceed by postulating a series representation for $m(x,\lambda)$, which is substituted into (3.1). We then choose the terms of the series, establish sufficient conditions for the validity of the representation, and investigate the existence of $\rho_0''(\lambda)$. Based on the known asymptotic behavior of $m(x,\lambda)$ as $\lambda \to \infty$ (cf. [10]; [18], Theorem 5.1), we seek a series representation in the form

$$m(x,\lambda) = i\sqrt{\lambda} + g(x,\lambda) \quad \text{where} \quad g(x,\lambda) := \sum_{n=0}^{\infty} m_n(x,\lambda) \tag{3.4}$$

is in $L_1([0,\infty); dx)$, and satisfies $g(x,\lambda) \to 0$ as $x \to \infty$. It may be shown that the Riccati equation has at most one solution of this form (see [7]), from which it follows that if the series representation in (3.4) is a valid solution of (3.1), then it does indeed represent the extension of the generalised Dirichlet m-function onto the real axis, as sought.

Substituting for $m(x,\lambda)$ from (3.4) into (3.1) and rearranging yields

$$m_1' + 2i\sqrt{\lambda}m_1 + m_2' + 2i\sqrt{\lambda}m_2 + \sum_{n=3}^{\infty}(m_n' + 2i\sqrt{\lambda}m_n)$$

$$= q - m_1^2 - \sum_{n=3}^{\infty}\left(m_{n-1}^2 + 2m_{n-1}\sum_{k=1}^{n-2} m_k\right)$$

where $'$ denotes differentiation with respect to x. Choosing the $\{m_n\}$ to satisfy

$$\begin{aligned} m_1' + 2i\sqrt{\lambda} m_1 &= q \\ m_2' + 2i\sqrt{\lambda} m_2 &= -m_1^2 \\ m_n' + 2i\sqrt{\lambda} m_n &= -\left(m_{n-1}^2 + 2m_{n-1} \sum_{k=1}^{n-2} m_k \right), \qquad n \geq 3, \end{aligned}$$

yields

$$\begin{aligned} m_1(x,\lambda) &= -e^{-2i\sqrt{\lambda}x} \int_x^\infty e^{2i\sqrt{\lambda}t} q(t)dt, \\ m_2(x,\lambda) &= e^{-2i\sqrt{\lambda}x} \int_x^\infty e^{2i\sqrt{\lambda}t} m_1^2(t,\lambda)dt, \\ m_n(x,\lambda) &= e^{-2i\sqrt{\lambda}x} \int_x^\infty e^{2i\sqrt{\lambda}t} \left(m_{n-1}^2 + 2m_{n-1} \sum_{k=1}^{n-2} m_k \right) dt, \; n \geq 3. \end{aligned} \tag{3.5}$$

In order to determine under what circumstances $\rho_0''(\lambda)$ exists for sufficiently large λ, we now define

$$w_n(x,\lambda) := \frac{\partial}{\partial\lambda} m_n(x,\lambda), \qquad n = 1, 2, 3, \ldots$$

for those λ for which the derivatives exist. The following lemma establishes some key properties of $\{m_n(x,\lambda)\}$ and $\{w_n(x,\lambda)\}$, and is proved in [7].

Lemma 3.1. *Let $q(x) \in L_1[0,\infty)$ and suppose that there exists $\Lambda_1 > 0$ such that for $x \geq 0$ and $\lambda > \Lambda_1$*

$$\left| \int_x^\infty e^{2i\sqrt{\lambda}t} q(t)dt \right| \leq a(x)\eta(\lambda),$$

where $a(x) \in L_1[0,\infty)$ is decreasing, $\eta(\lambda) \to 0$ as $\lambda \to \infty$, and $32\eta(\lambda)\int_0^\infty a(t)dt \leq 1$. Then for $x \geq 0$, $\lambda > \Lambda_1$ and $n = 1, 2, 3, \ldots$

$$\begin{aligned} |\, m_n(x,\lambda) \,| &\leq \frac{a(x)\eta(\lambda)}{2^{n-1}}, \\ |\, w_n(x,\lambda) \,| &\leq \frac{\eta(\lambda)}{2^{n-1}\sqrt{\lambda}} \int_x^\infty a(t)dt, \end{aligned}$$

so that the series $\sum_{n=1}^\infty m_n(x,\lambda)$ and $\sum_{n=1}^\infty w_n(x,\lambda)$ are uniformly absolutely convergent in x and λ.

We remark that the conditions of Lemma 3.1 are satisfied, for example, if $(1+x)q(x) \in L_1[0,\infty)$ or if $q(x) \in L_1[0,\infty)$ is monotonic [10]. The convergence and continuity properties of the series in Lemma 3.1 imply that

(i) $m(x, \lambda)$, as defined in (3.4) and (3.5), is a valid series representation of the generalised Dirichlet m-function for $x \geq 0$, $\lambda > \Lambda_1$, and
(ii) $\rho_0''(\lambda)$ exists for $\lambda > \Lambda_1$ and is given by

$$\rho_0''(\lambda) = \frac{1}{\pi}\left(\frac{1}{2\sqrt{\lambda}} + \sum_{n=1}^{\infty} \Im w_n(0, \lambda)\right).$$

This leads to the following result.

Theorem 3.2. *Let $q(x) \in L_1[0, \infty)$ and suppose that the hypothesis of Lemma 1 is satisfied. Then for all $\lambda > \Lambda_1$, $\rho_0''(\lambda)$ exists and satisfies*

$$\left|\rho_0''(\lambda) - \frac{1}{2\pi\sqrt{\lambda}}\right| \leq \frac{4}{\pi\sqrt{\lambda}}\eta(\lambda)\int_0^{\infty} a(t)dt,$$

so that $\rho_0''(\lambda) > 0$ for all $\lambda > \Lambda_1$. In particular, there are no points of spectral concentration of H_0 for $\lambda > \Lambda_1$.

Theorem 3.2 enables explicit upper bounds for points of spectral concentration to be calculated, as illustrated in the first example in Section 5. Details of the proof of this theorem are given in [7].

4. Long range potentials

In the case where $q(x) \to 0$ as $x \to \infty$, but $q(x) \notin L_1[0, \infty)$, the situation is more delicate. It is no longer true in general that $\rho_0'(\lambda)$ exists, is continuous and satisfies $\rho_0'(\lambda) > 0$ for all $\lambda > 0$. If $q(x)$ decays more slowly than the Coulomb potential, examples can be constructed where $\rho_0(\lambda)$ is discontinuous on a dense set of eigenvalues in $[0, \infty)$ [15]. If $q(x)$ fails to be in $L_2[0, \infty)$, then the absolutely continuous spectrum may be empty, in which case there is a dense set of points in $[0, \infty)$ on which $\rho_0'(\lambda)$ does not exist as a finite limit [4], [9], [20]. The situation is more fully understood in the case of von Neumann-Wigner type potentials, where the spectrum is purely absolutely continuous on $(0, \infty)$ apart from an at most countable set of isolated eigenvalues, known as resonances [1], [2]. Moreover, under fairly minimal smoothness conditions, classes of decaying, but non-integrable potentials do exist for which the spectrum is purely absolutely continuous on $(0, \infty)$ or on (M, ∞) for sufficiently large M [2], [6]. In such circumstances, we can investigate whether it is possible to extend $m(x, z)$ continuously onto part of the non-negative real axis in such a way that $\rho_0''(\lambda)$ exists and (3.1), (3.2) and (3.3) are satisfied for sufficiently large λ. If this can be achieved, then estimates of upper bounds for both embedded singular spectrum and points of spectral concentration can normally be obtained.

We proceed as before by postulating the existence of a series representation for the generalised Dirichlet m-function, $m(x, z)$, in this case for $\Re z > 0$, $\Im z \geq 0$, and $| z |$ sufficiently large. It is necessary to consider the m-function in the upper half-plane as well as on the real axis, so as to ensure that if $m(x, \lambda)$ exists and is continuous in x and λ for $|\lambda| > M$, then it is the unique continuous extension of $m(x,z)$, $z \in \mathbf{C}^+$, as $z \downarrow \lambda \in \mathbf{R}$. Also, in order to construct a series with the desired convergence properties, it is helpful to introduce an additional term, $R(x,z)$, into the expression for $m(x,z)$, so we now seek a representation in the form

$$m(x,z) = i\sqrt{z} + R(x,z) + g(x,z) \tag{4.1}$$

where

$$g(x,z) := \sum_{n=0}^{\infty} m_n(x,z)$$

is in $L_1([0,\infty);dx)$, and satisfies $g(x,z) \to 0$ as $x \to \infty$. The introduction of $R(x,z)$ enables the terms of the series to be generated iteratively with $Q(x,z) := q(x) - R' - R^2 - 2i\sqrt{\lambda}R \in L_1([0,\infty);dx)$ having an analogous role to that of $q(x)$ in the short range case. We observe that in general the choice of $R(x,z)$ is not unique.

Proceeding as before, we substitute for $m(x,z)$ from (4.1) into (2.3), and after rearrangement the $\{m_n\}$ are chosen to satisfy:

$$\begin{aligned} m_1' + (2i\sqrt{z} + 2R)m_1 &= Q \\ m_2' + (2i\sqrt{z} + 2R)m_2 &= -m_1^2 \\ m_n' + (2i\sqrt{z} + 2R)m_n &= -\left(m_{n-1}^2 + 2m_{n-1}\sum_{k=1}^{n-2} m_k\right), \quad n \geq 3, \end{aligned}$$

The solutions $\{m_n\}$ of these equations, and their derivatives, form the basis of our analysis of the long range case. This leads to the following theorem which is proved in [8].

Theorem 4.1. *Let $q(x)$ be continuously differentiable and satisfy $q(x) \to 0$ as $x \to \infty$, $q(x) \notin L_1[0,\infty)$. Define*

$$Q(x,z) := q(x) - R' - R^2 - 2i\sqrt{z}R$$

for $\Re z > 0$, $\Im z \geq 0$, where $R = R(x,z)$ is chosen so that $Q(\cdot,z) \in L_1[0,\infty)$, $'$ denotes differentiation with respect to x, and Q, R, $\frac{\partial Q}{\partial z}$, $\frac{\partial R}{\partial z}$ are continuous in x and z. Suppose that there exists $M > 0$ so that

(a) *for $\Re z > 0$, $\Im z \geq 0$, $|z| > M$,*

(i) *there exists $K \in \mathbf{R}$ so that for $0 \leq x < t$,*

$$\Re\left(2i\sqrt{z}(t-x) + 2\int_x^t R(s,z)ds\right) \leq K,$$

(ii) *for $0 \leq x < t$,*

$$\left|\int_x^{\infty} e^{2i\sqrt{z}(t-x)+2\int_x^t R(s,z)ds} Q(t,z)dt\right| \leq a(x)\eta(z),$$

where $a(x)$, $\eta(z)$, are real-valued functions with $a(x) \in L_1[0,\infty)$ and decreasing, $\eta(z) \to 0$ as $|z| \to \infty$ and $32\eta e^K \int_0^{\infty} a(t)dt \leq 1$,

(iii) $$\left|\frac{\partial}{\partial z}\int_x^t R(s,z)ds\right| \leq \text{const}(t-x) \quad \textit{for } 0 \leq x < t < \infty,$$

(b) *for* $\lambda = \Re z > M$, $\Im z = 0$, *there exists a decreasing function* $b(x)$ *such that for* $x \geq 0$,

$$e^K \int_x^\infty \left|\frac{\partial Q}{\partial \lambda}\right| + \left|\frac{i}{\sqrt{\lambda}} + 2\frac{\partial R}{\partial \lambda}\right| a(t)\eta(\lambda)dt \leq \frac{\eta(\lambda)}{\sqrt{\lambda}} b(x).$$

Then $\rho_0''(\lambda)$ *exists for* $\lambda > M$, *and satisfies*

$$\left|\rho_0''(\lambda) - \frac{1}{2\pi\sqrt{\lambda}} - \frac{1}{\pi}\Im R(0,\lambda)\right| \leq \frac{3}{\pi\sqrt{\lambda}}\eta(\lambda)b(0).$$

The following corollary may be inferred from the proof of Theorem 4.1 (see [8]).

Corollary 4.2. *Let* $q(x)$, $Q(x,z)$ *and* $R(x,z)$ *be as in Theorem* 4.1 *and suppose that* $\Lambda_0 > 0$ *exists such that for* $\Re z > 0$, $\Im z \geq 0$, $\mid z \mid > \Lambda_0$, *conditions (a)(i)and (ii) of Theorem* 4.1 *are satisfied. Then for* $\lambda = \Re z > \Lambda_0$, $\rho_0'(\lambda)$ *exists as a finite limit, and hence the spectrum of* H_0 *is purely absolutely continuous on* (Λ_0, ∞).

5. Applications

Example. Let $q(x) = (1+x)^{-\gamma}\sin(1+x)$, $\gamma > 1$. By expressing $\sin(1+x)$ in exponential form and integrating by parts we obtain

$$\left|\int_x^\infty e^{2i\sqrt{\lambda}t} q(t)dt\right| \leq \frac{2}{(2\sqrt{\lambda}-1)(1+x)^\gamma}$$

for $\lambda > \frac{1}{4}$, from which we may choose $\eta(\lambda) = 2(2\sqrt{\lambda}-1)^{-1}$ and $a(x) = (1+x)^{-\gamma}$. It is then straightforward to show from Theorem 3.2 that

$$\Lambda_1 = \left(\frac{1}{2} + \frac{32}{\gamma - 1}\right)^2$$

is an upper bound for points of spectral concentration of H_0.

Example. Let $q(x) = \sin(1+x)^{\frac{1}{2}}(1+x)^{-\frac{1}{2}}$. In this case $q(x)$, $q'(x)$ and $(q(x))^2$ are not in $L_1[0,\infty)$, so we take

$$R(x,z) = \frac{q}{2i\sqrt{z}} - \frac{q'}{(2i\sqrt{z})^2} - \frac{q^2}{(2i\sqrt{z})^3}$$

to give $Q(x,z) = O(1+x)^{-\frac{3}{2}} \in L_1[0,\infty)$. This leads to the choice

$$a(x) = \frac{1}{(1+x)^{\frac{3}{2}}}, \qquad \eta(z) = \frac{1}{5\mid z\mid}, \qquad b(x) = \frac{47}{5\sqrt{1+x}},$$

from which it follows by Theorem 4.1 that $\rho_0''(\lambda) > 0$ if $\lambda > 30$, so that $\Lambda_1 = 30$ is an upper bound for points of spectral concentration of H_0. Note that $\sigma(H_0)$ is known to be purely absolutely continuous on $(0,\infty)$ [2], so that the issue of embedded singular spectrum does not arise.

Example. We consider the von Neumann Wigner type potential

$$q(x) = \sum_{k=-M}^{M} h_k(x) e^{2ic_k x},$$

where for each $k = -M, \dots, M$, $c_k \in \mathbf{R}\backslash\{0\}$, $h_k(x) \to 0$ as $x \to \infty$, $h_k(x) \in C^L[0,\infty)$ and $h_k^{L+1}(x) \in AC[0,\infty)$. We suppose also that there exists a real-valued non-negative function $p(x)$ such that $x(p(x))^{L+2}$ is decreasing with $(p(x))^{L+2}$, $x(p(x))^{L+2} \in L_1[0,\infty)$, and that for $j = 0, \dots, L+1$, $k = -M, \dots, M$,

$$| h_k^{(j)}(x) | \le (p(x))^{j+1}.$$

Then $R(x,z)$ may be chosen so that, after successive integrations by parts,

$$| Q(x,z) | \le c \left(\frac{p(x)}{| z |^{\frac{1}{2}} - 2Lc_*} \right)^{L+2},$$

where c and c_* are real constants which are computable for given $q(x)$. We may then take

$$a(x) = \int_x^\infty (p(t))^{L+2} dt, \quad \eta(z) = \frac{c}{(| z |^{\frac{1}{2}} - 2Lc_*)^{L+2}}, \quad b(x) = \int_x^\infty a(t) dt,$$

from which the existence of computable upper bounds, Λ_0 and Λ_1, for resonances (embedded eigenvalues) and points of spectral concentration follows from Corollary 4.2 and Theorem 4.1 respectively. Further details may be found in [8].

References

[1] H. Behncke, Absolute continuity of Hamiltonians with von Neumann Wigner potentials, *Proc. AMS* **111** (1991), 373–384.

[2] H. Behncke, Absolute continuity of Hamiltonians with von Neumann Wigner potentials II, *Manuscripta Math.* **71** (1991), 163–181.

[3] B.M. Brown, M.S.P. Eastham and D.K.R. McCormack, Spectral concentration and perturbed discrete spectra, *J. Comp. Appl. Math.* **86** (1997), 415–425.

[4] P. Deift and R. Killip, On the absolutely continuous spectrum of one-dimensional Schrödinger operators with square summable potentials, *Comm. Math. Phys.* **203** (1999), 341–347.

[5] C.L. Dolph, J.B. Mcleod and D. Thoe, The analytic continuation of the resolvent kernel and scattering operator associated with the Schroedinger operator, *J. Math. Anal. Appl.* **16** (1966), 311–332.

[6] M.S.P. Eastham, On the convexity of the Sturm-Liouville spectral functions with slowly decaying potential, *Indian J. Math.* **42** (2000), 9–20.

[7] D.J. Gilbert and B.J. Harris, Bounds for the points of spectral concentration of Sturm-Liouville problems, *Mathematika* **47** (2000), 327–337 (2002).

[8] D.J. Gilbert, B.J. Harris and S. Riehl, The spectral function for Sturm-Liouville problems where the potential is of von Neumann-Wigner type or slowly decaying, *J. Differential Equations* **201** (2004), 139–159.

[9] D.J. Gilbert and D.B. Pearson, On subordinacy and analysis of the spectrum of one-dimensional Schrödinger operators, *J. Math. Anal. Appl.* **128** (1987), 30–56.

[10] B.J. Harris, The form of the spectral functions associated with Sturm- Liouville problems with continuous spectrum, *Mathematika* **44** (1997), 149–161.

[11] D.B. Hinton and J.K. Shaw, Spectral concentration for Hamiltonian systems, *Helv. Phys. Acta* **58** (1985), 982–994.

[12] P.D. Hislop and I.M. Segal, *Introduction to Spectral Theory, with Applications to Schrödinger Operators* , Springer-Verlag, New York, 1996.

[13] K. Kodaira, The eigenvalue problem for ordinary differential equations of the second order, and Heisenberg's theory of S-matrices, *Amer. J. Math.* **71** (1949), 921–945.

[14] J. von Neumann and E. Wigner, Über merkwürdige diskrete Eigenwerte, *Phys. Z.* **30** (1929), 465–467.

[15] S. Naboko, On the dense point spectrum of Schrödinger and Dirac operators, *Teoret. Mat. Fiz.* **68** (1986), 18–28.

[16] J.R. Oppenheimer, Three notes on the quantum theory of aperiodic effects, *Phys. Rev.* **31** (1928), 66–81.

[17] M. Reed and B. Simon, *Methods of Modern Mathematical Physics, IV: Analysis of Operators,* Academic Press, 1978.

[18] A. Rybkin, Some new and old asymptotic representations of the Jost solution and the Weyl m-function for Schrödinger operators on the line, *Bull. London Math. Soc.* **34** (2002), 61–72.

[19] E. Schrödinger, Quantisierung als Eigenwertproblem III, *Ann. der Physik* **80** (1926), 457–490.

[20] B. Simon, Operators with singular continuous spectrum, I. General operators, *Annals Math.* **141** (1995), 131–145.

[21] E.C. Titchmarsh, *Eigenfunction Expansions I*, 2nd edition, Oxford University Press, 1962.

[22] E.C. Titchmarsh, *Eigenfunction Expansions II*, Oxford University Press, 1958.

Daphne J. Gilbert
School of Mathematical Sciences
Dublin Institute of Technology
Ireland

Bernard J. Harris
Department of Mathematical Sciences
Northern Illinois University, USA

Suzanne M. Riehl
Department of Mathematics
University of Northern Iowa, USA

Operator Theory:
Advances and Applications, Vol. 154, 151–154

Reconstructing Jacobi Matrices from Three Spectra

Johanna Michor and Gerald Teschl

Abstract. Cut a Jacobi matrix into two pieces by removing the n-th column and n-th row. We give necessary and sufficient conditions for the spectra of the original matrix plus the spectra of the two submatrices to uniquely determine the original matrix. Our result contains Hochstadt's theorem as a special case.

Mathematics Subject Classification (2000). Primary 36A10, 39A70; Secondary 34B24, 34L05.

Keywords. Jacobi matrices, spectral theory, trace formulas, Hochstadt's theorem.

1. Introduction

The topic of this paper is inverse spectral theory for Jacobi matrices, that is, matrices of the form

$$H = \begin{pmatrix} b_1 & a_1 & & & \\ a_1 & b_2 & a_2 & & \\ & \ddots & \ddots & \ddots & \\ & & a_{N-2} & b_{N-1} & a_{N-1} \\ & & & a_{N-1} & b_N \end{pmatrix}. \tag{1.1}$$

This is an old problem closely related to the moment problem (see [9] and the references therein), which has attracted considerable interest recently (see, e.g., [1] and the references therein, [3], [4], [8]). For analogous results in the case of Sturm-Liouville operators see [2], [6], and [7]. In this note we want to investigate the following question: Remove the n-th row and the n-th column from H and denote the resulting submatrices by H_- (from b_1 to b_{n-1}) respectively H_+ (from b_{n+1} to b_N). When do the spectra of these three matrices determine the original matrix H? We will show that this is the case if and only if H_- and H_+ have no eigenvalues in common.

From a physical point of view such a model describes a chain of N particles coupled via springs and fixed at both end points (see [11], Section 1.5). Determining the eigenfrequencies of this system and the one obtained by keeping one particle fixed, one can uniquely reconstruct the masses and spring constants. Moreover, these results can be applied to completely integrable systems, in particular the Toda lattice (see, e.g., [11]).

2. Main result

To set the stage let us introduce some further notation. We denote the spectra of the matrices introduced in the previous section by

$$\sigma(H) = \{\lambda_j\}_{j=1}^N, \quad \sigma(H_-) = \{\mu_k^-\}_{k=1}^{n-1}, \quad \sigma(H_+) = \{\mu_l^+\}_{l=1}^{N-n}. \tag{2.1}$$

Moreover, we denote by $(\mu_j)_{j=1}^{N-1}$ the ordered eigenvalues of H_- and H_+ (listing common eigenvalues twice) and recall the well-known formula (see [1], Theorem 2.4 and Theorem 2.8)

$$g(z,n) = -\frac{\prod_{j=1}^{N-1}(z-\mu_j)}{\prod_{j=1}^{N}(z-\lambda_j)} = \frac{-1}{z - b_n + a_n^2 m_+(z,n) + a_{n-1}^2 m_-(z,n)}, \tag{2.2}$$

where $g(z,n)$ are the diagonal entries of the resolvent $(H-z)^{-1}$ and $m_\pm(z,n)$ are the Weyl m-functions corresponding to H_- and H_+. The Weyl functions $m_\pm(z,n)$ are Herglotz and hence have a representation of the following form

$$m_-(z,n) = \sum_{k=1}^{n-1} \frac{\alpha_k^-}{\mu_k^- - z}, \qquad \alpha_k^- > 0, \quad \sum_{k=1}^{n-1} \alpha_k^- = 1, \tag{2.3}$$

$$m_+(z,n) = \sum_{l=1}^{N-n} \frac{\alpha_l^+}{\mu_l^+ - z}, \qquad \alpha_l^+ > 0, \quad \sum_{l=1}^{N-n} \alpha_l^+ = 1. \tag{2.4}$$

With this notation our main result reads as follows

Theorem 2.1. *To each Jacobi matrix H we can associate spectral data*

$$\{\lambda_j\}_{j=1}^N, \quad (\mu_j, \sigma_j)_{j=1}^{N-1}, \tag{2.5}$$

where $\sigma_j = +1$ if $\mu_j \in \sigma(H_+)\backslash\sigma(H_-)$, $\sigma_j = -1$ if $\mu_j \in \sigma(H_-)\backslash\sigma(H_+)$, and

$$\sigma_j = \frac{a_n^2\alpha_l^+ - a_{n-1}^2\alpha_k^-}{a_n^2\alpha_l^+ + a_{n-1}^2\alpha_k^-} \tag{2.6}$$

if $\mu_j = \mu_k^- = \mu_l^+$.

Then these spectral data satisfy

(i) $\lambda_1 < \mu_1 \le \lambda_2 \le \mu_2 \le \cdots < \lambda_N$,

(ii) $\sigma_j = \sigma_{j+1} \in (-1,1)$ *if* $\mu_j = \mu_{j+1}$ *and* $\sigma_j \in \{\pm 1\}$ *if* $\mu_j \neq \mu_i$ *for* $i \neq j$

and uniquely determine H. Conversely, for every given set of spectral data satisfying (i) and (ii), there is a corresponding Jacobi matrix H.

Proof. We first consider the case where H_- and H_+ have no eigenvalues in common. The interlacing property (i) is equivalent to the Herglotz property of $g(z,n)$. Furthermore, the residues α_i^- can be computed from (2.2)

$$\frac{\prod_{j=1}^{N}(z-\lambda_j)}{\prod_{k=1}^{n-1}(z-\mu_k^-)\prod_{l=1}^{N-n}(z-\mu_l^+)} = z - b_n + a_n^2 \sum_{l=1}^{N-n} \frac{\alpha_l^+}{z-\mu_l^+} + a_{n-1}^2 \sum_{k=1}^{n-1} \frac{\alpha_k^-}{z-\mu_k^-}. \tag{2.7}$$

and are given by $\alpha_i^- = a_{n-1}^{-2}\beta_i^-$, where

$$\beta_i^- = -\frac{\prod_{j=1}^{N}(\mu_i^- - \lambda_j)}{\prod_{l\neq i}(\mu_i^- - \mu_l^-)\prod_{l=1}^{N-n}(\mu_i^- - \mu_l^+)}, \qquad a_{n-1}^2 = \sum_{i=1}^{n-1}\beta_i^-. \tag{2.8}$$

Similarly, $\alpha_l^+ = a_n^{-2}\beta_l^+$, where

$$\beta_l^+ = -\frac{\prod_{j=1}^{N}(\mu_l^+ - \lambda_j)}{\prod_{k=1}^{n-1}(\mu_l^+ - \mu_k^-)\prod_{p\neq l}(\mu_l^+ - \mu_p^+)}, \qquad a_n^2 = \sum_{l=1}^{N-n}\beta_l^+. \tag{2.9}$$

Hence $m_\pm(z,n)$ are uniquely determined and thus $H_\pm$ by standard results from the moment problem. The only remaining coefficient b_n follows from the well-known trace formula

$$b_n = \operatorname{tr}(H) - \operatorname{tr}(H_-) - \operatorname{tr}(H_+) = \sum_{j=1}^{N}\lambda_j - \sum_{k=1}^{n-1}\mu_k^- - \sum_{l=1}^{N-n}\mu_l^+. \tag{2.10}$$

Conversely, suppose we have the spectral data given. Then we can define a_n, a_{n-1}, b_n, α_k^-, α_l^+ as above. By (i), α_k^- and α_l^+ are positive and hence give rise to $H_\pm$. Together with a_n, a_{n-1}, b_n we have thus defined a Jacobi matrix H. By construction, the eigenvalues μ_k^-, μ_l^+ are the right ones and also (2.2) holds for H. Thus λ_j are the eigenvalues of H, since they are the poles of $g(z,n)$.

Next we come to the general case where $\mu_{j_0} = \mu_{k_0}^- = \mu_{l_0}^+$ $(=\lambda_{j_0})$ at least for one j_0. Now some factors in the left-hand side of (2.7) will cancel and we can no longer compute $\beta_{k_0}^-$, $\beta_{l_0}^+$, but only $\gamma_{j_0} = \beta_{k_0}^- + \beta_{l_0}^+$. However, by definition of σ_{j_0} we have

$$\beta_{k_0}^- = \frac{1-\sigma_{j_0}}{2}\gamma_{j_0}, \qquad \beta_{l_0}^+ = \frac{1+\sigma_{j_0}}{2}\gamma_{j_0}. \tag{2.11}$$

Now we can proceed as before to see that H is uniquely determined by the spectral data.

Conversely, we can also construct a matrix H from given spectral data, but it is no longer clear that λ_j is an eigenvalue of H unless it is a pole of $g(z,n)$. However, in the case $\lambda_{j_0} = \mu_{k_0}^- = \mu_{l_0}^+$ we can glue the eigenvectors u_- of H_- and u_+ of H_+ to give an eigenvector $(u_-, 0, u_+)$ corresponding to λ_{j_0} of H. □

The special case where we remove the first row and the first column (in which case H_- is not present) corresponds to Hochstadt's theorem [5]. Similar results for (quasi-)periodic Jacobi operators can be found in [10].

Acknowledgments

We thank the referee for useful suggestions with respect to the literature.

References

[1] F. Gesztesy and B. Simon, *m-functions and inverse spectral analysis for finite and semi-infinite Jacobi matrices*, J. Anal. Math. **73**, 267–297 (1997).

[2] F. Gesztesy and B. Simon, On the determination of a potential from three spectra, in *Differential operators and spectral theory*, 85–92, *Amer. Math. Soc. Transl. Ser. 2*, **189**, Amer. Math. Soc., Providence, RI, 1999.

[3] G.M.L. Gladwell, *On isospectral spring-mass systems*, Inverse Probl. **11**, 591–602 (1995).

[4] P. C. Gibson, *Inverse spectral theory of finite Jacobi matrices*, to appear in Memoires of the Amer. Math. Soc.

[5] H. Hochstadt, *On the construction of a Jacobi matrix from spectral data*, Lin. Algebra Appl. **8**, 435–446, 1974.

[6] R. Hryniv and Ya. Mykytyuk, *Inverse spectral problems for Sturm-Liouville operators with singular potentials. Part III: Reconstruction by three spectra*, J. Math. Anal. Appl. **248**, 626–646, (2003)

[7] V. Pivovarchik, *An inverse Sturm-Liouville problem by three spectra*, Integral Equations Operator Theory **34** no. 2, 234–243, (1999).

[8] C.-T. Shieh, *Reconstruction of a Jacobi matrix with mixed data*, preprint.

[9] B. Simon, *The classical moment problem as a self-adjoint finite difference operator*, Advances in Math. **137**, 82–203 (1998).

[10] G. Teschl, *Trace Formulas and Inverse Spectral Theory for Jacobi Operators*, Comm. Math. Phys. **196**, 175–202 (1998).

[11] G. Teschl, *Jacobi Operators and Completely Integrable Nonlinear Lattices*, Math. Surv. and Mon. **72**, Amer. Math. Soc., Rhode Island, 2000.

Johanna Michor and Gerald Teschl
Institut für Mathematik
Strudlhofgasse 4
A-1090 Wien, Austria
and
International Erwin Schrödinger Institute for Mathematical Physics
Boltzmanngasse 9
A-1090 Wien, Austria
e-mail: `Johanna.Michor@esi.ac.at`
e-mail: `Gerald.Teschl@univie.ac.at`
URL: `http://www.mat.univie.ac.at/~gerald/`

Operator Theory:
Advances and Applications, Vol. 154, 155–162

On Some Asymptotic Properties of Solutions for a Particular Class of Finite Difference Equations

Andrey S. Osipov

Abstract. For one class of linear difference equations of arbitrary order we obtain some results about the asymptotic behavior of its solutions, applying a method used in the spectral analysis of discrete Sturm-Liouville operators. We apply this asymptotic method to show, that the analog of the Hellinger-Wall theorem of invariance in l^2 is not valid in l^p for $p > 2$.

Mathematics Subject Classification (2000). 47B36;39A11.

Keywords. Band operators; asymptotic methods; l^p-property; Hellinger-Wall theorem.

One of the main methods of spectral analysis of nonsymmetric difference operators generated by infinite band matrices (band operators) is based on the study of asymptotical behavior of polynomials defined by systems of orthogonality relations [1], [2] . Such polynomials satisfy the difference equation, associated with the studied operator. Although some asymptotic methods for such sort of linear difference equations are known [3], [4] either they are not giving the explicit formula, or require some special conditions to be fulfilled (e.g., dichotomy-type condition). In the present paper, we consider the asymptotic method, which was used for the analysis of absolutely continuous spectrum of second order difference operators in [5]. This method is based on direct analysis of the transfer matrices, corresponding to the difference equation, and under some simple assumptions it gives an explicit asymptotic formula. First for the studied system of finite-difference equations of an arbitrary order we obtain some asymptotic formulas similar to [5]. Then we apply the method to study an analog of the Hellinger-Wall theorem for the difference operator of a second order. Note that the continuous analog of the Hellinger-Wall theorem is a known Weyl theorem, which states that for the Sturm-Liouville operator on a semi-axis, the property of all solutions of the corresponding differential

This work was supported by the Academy of Finland and RFBR: projects 02-01-00790 and 00-15-96100.

equation to belong to the space L^2, is invariant with respect to the spectral parameter [6]. We show that the claim of the Hellinger-Wall theorem is not true for the spaces l^p, when $p > 2$.

1. The asymptotic method

We consider the following infinite system of finite difference equations

$$\mu_{n-q}u_{n-q} + a_{n,n-q+1}u_{n-q+1} + \cdots + a_{n,n+q-1}u_{n+q-1} + \lambda_n u_{n+q} = zu_n, \quad (1)$$
$$n > q, \quad \mu_n \neq 0, \quad \lambda_n \neq 0,$$

(for some fixed index $q \geq 1$ and parameter $z \in \mathbb{C}$) with the complex coefficients, satisfying the following conditions:

$$\mu_n = n^{\gamma(n)}(1+\xi_n),\ \lambda_n = n^{\alpha(n)}(1+\delta_n);\ \{\delta_n\}, \{\xi_n\} \in l^2, \quad (2)$$
$$\text{for all } n > q,\ \alpha(n), \gamma(n) \in (1/2, \infty),$$

$$\sum_{n=q+1}^{\infty} |a_{n,n-i}| n^{-\alpha(n)} < \infty,\ i = -q+1, \ldots, q-1. \quad (3)$$

First consider the case when for all $n > q$ $\alpha(n) = \gamma(n) = \alpha$ and assume for simplicity that all the coefficients $a_{n,n-i}$, $(i = -q+1, \ldots, q-1)$ for $n > q$ are equal to zero. Denote by $I_k(a_1, \ldots, a_{2q-k})$ (or $I_{-k}(a_1, \ldots, a_{2q-k})$) the matrix of order $2q$ where the only nontrivial diagonal, consisting of the elements $a_1, \ldots, a_{2q-k}$ is shifted k steps to the right (or k steps to the left) with respect to the main diagonal; $I_k \equiv I_k(1, \ldots, 1)$. Then system (1) admits the following representation:

$$\mathbf{u}_{n+1} = B(n)\mathbf{u}_n, \quad n > q, \quad \text{where } \mathbf{u}_n = \begin{pmatrix} u_{n-q} \\ \vdots \\ u_{n+q-1} \end{pmatrix}, \quad \text{and} \quad (4)$$

$$B(n) = (B(n)_{i,j})_{i,j=1}^{2q} = \begin{pmatrix} 0 & 1 & 0 & 0 & 0 & \ldots & 0 & 0 \\ 0 & 0 & 1 & 0 & 0 & \ldots & 0 & 0 \\ 0 & 0 & 0 & 1 & 0 & \ldots & 0 & 0 \\ \vdots & & & & & & & \vdots \\ 0 & 0 & 0 & 0 & 0 & \ldots & 0 & 1 \\ -b_n & 0 & \ldots & 0_{2q,q} & \rho_n & 0_{2q,q+2} & \ldots & 0 \end{pmatrix}$$
$$= I_1 + I_{-q+1}(0, \ldots, 0, \rho_n) - I_{-2q+1}(b_n),\ b_n = \frac{\mu_{n-q}}{\lambda_n}, \rho_n = \frac{z}{\lambda_n}.$$

Matrix $B(n)$ is a so-called transfer matrix corresponding to (1) with the restriction that all coefficients $a_{n,n-i}$, $(i = -q+1, \ldots, q-1)$ for $n > q$ are equal to zero. If $q = 1$,

$$B(n) = \begin{pmatrix} 0 & 1 \\ -b_n & \rho_n \end{pmatrix}.$$

One can easily check that for $n-2q+1>q$

$$B(n,n-2q+1)\equiv B(n)B(n-1)\dots B(n-2q+1)$$
$$=B(n,n-q+1)B(n-q,n-2q+1)$$
$$=[I_q+\operatorname{diag}(\overbrace{0,\dots,0}^{q\text{ times}},\rho_{n-q+1},\rho_{n-q+2},\dots,\rho_n)-I_{-q}(b_{n-q+1},\dots,b_n)]$$
$$\times[I_q+\operatorname{diag}(\overbrace{0,\dots,0}^{q\text{ times}}\rho_{n-2q+1},\rho_{n-2q+2},\dots\rho_{n-q})-I_{-q}(b_{n-2q+1},\dots,b_{n-q})]$$
$$=\operatorname{diag}(-b_{n-2q+1},\dots,-b_{n-q},-\tilde b_{n-q+1},\dots,-\tilde b_n)+I_q(\rho_{n-2q+1},\dots,\rho_{n-q})$$
$$-I_{-q}(\rho_{n-q+1}b_{n-2q+1},\rho_{n-q+2}b_{n-2q+2},\dots,\rho_n b_{n-q}),$$

where $\tilde b_i=b_i-\rho_i\rho_{i-q}$. Since $\{\xi\},\{\delta\}\in l^2$, we get

$$b_n=\frac{\mu_{n-q}}{\lambda_n}=(1-\frac{q\alpha}{n}+O(\frac{1}{n^2}))(1-\epsilon_n+\frac{\epsilon_n\delta_n}{1+\delta_n})=1-\beta_n+r_n,$$

where $\epsilon_n=\delta_n-\xi_{n-q},\beta_n=q\alpha/n+\epsilon_n,\{r_n\}\in l^1$. We also have

$$\{\rho_n b_{n-q}-\rho_n\},\ \{\tilde b_n-b_n\}\in l^1,$$

because $\alpha>1/2$.

In view of the above, we can write

$$B(n,n-2q+1)=\tilde B(n,n-2q+1)+R_n,\quad \text{where}$$
$$\tilde B(n,n-2q+1)=\operatorname{diag}(-b_{n-2q+1},\dots,-b_n)$$
$$+I_q(\rho_{n-2q+1},\dots,\rho_{n-q})-I_{-q}(\rho_{n-q+1},\dots,\rho_n),$$

and $\{\|R_n\|\}\in l^1$. Again, using the condition $\alpha>1/2$ we obtain that $\{\rho_n-\rho_{n-i}\}\in l^1$ for $i=1,\dots,2q-1$ and $\{\rho_n\}\in l^2$, so the formula for $B(n,n-2q+1)$ can be written as

$$-\exp(-\rho_n P)[I-\operatorname{diag}(\beta_{n-2q+1},\dots,\beta_n)][I+S_n],$$

where $P=I_q-I_{-q},\{\|S_n\|\}\in l^1$, and I is the identity matrix.

Now assume that for a fixed index $i=1,\dots,2q-1$, $\{\epsilon_n-\epsilon_{n-i}\}\in l^1$ for $n=q(2k+1),k\in\mathbb{N}$. Then $\{\beta_n-\beta_{n-i}\}\in l^1$, and we get

$$B(n,n-2q+1)=-(1-\beta_n)\exp(-\rho_n P)[I+\tilde S_n],\text{where }\{\|\tilde S_n\|\}\in l^1.$$

Thus, for $n=q(2k+1)$ we are coming to the formula

$$\mathbf{u}_{n+1}=(-1)^{\frac{n-q}{2q}}\prod_{j=1}^{\frac{n-q}{2q}}(1-\beta_{q(2j+1)})\exp(-\rho_n P)(I+\tilde S_n)\tag{5}$$
$$\times\exp(-\rho_{n-2q}P)(I+\tilde S_{n-2q})\dots\exp(-\rho_{3q}P)(I+\tilde S_{3q})\mathbf{u}_{q+1}.$$

We have

$$\exp(-\rho_n P)=\cos(\rho_n)I-\sin(\rho_n)I_q+\sin(\rho_n)I_{-q}.$$

Hence, for real ρ_n this is a unitary matrix.

Theorem 1. *Suppose that for the system* (1–3) *the following conditions are fulfilled*

(i) $\alpha(n) = \gamma(n) = \alpha$ *for all* $n > q$.
(ii) *The series* $\sum_k \epsilon_k$ *is convergent.*
(iii) *There exists an integer* $m, 0 \le m \le 2q-1$ *such that* $\{\epsilon_n - \epsilon_{n-i}\} \in l^1$, $n = 2qk + m,\ k \in \mathbb{N}, i = 1, \ldots, 2q-1$.
(iv) *For all* $k > q$, ρ_k *are real.*

If $u = \{u_n\}_{n=q+1}^{\infty}$ *is a non-zero solution of* (1–3), *then for* $\mathbf{u}_n$ *defined by* (4)

$$\mathbf{u}_n = P^n n^{-\frac{\alpha}{2}} (\exp(-\frac{1}{2q} \sum_{k=q+1}^{n} \rho_k P)(f + (\mathbf{o(1)})),\ n \to \infty \tag{6}$$

for a certain vector $f \in \mathbb{C}^{2q}$ *and a vector* $\mathbf{o(1)}$ *of norm* $o(1)$.

The method of proof is the same as for the Theorem 3.2 from [5] (see also the example in the next section). If $\lambda_n = \mu_n,\ n = q+1, \ldots, \infty$ (as in [5] for $q = 1$), then

$$\sum_{k=q+1}^{N} \epsilon_k = \sum_{k=N-q+1}^{N} \xi_k - \sum_{k=1}^{q} \xi_k,$$

so the condition (ii) is automatically fulfilled (since $\{\xi_k\} \in l^2$). Note that formula (5) used in the proof has been derived from the additional assumption that $a_{n,n-i} = 0$ for all $i = -q+1, \ldots, q-1,\ n > q$. But, if in case of nonzero $a_{n,n-i}$ the condition (3) is fulfilled then the corresponding additional terms in the transfer matrix $B(n)$ give a summable error, and the formula (5) is also valid for this case.

Theorem 2. *Under the assumptions of the Theorem* 1 *for any non-zero solution* u *of* (1) *one obtains the asymptotic formula as* $N \to \infty$

$$\sum_{n=1}^{N} |u_n|^2 = \begin{cases} O(N^{1-\alpha}), & \alpha \neq 1, \\ O(ln(N)), & \textit{otherwise}. \end{cases}$$

Proof. Let the conditions of the previous theorem be fulfilled for $m = q$. Then, for $N = q(2k+1)$, $\sum_{n=1}^{N} |u_n|^2 \le \sum_{p=0}^{\frac{N-q}{2q}} \|\mathbf{u}_{q(2p+1)+1}\|^2$. Applying (5) and using, that $\|\exp(-\rho_n P)\| = 1$ we get

$$\begin{aligned}
\textstyle\sum_{n=1}^{q(2k+1)} |u_n|^2 &\le \textstyle\sum_{l=1}^{k} \prod_{j=1}^{l} |1 - \beta_{q(2j+1)}|^2 (1 + \|\tilde{S}_{q(2j+1)}\|^2) \|\mathbf{u}_{q+1}\|^2 \\
&\le C_1 \textstyle\sum_{l=1}^{k} \prod_{j=1}^{l} |1 - \beta_{q(2j+1)}|^2 \|\mathbf{u}_{q+1}\|^2 \\
&\le C_2 \textstyle\sum_{l=1}^{k} \exp(-2 \sum_{j=1}^{l} |\beta_{q(2j+1)}|) \|\mathbf{u}_{q+1}\|^2 \\
&\le C_3 \textstyle\sum_{l=1}^{k} \exp(-\alpha \sum_{j=1}^{l} \frac{1}{j}) \|\mathbf{u}_{q+1}\|^2 \le C_4 \sum_{l=1}^{k} l^{-\alpha},
\end{aligned}$$

for some suitable positive constants $C_m, m = 1, \ldots, 4$. Here we used that $\{\beta\}, \{\epsilon\} \in l^2$ and the condition (i) of the Theorem 1.

On the other hand, there exists N_0 so large, that $\|S_{q(2j+1)}\| < 1$ and $|\beta_{q(2j+1)}| < 1$ for $j \geq N_0$. So we obtain

$$\sum_{n=1}^{q(2k+1)} |u_n|^2 \geq C_5 \sum_{l=N_0}^{k} \prod_{j=N_0}^{l} |(1-\beta_{q(2j+1)})|^2 (1 - \|\tilde{S}_{q(2j+1)}\|^2) \|\mathbf{u}_{q+1}\|^2$$

$$\geq C_6 \sum_{l=N_0}^{k} l^{-\alpha},$$

for some C_5 and C_6. Thus the claim of the theorem is proved for $N = q(2k+1)$ and for the other N it follows from the fact, that $\|B(n)^{\pm 1}\| \to 1$ as $n \to \infty$. □

This theorem can be applied to study the subordinated solutions of (1) for $q = 1$. A solution $u \neq 0$ is called a subordinate solution of (1) if for every solution v linearly independent with u we have $\lim_{n\to\infty} \sum_{k=1}^{n} |u_k|^2 / \sum_{k=1}^{n} |v_k|^2 = 0$. We mention here that one of the main tools in the spectral analysis of infinite Jacobi matrices is the study of subordinated solutions of the systems (1) related to such matrices [7], [8].

Let $u = \{u_n\}_{n=1}^{\infty}$ and $v = \{v_n\}_{n=1}^{\infty}$ be two linearly independent solutions of (1) for $q = 1$. One can easily check that for $n \geq 2$

$$u_n v_{n+1} - v_n u_{n+1} = \frac{\mu_{n-1}}{\lambda_n}(u_{n-1}v_n - v_{n-1}u_n)$$

$$= \frac{\mu_{n-1}\cdots\mu_1}{\lambda_n\cdots\lambda_2}(u_1 v_2 - v_1 u_2) = \frac{\delta_{n-1} w_0}{\lambda_n},$$

where $\delta_n = \frac{\mu_n \cdots \mu_1}{\lambda_n \cdots \lambda_2}$ and $w_0 = u_1 v_2 - v_1 u_2$. Assuming that $\inf_n |\delta_n| \geq \alpha > 0$ we get

$$\sum_{n=1}^{N} \frac{|w_0|\alpha}{|\lambda_n|} \leq 2\Big(\sum_{n=1}^{N+1} |u_n|^2\Big)^{\frac{1}{2}} \Big(\sum_{n=1}^{N+1} |v_n|^2\Big)^{\frac{1}{2}}.$$

Applying the above theorem and using the same arguments as in [5] page 222, we finally obtain

$$\frac{\sum_{n=1}^{N+1} |u_n|^2}{\sum_{n=1}^{N+1} |v_n|^2} \geq c_0 > 0.$$

Thus the following theorem is proved.

Theorem 3. *If the system (1–3), with $q = 1$, satisfies the assumptions of the Theorem 1 and the additional assumption $inf_n|\delta_n| > 0$, then it has no subordinated solutions.*

2. Application

The above asymptotic formulas were established under the assumption that ρ_k are real, so $\exp(-\rho_k P)$ are unitary. To establish (6) for the complex ρ_k we have to additionally assume, that $\exp(\sum_{k=1}^{\infty} \rho_k P)$ is convergent, which is fulfilled, e.g., for

$\alpha > 1$. For such α, as it follows from (6), all non-zero solutions of (1) are decreasing not slower than $n^{-\alpha/2}$. Nevertheless, for some cases when $\exp(\sum_{k=1}^{\infty} \rho_k P)$ is divergent, it is also possible to establish the analog of (6).

As an example, we consider the application of the studied asymptotic method to the analysis of invariance of l^p property for (1), with respect to z. Namely, the following generalization of the Hellinger-Wall theorem [9] takes place.

Theorem 4. *Suppose that the coefficients of the system*

$$\mu_{n-q}u_{n-q} + a_{n,n-q+1}u_{n-q+1} + \cdots + a_{n,n+r-1}u_{n+r-1} + \lambda_n u_{n+r} = zu_n, \quad (7)$$
$$n > q, \quad q, r \geq 1, \quad \mu_{n-q} \neq 0, \quad \lambda_n \neq 0, \quad z \in \mathbb{C},$$

satisfy the condition

$$\inf_i |\delta_i| > \delta > 0, \quad (8)$$

where

$$\delta_1 = \mu_1, \; \delta_i = \frac{\mu_1\mu_2 \cdots \mu_i}{\lambda_{q+1}\lambda_{q+2} \cdots \lambda_{q+i-1}}, \; i \geq 2.$$

Suppose also that for some p, $1 \leq p \leq 2$, and $z = z_0 \in \mathbb{C}$, the series $\sum_{i=1}^{\infty} |u_i(z_0)|^p$ is convergent for every solution $u = u(z) = \{u_i(z)\}_{i=1}^{\infty}$ of (7),and, therefore $u(z_0) \in l^p$. Then, for any $M > 0$, the series $\sum_{i=1}^{\infty} |u_i(z)|^p$ is uniformly convergent for $|z - z_0| < M$.

The proof is contained in [10]. Thus, the property of all solutions of (7) to belong to l^p (the l^p property) is invariant with respect to $z \in \mathbb{C}$ for $1 \leq p \leq 2$.

To check whether or not the statement of this theorem is true for l^p, $p > 2$ we consider the system:

$$\mu_{n-1}u_{n-1} + \lambda_n u_{n+1} = zu_n, \; n > 1, \quad (9)$$

where

$$\mu_n = \begin{cases} n(n+1), & n = 2k, \\ n+1, & n = 2k+1, \end{cases} \quad \lambda_n = \begin{cases} n+1, & n = 2k, \\ n^2, & n = 2k+1, \end{cases} \quad k \in \mathbb{N}, \mu_1 = 2.$$

Or, in other words,

$$nu_{n-1} + (n+1)u_{n+1} = zu_n \quad \text{for even } n,$$
$$(n-1)nu_{n-1} + n^2 u_n = zu_n \quad \text{otherwise.}$$

It satisfies the condition (8) of Theorem 4. For even n $\alpha(n) = \gamma(n-1) = 1$, $\delta_n = 1/n, \xi_{n-1} = 1/(n-1)$; for odd n $\alpha(n) = \gamma(n-1) = 2, \delta_n = 0, \xi_{n-1} = 1/(n-1)$, and for all n $\mu_{n-1}/\lambda_n = 1 - 1/n$. Our purpose is to obtain an asymptotic formula for the solutions of (9). Since $\alpha(n) \neq \gamma(n)$ we cannot apply here Theorem 1 (the conditions (ii)–(iii) are also not fulfilled), but we can use similar techniques. Namely, consider the case of odd n. Then

$$\begin{aligned} B(n)B(n-1) &= \begin{pmatrix} -1 + \frac{1}{n} & \frac{z}{n} \\ 0 & -1 + \frac{1}{n} \end{pmatrix} + \begin{pmatrix} 0 & 0 \\ \frac{z}{n^3} - \frac{z}{n^2} & \frac{z^2}{n^3} \end{pmatrix} \\ &= -(1 - \frac{1}{n}) \exp(-\tilde{\rho}_n \tilde{P})(I + \tilde{S}_n), \end{aligned}$$

where $\tilde{\rho}_n = \frac{z}{n-1}$,

$$\tilde{P} = \begin{pmatrix} 0 & 1 \\ 0 & 0 \end{pmatrix}, \quad \tilde{S}_n = \begin{pmatrix} -\frac{z^2}{n^2(n-1)^2} + \frac{z^2}{n(n-1)^2} & -\frac{z^3}{n^2(n-1)^2} \\ \frac{z}{n(n-1)} - \frac{z}{n^2(n-1)} & -\frac{z^2}{n^2(n-1)} \end{pmatrix}.$$

Hence for odd n we have:

$$\mathbf{u}_{n+1} = B(n)B(n-1)\dots B(2)\mathbf{u}_2 = (-1)^{\frac{n-1}{2}} \prod_{k=1}^{\frac{n-1}{2}} (1 - \tfrac{1}{2k+1})$$
$$\times \exp(-\tilde{\rho}_n\tilde{P})(I+\tilde{S}_n)\exp(-\tilde{\rho}_{n-2}\tilde{P})(I+\tilde{S}_{n-2})\dots\exp(-\tilde{\rho}_3\tilde{P})(I+\tilde{S}_3)\mathbf{u}_2$$
$$= (-1)^{\frac{n-1}{2}} \prod_{k=1}^{\frac{n-1}{2}} (1 - \tfrac{1}{2k+1}) \exp(-\textstyle\sum_{k=1}^{\frac{n-1}{2}} \tilde{\rho}_k\tilde{P}) \prod_{k=1}^{\frac{n-1}{2}} (I + \tilde{R}_{2k+1})\mathbf{u}_2,$$

where $\tilde{R}_3 = \tilde{S}_3$, and for $k > 1$

$$\tilde{R}_{2k+1} = \exp(\sum_{i=1}^{k-1} \tilde{\rho}_{2i+1}\tilde{P})\tilde{S}_{2k+1}\exp(\sum_{i=1}^{k-1} -\tilde{\rho}_{2i+1}\tilde{P})$$
$$= \begin{pmatrix} 1 & \sum_{i=1}^{k-1} \frac{z}{2i+1} \\ 0 & 1 \end{pmatrix} \begin{pmatrix} \frac{z^2}{(2k)^2(2k+1)} - \frac{z^2}{(2k)^2(2k+1)^2} & \frac{-z^3}{(2k)^2(2k+1)^2} \\ \frac{z}{2k(2k+1)} - \frac{z}{(2k)(2k+1)^2} & \frac{-z^2}{(2k)(2k+1)^2} \end{pmatrix} \begin{pmatrix} 1 & \sum_{i=1}^{k-1} \frac{-z}{2i+1} \\ 0 & 1 \end{pmatrix}.$$

Thus the elements of $\tilde{R}_{2k+1}$ are of order $O(\frac{\ln^2(k)}{k^2})$ and therefore $\{\|\tilde{R}_{2k+1}\|\} \in l^1$. Then, using that

$$\prod_{k=1}^{\frac{n-1}{2}} (1 - \frac{1}{2k+1}) \asymp \exp(-\sum_{k=1}^{\frac{n-1}{2}} \frac{1}{2k+1}), \quad P^2 = -I, \quad (\text{where } P = \begin{pmatrix} 0 & 1 \\ -1 & 0 \end{pmatrix});$$

$$\sum_{k=1}^{\frac{n-1}{2}} \frac{1}{2k+1} = \frac{1}{2}\sum_{k=2}^{n} \frac{1}{k} - \frac{1}{2}\sum_{k=1}^{\frac{n-1}{2}} (\frac{1}{2k} - \frac{1}{2k+1}),$$

$$\sum_{k=1}^{\frac{n-1}{2}} \tilde{\rho}_{2k+1} = \frac{1}{2}\sum_{k=1}^{n} \tilde{\rho}_k - \frac{1}{2}\sum_{k=1}^{\frac{n-1}{2}} (\tilde{\rho}_{2k} - \tilde{\rho}_{2k+1}), \quad (\tilde{\rho}_1 = 0),$$

$$\left\{\frac{1}{2k} - \frac{1}{2k+1}\right\} \quad \text{and} \quad \{\tilde{\rho}_{2k} - \tilde{\rho}_{2k+1}\} \in l^1,$$

for odd n we obtain

$$\mathbf{u}_{n+1} = P^{n-1}\exp(-\textstyle\sum_{k=1}^{\frac{n-1}{2}} \frac{1}{2k+1})\exp(-\sum_{k=1}^{\frac{n-1}{2}} \tilde{\rho}_{2k+1}\tilde{P}) \prod_{k=1}^{\frac{n-1}{2}} (I + \tilde{R}_{2k+1})\mathbf{u}_2$$
$$= P^{n-1}\exp(-\tfrac{1}{2}\textstyle\sum_{k=1}^{n+1} \frac{1}{k})\exp(-\frac{1}{2}\sum_{k=1}^{n+1} \tilde{\rho}_k\tilde{P})(e + \mathbf{o}(\mathbf{1}))$$
$$= P^{n+1}(n+1)^{-1/2}\exp(-\tfrac{1}{2}\textstyle\sum_{k=1}^{n+1} \tilde{\rho}_k\tilde{P})(f + \mathbf{o}(\mathbf{1})), \ n \to \infty,$$

where e and f are some vectors in $\mathbb{C}^2$ and $\mathbf{o}(\mathbf{1})$ is a vector of norm $o(1)$. Since $B(n) \to P$ as $n \to \infty$, the asymptotics for even n is the same.

Hence for all n

$$\mathbf{u}_n = P^n n^{-\frac{1}{2}} \begin{pmatrix} 1 & n^{\frac{-z}{2}} \\ 0 & 1 \end{pmatrix} (g + \mathbf{o}(\mathbf{1})), \ n \to \infty,$$

where g is a certain vector in $\mathbb{C}^2$. From this, one can immediately see that for $z = 0$ or $z = i$ all solutions of the equation are of order $O(n^{-1/2})$ and therefore belong to the space $l^{2+\epsilon}$ for any $\epsilon > 0$. However, for $z = -1$ there is a one-dimensional subspace of solutions which are bounded but do not tend to zero, as $n \to \infty$. Thus the following proposition is proved:

Proposition. *The statement of the Theorem 4 is not true for $p > 2$.*

Acknowledgment
The author is thankful to Bernhard Beckermann for important suggestions and remarks and to the anonymous referee for valuable comments.

References

[1] A.I. Aptekarev V. Kaliaguine and W. Van Assche, Criterion for the resolvent set of nonsymmetric tridiagonal matrix, Proc. Amer. Math Soc., vol. 123 no 8, (1995), 2423–2430.

[2] V.A. Kaliaguine, Hermite-Padé approximants and spectral analysis of nonsymmetric operators, Mat. Sb., vol. 185 (1994) 79–100 (In Russian). English transl. in Russian Acad. Sci. Sb. Math., vol. 82 (1995) 199–216.

[3] Yu.V. Sidorov, M.V. Fedoryuk, M.A. Shabunin, Lectures on the theory of functions of a complex variable. Textbook, Third ed.(Russian) Moscow, Nauka, 478 p. (1989).

[4] Z. Benzaid and D.A. Lutz, Asymptotic representation of solutions of perturbed systems of linear difference equations, Studies in Applied Mathematics, Vol. 77, (1987) 195–221.

[5] J. Janas, S. Naboko, Jacobi matrices with power-like weights – grouping in blocks approach, Journal of Functional Analysis, vol. 166., (1999), 218–243.

[6] E.A. Coddington, N. Levinson, Theory of ordinary differential equations, McGraw-Hill Book Company, NY, 1955.

[7] S. Khan and D.B. Pearson, Subordinacy and spectral theory for infinite matrices, Helv. Phys. Acta, vol. 65, no. 4, (1992), 505–527.

[8] G. Stolz, Spectral theory for slowly oscillating potentials. I. Jacobi matrices, Manuscripta Math., vol. 84, no. 3–4, (1994), 245–260.

[9] H. Wall,Analytic Theory of Continued Fractions, Chelsea, Bronx, NY, 1973.

[10] A.S. Osipov, On the Hellinger theorem and l^p properties of solutions of difference equations,Journal of Difference Equations and Applications, vol. 9, no. 9, (2003), 841–851.

Andrey S. Osipov
Scientific Research Institute for System Studies
Russian Academy of Sciences
Nakhimovskii pr. 36-1
Moscow, 117218, Russia
e-mail: andrei@sgo.fi

Operator Theory:
Advances and Applications, Vol. 154, 163–177

A Remark on Spectral Meaning of the Symmetric Functional Model

Boris Pavlov

Abstract. The imaginary part of a dissipative operator L is weak if it is presented by a positive operator $\mathcal{T}$ such that the square $\mathcal{T}^2$ of it is a product of an operator with a finite trace and an operator from Macaev class. For a dissipative operator with a weak imaginary part the families of incoming and outgoing scattered waves form a non-orthogonal and often even over-complete system $\{\Psi_{\rm in}, \Psi_{\rm out}\}$ of eigenfunctions of the corresponding self-adjoint dilation $\mathcal{L}$. The rescription of L in the spectral representation associated with $\{\Psi_{\rm in}, \Psi_{\rm out}\}$ gives the Symmetric Functional Model of L, and the characteristic function S of L coincides with the transmission coefficient of the outgoing waves. A general construction based on the self-adjoint delation and an example of the Lax-Phillips Semigroup for the 1-D wave equation on the infinite string with a bounded non-negative potential supported by semi-axis are considered.

Mathematics Subject Classification (2000). 47A45.
Keywords. Symmetric functional model.

1. Introduction

The classical Nagy-Foias Functional Model of a dissipative operator is obtained as an image of the dissipative operator in incoming or outgoing spectral representation of it's self-adjoint dilation, see [10]. The symmetric rescription of the Nagy-Foias Functional Model (we will call it further just a Symmetric Model) was suggested in [12, 13] and developed in [6, 8, 15, 11, 17, 18], see complete bibliography in [15, 11, 18]. The symmetric rescription of the Nagy-Foias functional model gives most simple algebraic formulae for spectral objects of the model (the eigenfunctions, the spectral density and others) and permits to calculate directly important spectral characteristics, see [15, 11, 18]. The Symmetric Model can be

The author recognizes support from the grant of the Russian Academy of Sciences, RFBR 03-01-00090.

obtained formally from the original Nagy-Foias model just by a non-degenerate transformation of the space, see [12], which transforms the spectral density matrix of the dilation into the spectral density matrix of the symmetric model:

$$\begin{pmatrix} I & 0 \\ 0 & I - S^+S \end{pmatrix} \longrightarrow \begin{pmatrix} S^+ & I \\ I & 0 \end{pmatrix} \begin{pmatrix} I & 0 \\ 0 & I - S^+S \end{pmatrix} \begin{pmatrix} S & I \\ I & 0 \end{pmatrix}$$

$$= \begin{pmatrix} I & S^+ \\ S & I \end{pmatrix} := \rho(\lambda), \tag{1.1}$$

where $S = S(k + i0)$ is the limit value of the characteristic function of the original dissipative operator on the real axis from the upper half-plane and S^+ are the limit values of $S^+(k - i0)$ on the real axis from the lower half-plane. The above transformation of the spectral data connects actually different kinds of eigenfunctions which may be used for construction of the spectral representation. The standard Nagy-Foias model was connected with the orthogonal system of eigenfunctions of incoming and complementary ("radiating") components of the dilation, see [13, 14]. We denote them by $\{\psi_-, \psi^<\}$ respectively. An equivalent model may be connected with outgoing $\{\psi_+\}$ and "absorbing" eigenfunction $\{\psi^>\}$ of the dilation. Then the symmetric model is formally connected by the above formula (1.1) to the orthogonal system $\{\psi_<, \psi_- - S^+\psi^<\}$ or the dual system $\{\psi_>, \psi_+ - S\psi^>\}$. Both systems look rather ugly, and hence the recipe of the construction of the functional model as an image of the original dissipative operator in the spectral representation associated with this system seems neither elegant nor persuading. Besides the suggested transformation of the classical Nagy-Foias model to the symmetric one is not unique. Hence there exists an extended set of systems of eigenfunctions of the dilation which can be used for construction of the spectral representation relevant to the symmetric model.

We will reveal the spectral meaning of the symmetric model via finding *the most natural spectral representation* of the dilation, such that the rescription of the original dissipative operator in terms of it coincides with the symmetric functional model, similarly to the classical case. We show that the symmetric functional model is associated with a certain set of eigenfunctions of the absolutely-continuous spectrum of the model: the incoming and outgoing scattered waves $\{\psi_-, \psi_+\}$. Generally, this set of eigenfunctions is not orthogonal. Hence the corresponding matrix ρ of the spectral density is non-diagonal and may degenerate. Nevertheless the spectral representation is properly defined, up to non-essential addenda, as stated in [13, 11].

The plan of actual paper is the following: in the second section we revisit the construction of the functional model in [12] for the scalar wave-equation in R_1, paying a special attention to the calculation of the symmetric non-orthogonal spectral representation. Then in the third section we study an abstract dissipative operator B with finite-dimensional defect dim $(B - B^+) < \infty$ [1] and square characteristic matrix-function. We obtain the corresponding eigenfunctions of the

[1] See the remark after Theorem 3.2.

self-adjoint dilation as limits of the resolvent in the properly rigged space. Though the spectral components are, generally, non-uniquelly defined, we calculate them with the help of the wave-operators. The characteristic function is interpreted both as a stationary transmission coefficient and as a (non-stationary) limit, in full agreement with the classical Adamyan-Arov results.

2. Symmetric representation for Lax-Phillips semigroup

Assuming that q is a piece-wise continuous non-negative locally bounded function supported by the left semi-axis $R_- = (-\infty, 0)$, consider the wave-equation on $R = (-\infty, \infty)$

$$u_{tt} - u_{xx} + q(x)u = 0, \ -\infty < x < \infty, \tag{2.1}$$

with proper initial data with finite energy:

$$u(x,0) = u_0, \ u_t(x,0) = u_1.$$

There exists a unique generalized solution of the above equation from the domain of the quadratic form of the corresponding Schrödinger operator $L, Lu = -u_{xx} + q(x)u$, defined as a Friedrichs extension of the corresponding symmetric operator defined on smooth compactly-supported functions. The energy of the solution $u(x,t)$ is conserved in course of evolution and is equal to the energy of the initial data:

$$\mathcal{E}_u = \frac{1}{2}\int_{-\infty}^{\infty} \left[|\nabla u_0|^2 + q(x)|u_0|^2 + |u_1|\right] dx.$$

The evolution operator of Cauchy data

$$\begin{pmatrix} u_0 \\ u_1 \end{pmatrix} \longrightarrow \begin{pmatrix} u(x,t) \\ u_t(x,t) \end{pmatrix} = \mathbf{u}(t)$$

is unitary in the space $\mathcal{E}$ of all Cauchy data $\mathbf{u}$ supplied with the energy dot-product:

$$\langle \mathbf{u}, \mathbf{v}\rangle_{\mathcal{E}} = \frac{1}{2}\int_{-\infty}^{\infty} \left[\langle \nabla u, \nabla \bar{v}\rangle + q(x)u\bar{v} + u_t\bar{v}_t\right] dx.$$

The generator of the evolution of Cauchy data is a self-adjoint operator $\mathcal{L}$ in $\mathcal{E}$ which appears in the right side of the wave-equation represented in Schrödinger form:

$$\frac{1}{i}\frac{\partial \mathbf{u}}{\partial t} = i\begin{pmatrix} 0 & -1 \\ L & 0 \end{pmatrix}\mathbf{u} := \mathcal{L}\mathbf{u}.$$

The incoming and outgoing subspaces $D_{\text{in,out}} \in \mathcal{E}$ consist of all Cauchy data of incoming and outgoing waves $u(x \pm t)$ supported by the right semi-axis and are mutually orthogonal in $\mathcal{E}$, see [7].

The spectrum of the operator L is absolutely continuous on the interval $R_+ = [0, \infty)$, possibly with varying multiplicity. We assume that the spectral analysis of the Schrödinger operator L is done and a complete orthogonal set of eigenfunction of the absolutely-continuous spectrum – the scattered waves $\psi_\pm$ – is constructed as

$$\psi_+(x,k) = \begin{cases} e^{-ikx} + Se^{ikx} & \text{if} \quad x > 0 \\ \alpha_+\chi_+(x,k) & \text{if} \quad x < 0 \end{cases}$$

and

$$\psi_-(x,k) = \begin{cases} e^{ikx} + \bar{S}e^{-ikx} & \text{if} \quad x > 0 \\ \alpha_-\chi_-(x,k) & \text{if} \quad x < 0 \end{cases}$$

where $\alpha_\pm$ are corresponding transmission coefficients and $\chi_\pm(x,k) = \varphi(x,k) + m_\pm(k)\theta(x,k)$ are the limit values (from the upper and lower half-planes) of Weyl-solutions of the homogeneous equation $Lu = k^2 u$ which are square integrable on $(-\infty, 0]$ for k in upper or lower half-plane $\Im k > 0, \Im k < 0$ respectively. Here $\varphi(x,k), \theta(x,k)$ are the standard solutions of the homogeneous equation with initial data $(0,1)$, $(1,0)$ and $m_\pm(k)$ are the corresponding Weyl-Titchmarsh functions, see [19], selected with the condition of analyticity in upper or lower half-planes respectively. The role of the spectral parameter is played by k^2 with the branch of the square root $\sqrt{k^2} = k$ defined for S by the condition $\Im k > 0$ and for $\bar{S}$ by $\Im k < 0$.

Theorem 2.1. *The reflection coefficients on the real axis $\Im k = 0$ are defined by the limit values of the corresponding Weyl-Titchmarsh functions on real axis k from the upper (lower) half-planes respectively:*

$$S(k) = -\frac{m_+(k) + ik}{m_+(k) - ik}, \ \bar{S}(k) = -\frac{m_-(k) - ik}{m_-(k) + ik}$$

and admit an analytic continuation into the upper and lower half-planes respectively. The incoming and outgoing solutions of the wave-equation on the positive semi-axis $0 < x < \infty$ are parametrized by elements of the Hardy classes $h_\pm \in H^2_\pm$ as

$$u_{\text{in,out}}(x,t) = \int_{-\infty}^{\infty} e^{ikt}\psi_\pm(x,k)\frac{1}{ik}h_\pm(k)dk.$$

The incoming and outgoing subspaces $\mathcal{D}_{\text{in}}$, $\mathcal{D}_{\text{out}}$ of the energy-normed space $\mathcal{E}$ are images of the corresponding Hardy-classes in spectral representations associated with proper eigenfunctions

$$\Psi_\pm(x,k) = \begin{pmatrix} \frac{1}{ik}\psi_\pm(x,k) \\ \psi_\pm(x,k) \end{pmatrix}$$

of the generator $\mathcal{L}$:

$$\mathcal{D}_{\text{in}} = \left\{ \int_{-\infty}^{\infty} e^{ikt} \begin{pmatrix} \frac{1}{ik}\psi_-(x,k) \\ \psi_-(x,k) \end{pmatrix} h_-(k)dk, \ h_- \in H^2_- \right\},$$

$$\mathcal{D}_{\text{out}} = \left\{ \int_{-\infty}^{\infty} e^{ikt} \begin{pmatrix} \frac{1}{ik}\psi_+(x,k) \\ \psi_+(x,k) \end{pmatrix} h_+(k)dp, \ h_+ \in H^2_+ \right\} \tag{2.2}$$

and the correspondence between the functional parameters $h_\pm$ and the relevant Cauchy data is isometric in energy-normed space.

The invariant subspaces $\mathcal{E}_\pm$ of the generator $\mathcal{L}$ of the evolution developed from $\mathcal{D}_{\text{in, out}}$ are isometrical images of $L_2(R)$ defined by the spectral map

$$\mathcal{J}_\pm : h \to \frac{1}{\sqrt{2\pi}} \int \Psi_\pm(x,p)h(p)dp = \mathcal{J}_\pm h, \ , \ h \in L_2$$

$$\mathcal{E}_\pm = \left\{ \int \Psi_\pm(x,k)h(k)dk, \ h \in L_2(R) \right\}, \ |\mathcal{J}_\pm h|_\mathcal{E} = |h|_{L_2}.$$

In particular, if $\mathbf{f}, \mathbf{g}$ *are elements of the energy-normed space presented as*

$$\mathbf{f} = \mathcal{J}_+ f_+ + \mathcal{J}_- f_- := \mathcal{J} \begin{pmatrix} f_+ \\ f_- \end{pmatrix}, \ \mathbf{g} = \mathcal{J}_+ g_+ + \mathcal{J}_- g_- := \mathcal{J} \begin{pmatrix} g_+ \\ g_- \end{pmatrix},$$

then

$$\langle \mathcal{J} \begin{pmatrix} f_+ \\ f_- \end{pmatrix}, \mathcal{J} \begin{pmatrix} g_+ \\ g_- \end{pmatrix} \rangle_\mathcal{E} = \int_{-\infty}^{\infty} \langle \begin{pmatrix} 1 & \bar{S} \\ S & 1 \end{pmatrix} \begin{pmatrix} f_+ \\ f_- \end{pmatrix}, \begin{pmatrix} g_+ \\ g_- \end{pmatrix} \rangle_{C_2}. \qquad (2.3)$$

Each of systems of eigenfunctions $\Psi_\pm$ *of the generator* $\mathcal{L}$ *is non-complete, if the modulo* $|S|$ *of the Scattering matrix is not equal to one identically:*

$$\mathcal{E}_\pm \neq \mathcal{E}, \ \mathcal{E}_+^\perp = \mathcal{E}^> \neq 0, \ \mathcal{E}_-^\perp = \mathcal{E}^< \neq 0,$$

but the joining of them $\{\Psi_\pm\}$ *is complete* $\mathcal{E}_+ + \mathcal{E}_- = \mathcal{E}$ *and the corresponding Parseval identity* (2.3) *is true with proper spectral density matrix.*

Proof. The proof of the statement is obtained by straightforward verification of the corresponding algebra, see, for instance [13]. The only analytical question appears when deriving the above Parseval identity (2.3). It is actually equivalent to the following equations:

$$\begin{array}{ll} \langle \mathcal{J}_+ h, \mathcal{J}_+ g \rangle_\mathcal{E} = \langle h, g \rangle_{L_2}, & \langle \mathcal{J}_+ h, \mathcal{J}_- g \rangle_\mathcal{E} = \langle \bar{S} h, g \rangle_{L_2} \\ \langle \mathcal{J}_- h, \mathcal{J}_+ g \rangle_\mathcal{E} = \langle S h, g \rangle_{L_2}, & \langle \mathcal{J}_- h, \mathcal{J}_- g \rangle_\mathcal{E} = \langle h, g \rangle_{L_2}. \end{array} \qquad (2.4)$$

We will verify only one of them assuming that $h = h_+ \in H_+^2$ and $g = e^{ikt} g_-$, $g_- \in H_-^2$, $t > 0$. Then the integration in the energy dot-product is extended on the positive semi-axis $0 < x < \infty$ where the eigenfunctions $\psi_\pm$ are presented just as linear combinations of exponentials. Then assuming that the elements h_+, g_- are smooth and rapidly decreasing we may accomplish the integration over the space variable x, acquiring proper delta-functions. Then the integral over wave number k is reduced, for any positive t, to:

$$\langle \mathcal{J}_+ h, \mathcal{J}_- g \rangle_\mathcal{E} = \frac{1}{4\pi} \int_0^\infty dx \int dk \int d\hat{k} \left[-e^{-ikx} + S(k) e^{ikx} \right] h_+(k)$$

$$\overline{\left[e^{i\hat{k}x} - \bar{S}(\hat{k}) e^{-i\hat{k}x} \right]} e^{-i\hat{k}t} \bar{h}_-(\hat{k})$$

$$+ \frac{1}{4\pi} \int_0^\infty dx \int dk \int d\hat{k} \left[e^{-ikx} + S(k) e^{ikx} \right] h_+(k) \overline{\left[e^{i\hat{k}x} - \bar{S}(\hat{k}) e^{-i\hat{k}x} \right]} e^{-i\hat{k}t} \bar{h}_-(\hat{k}).$$

The ultimate expression may be presented as an integral over the semi-axis $0 < x < \infty$ and, after cancellation of few terms which do not contain S, results in:

$$\frac{1}{2\pi}\int_0^\infty dx \int dk \int d\hat{k} e^{i(k-\hat{k})x} S(k) h_+(k) \bar{h}_-(\hat{k}) e^{-i\hat{k}t}$$

$$+\frac{1}{2\pi}\int_0^\infty dx \int dk \int d\hat{k} e^{-i(k-\hat{k})x} S(k) h_+(k) \bar{h}_-(\hat{k}) e^{-i\hat{k}t}$$

$$\int_{-\infty}^{\infty} e^{-ikt}\left[S h_+ \bar{g}_-\right] dk = \int_{-\infty}^{\infty} \left[S h_+ e^{-ikt} \bar{g}_-\right] dk.$$

Note that the integral with negative t is equal to zero because of orthogonality of incoming and outgoing subspaces.

The obtained result may be extended to the dense set of elements via considering the products $h = e^{ikt_1} h_+$, $g = e^{ikt_2} g_-$ for any finite t_1, t_2. Taking a limit $t_1, t_2 \to \pm\infty$ we obtain the statement

$$\langle \mathcal{J}_+ h, \mathcal{J}_- g \rangle_{\mathcal{E}} = \langle S h, g \rangle$$

for any $h, g \in L_2(R)$, that is for any elements $\mathcal{J}_+ h \in \mathcal{E}_+$, $\mathcal{J}_- g \in \mathcal{E}_-$ from the invariant subspaces of the generator $\mathcal{L}$ produced by the development of the incoming and outgoing subspaces. Other announced statements may be proved in a similar way. □

Note that the column of "symmetric" coordinates $\begin{pmatrix} f_+ \\ f_- \end{pmatrix}$ of an element $\mathbf{f} = \mathcal{J}_+ f_+ + \mathcal{J}_- f_-$ in the symmetric spectral representation is not defined uniquely. Nevertheless, one may suggest a natural procedure of recovering the symmetric coordinates of an element (u, u_t) from the energy-normed space of Cauchy data of the wave equation based on Adamyan-Arov wave-operators.

Really, consider the non-perturbed wave equation

$$u_{tt} - u_{xx} = 0$$

on the whole axis $(-\infty, \infty)$. The corresponding unitary evolution group $e^{i\mathcal{L}_0 t}$ in proper "non-perturbed" energy-normed space $\mathcal{E}_0$ has a common pair $\mathcal{D}_{\text{out, in}}$ of incoming and outgoing subspaces with the evolution group generated by the equation (2.1) in $\mathcal{E}$. Denote by $P_{\text{in, out}}$ the orthogonal projections in the energy-normed space $\mathcal{E}$ onto incoming and outgoing subspaces $\mathcal{D}_{\text{in, out}}$.

Theorem 2.2. *The Adamyan-Arov wave operators, see* [1]

$$W_+ = s - \lim_{t\to\infty} e^{-i\mathcal{L}_0 t} P_{\text{out}} e^{-i\mathcal{L}t},\ W_- = s - \lim_{t\to-\infty} e^{-i\mathcal{L}_0 t} P_{\text{in}} e^{-i\mathcal{L}t}$$

exist as strong limits and are isometrical from the invariant subspaces $\mathcal{E}_{\text{in, out}}$ *obtained via development of* $\mathcal{D}_{\text{in, out}}$ *with perturbed evolution – into the corresponding invariant subspaces* $\overset{0}{\mathcal{D}}_{\text{in, out}}$ *of the unitary evolution group* $e^{i\mathcal{L}_0 t}$ *in the "non-perturbed" energy-normed space* $\mathcal{E}_0$ *spanned by Cauchy data of incoming "from the right" and outgoing "to the right" waves*

$$h(x \pm t) = \frac{1}{\sqrt{2\pi}} \int_{-\infty}^{\infty} e^{ik(t \pm x)} \frac{i}{k}\, h_{\text{in, out}}(k)\, dk, \quad \frac{i}{k} h_{\text{in, out}} := \mathcal{F} W_{\mp} \mathbf{h}(0), \tag{2.5}$$

where

$$\mathbf{h}(0) = \begin{pmatrix} f(x,0) \\ f_t(x,0) \end{pmatrix}$$

is the column of Cauchy data of the original perturbed problem and $\mathcal{F}$

$$f(x) \to \frac{1}{\sqrt{\pi}} \int_{-\infty}^{\infty} e^{ikx}\, dx : (\mathcal{F} f)(k)$$

is a standard Fourier transform in L_2.

Corollary. *The column of coordinates* $\begin{pmatrix} f_- \\ f_+ \end{pmatrix}$ *of an element* $\mathbf{f} = \mathcal{J}_- f_- + \mathcal{J}_+ f_+$ *in the symmetric spectral representation may be chosen as* $\begin{pmatrix} f_{\text{in}} \\ f_{\text{out}} \end{pmatrix}$, *where the components* $f_{\text{in, out}} := -ik\mathcal{F} W_{\mp} F(0)$ *are defined by the initial data* $F(0) = \begin{pmatrix} f_0 \\ f_1 \end{pmatrix}$ *of the perturbed wave equation.*

3. Self-adjoint dilation and symmetric model

Consider a dissipative operator in a Hilbert space K

$$L = A + \frac{i}{2} \Gamma^+ \Gamma$$

with a real part $A = A^+$ and a finite-dimensional [2] positive imaginary part. Assuming that $\Gamma K = E$ contains a generating subspace of the operator A, consider the extended space $\mathcal{E} = L_2(R_-, E) \oplus K \oplus L_2(R_+, E)$ of vector functions

$$\vec{u} = \begin{pmatrix} u_- \\ u \\ u_+, \end{pmatrix}$$

and the operator $\mathcal{L}$ in $\mathcal{E}$ defined on vector functions $\vec{u}, u \in D(A), u_\pm \in W_2^1(R_\pm, E)$ submitted to the condition $u_-(0) - u_+(0) = i\Gamma u$:

$$\mathcal{L}\vec{u} = \begin{pmatrix} i\frac{du_-}{dx} \\ Au + \frac{\Gamma^+}{2}[u_-(0) + u_+(0)] \\ i\frac{du_+}{dx} \end{pmatrix}.$$

Without loss of generality we can assume that the operator Γ is self-adjoint $\Gamma = \Gamma^+$ and positive, but still we will use both Γ, Γ^+ in further formulae, if not specified otherwise. The elements from the domain of $\mathcal{L}$, which vanish on either of semi-axes

[2]This condition can be essentially relaxed, see the remark after Theorem 3.2.

$R_\pm$, satisfy the corresponding homogeneous boundary conditions $u_-(0) = i\Gamma u$, or $-u_+(0) = i\Gamma u$. These conditions are fulfilled for the absorbing and radiating eigenvectors of the dilation, see (3.4) below.

Theorem 3.1. *The operator $\mathcal{L}$ is a self-adjoint operator in $\mathcal{E}$. Moreover, the compression of the resolvent $\mathcal{R}_\lambda$, $\Im\lambda < 0$ of the operator $\mathcal{L}$ onto the subspace K coincides with the resolvent of the operator L:*

$$P_K [\mathcal{L} - \lambda I]^{-1} P_K = (L - \lambda I)^{-1}, \ \Im\lambda < 0. \tag{3.1}$$

Proof of the corresponding statement for the dissipative Schrödinger operator with complex potential is given in [14]. Proof of the announced abstract result follows the same pattern: to obtain the first statement we have to verify the symmetry of $\mathcal{L}$ and the symmetry of the adjoint operator; to prove the second statement we may use a simple algebra and the basic fact of existence of limits of R-function on the real axis from the upper half-plane. Then using Riesz-integral of the resolvent one may derive from (3.1) that for any bounded analytic function $\Phi(\lambda)$ in upper half-plane $\Im\lambda > -2\epsilon$

$$P_K \Phi(\mathcal{L}) P_K =$$

$$-\frac{1}{2\pi i}\int_{-\infty-i\epsilon}^{\infty-i\epsilon} P_K \Phi(\lambda)\,[\mathcal{L} - \lambda I]^{-1} P_K \, d\lambda =$$

$$-\frac{1}{2\pi i}\int_{-\infty-i\epsilon}^{\infty-i\epsilon} \Phi(\lambda)[L - \lambda I]^{-1} d\lambda = \Phi(L)$$

which means, in particular, that for $t > 0$: $e^{iLt} = P_K e^{i\mathcal{L}t} P_K$. □

The unitary group $e^{i\mathcal{L}t}$ is a *unitary dilation* [10] of the contracting semigroup e^{iLt} and the operator $\mathcal{L}$ is the *self-adjoint dilation* of the dissipative operator L. The constructed dilation is *minimal* – i.e., it does not have proper self-adjoint parts – if the subspace E is a generating subspace of the operator A.

We can construct the symmetric spectral representation for the original dissipative operator following the pattern of the previous section. We begin with description of eigenfunctions of the dilation.

The space $\mathcal{E}$ of the dilation $\mathcal{L}$ may be decomposed into orthogonal sum of invariant subspaces generated by incoming and outgoing waves and corresponding complementary ("radiating" and "absorbing") components $\mathcal{E} = \mathcal{E}_- \oplus \mathcal{E}^< = \mathcal{E}_+ \oplus \mathcal{E}^>$. The spectrum of $\mathcal{L}_\pm$ in each of components $\mathcal{E}_\pm$ is absolutely continuous with the constant multiplicity $\dim\Gamma$ on the whole real axis R. The spectrum of the "absorbing" and "radiating" components $\mathcal{L}^>$, $\mathcal{L}^<$ in subspaces $\mathcal{E}^>$, $\mathcal{E}^<$ consists of intervals of real axis where nonzero generalized solutions of the homogeneous equation $\mathcal{L}\psi - \lambda\psi = 0$ exist which vanish on $L_2(R_+)$ (for $\psi^>$) or vanish on $L_2(R_-)$ (for $\psi^<$). The corresponding eigen-functions of the dilation in each component $\mathcal{E}_\pm$ can be found, according to philosophy developed in [5, 4, 3] as elements of some rigged space constructed with a help of some Hilbert-Schmidt operator $\mathcal{T}$ which has a dense range, i.e., with all non-zero eigenvalues. This general statement can be specified in our case by selection of a special class of eigenfunctions which play a role of Scattered waves. This result can be obtained via selection of a special rigging (i.e., the operator $\mathcal{T}$) correlated with the imaginary part of the considered dissipative operator. Without loss of generality we can assume that the operator $\Gamma = \Gamma^{+}$ is a part of the positive Hilbert-Schmidt operator $\mathcal{T}$ acting in K, $\langle Ku, u\rangle > 0$. Moreover we can assume that the operator $\mathcal{T}^{2} > 0$ is presented as a product of an operator of the trace class and an operator from Matsaev class, so that its eigenvalues are $s_n(\mathcal{T}^{2}) = O(\alpha_n \beta_n)$ with $\sum_n |\alpha_n| < \infty$, $\sum_n \beta_n/n < \infty$ and β_n tend to zero monotonically. Consider the Gelfand triple [5] associated with the operator $\mathcal{T}$ as $\mathcal{T}K = \mathcal{K}_1 \subset K \subset \mathcal{K}^{1} = \mathcal{T}^{-1}K$. Then the following statement is true:

Theorem 3.2. *The incoming and outgoing eigen-functions of the ditation $\mathcal{L}$ can be presented as generalized solutions of the corresponding homogeneous equation with exponential behavior in $L_2(R_\pm, E)$:*

$$\psi_-(e) = \begin{cases} e^{-ikx}e, & e \in E, \quad x \in R_-, \\ u_-(e) & \text{in} \quad \mathcal{K}^{1}, \\ e^{-ikx}\mathbf{S}^+e, & e \in E, \quad x \in R_+, \end{cases}$$

$$\psi_+(e) = \begin{cases} e^{-ikx}\mathbf{S}e, & e \in E, \quad x \in R_-, \\ u_+(e) & \text{in} \quad \mathcal{K}^{1}, \\ e^{-ikx}e, & e \in E, \quad x \in R_+. \end{cases} \tag{3.2}$$

These eigenfunctions are labelled by the "direction vectors" [3] *$e \in E$. The mid-components $u_\mp$ are generalized solutions of the non-homogeneous equation in complex plane and are uniquely defined by the direction vectors $e \in E$, see (3.4) below, as $\mathcal{T}^{-1}$ images of strong limits of properly framed resolvent of the self-adjoint operator A or the resolvent of L, L^{+} on the real axis from the lower (upper) half-planes. The transmission coefficients $\mathbf{S}$, $\mathbf{S}^+$ are also uniquely defined from the homogeneous equation. In particular, $\mathbf{S}$, $\mathbf{S}^+$ are analytic matrix-function in upper and lower half-planes $\Im k > 0$, $\Im k < 0$*

$$\mathbf{S}^+(k - i0) = I - i \lim_{\lambda \to k - i0} \Gamma \frac{I}{L - \lambda I} \Gamma^+ = \lim_{\lambda \to k - i0} \frac{I - \frac{i}{2}\Gamma \frac{I}{A - \lambda I}\Gamma^+}{I + \frac{i}{2}\Gamma \frac{I}{A - \lambda I}\Gamma^+},$$

$$\mathbf{S}(k + i0) = I + i \lim_{\lambda \to k + i0} \Gamma \frac{I}{L^+ - \lambda I} \Gamma^+. \tag{3.3}$$

$$u_-(e) = -\frac{1}{2}\frac{1}{A - (k - i0)}\left(I + \mathbf{S}^+(k - i0)\right)e = -\frac{I}{L - (k - i0)}\Gamma^+ e, \; e \in E.$$

$$u_+(e) = -\frac{1}{2}\frac{I}{A - (k + i0)}\left(I + \mathbf{S}(k + i0)\right)e = -\frac{I}{L^+ - (k + i0)}\Gamma^+ e, \; e \in E. \tag{3.4}$$

[3] The term is borrowed from [2].

The eigenfunctions $\psi^{>}$, $\psi^{<}$ of components of the dilation in complementary subspaces $\mathcal{E}\ominus\mathcal{E}_{-}=\mathcal{E}^{<}$ and $\mathcal{E}\ominus\mathcal{E}_{+}=\mathcal{E}^{>}$ have a form:

$$\psi^{<}=\left(\begin{array}{c}0\\ u^{<}\\ e^{-ikx}e^{<}\end{array}\right),\ \psi^{>}=\left(\begin{array}{c}e^{-ikx}e^{>}\\ u^{>}\\ 0\end{array}\right), \tag{3.5}$$

when choosing vectors $e^{>}$, $e^{<}$ as eigenvectors of operators $\Delta^{>}=I-\mathbf{S}^{+}\mathbf{S}$, $\Delta^{<}=I-\mathbf{S}\mathbf{S}^{+}$ with non-zero eigenvalues $\delta^{>}$, $\delta^{<}$ respectively, we obtain:

$$u^{>}(e^{>})=\frac{1}{\delta^{>}}\left[u_{-}(e^{>})-u_{+}(\mathbf{S}^{+}e^{>})\right],$$

$$u^{<}(e^{<})=\frac{1}{\delta^{<}}\left[u_{+}(e^{<})-u_{-}(\mathbf{S}e^{<})\right].$$

Proof. It is easy to verify the above formulae for the eigenfunctions on a formal level, see for instance [14]. Note that analysis of the absolutely-continuous spectrum of the symmetric model in the rigged space is presented in [17, 18, 15]. We suggest below only the sketch of the proof of existence of the scattered wave of the dilation (actually, the proof of existence of their mid-components) based on the Theorem 7 from [9]. It is proved in that theorem, in particular, that the non-tangential limits exist on the real axis for the operator-valued R-function presented by the properly framed resolvent of a self-adjoint operator. The remark attached to the theorem shows that the statement remains true for the limit of the R-function

$$\mathcal{T}\,P_{K}\frac{I}{\mathcal{L}-\lambda I}P_{K}\,\mathcal{T}=$$

$$\mathcal{T}\,\frac{I}{L^{+}-\lambda I}\,\mathcal{T}$$

from the upper half-plane $\lambda\to k+i0$ and for the adjoint function $\mathcal{T}\,\frac{I}{L-\lambda I}\,\mathcal{T}$ from the lower half-plane, $\lambda\to k-i0$. Now it is easy to verify the existence of the mid-component u_{+} of the outgoing scattered wave. Really, the limit

$$\lim_{\lambda\to k-i0}\ \mathcal{T}\frac{I}{L^{+}-\lambda I}\mathcal{T}$$

exists in the trace class, hence the mid-component can be presented as

$$u_{+}=\mathcal{T}^{-1}\lim_{\lambda\to k-i0}\ \mathcal{T}\frac{I}{L^{+}-\lambda I}\mathcal{T}e\in\mathcal{K}^{1}.$$

Similarly the mid-component of the scattered wave ψ_{-} can be obtained.The mid-components $u^{<}$, $u^{>}$ of the eigenfunctions in the complementary subspaces can be obtained as linear combinations of them with proper coefficients and properly chosen direction vectors. □

Remark. One can see that the above calculation can be applied to the situation when the imaginary part of the dissipative operator is a positive operator presented as a part in E of the positive operator $\mathcal{T}$ such that the square $\mathcal{T}^{2}$ of it is product of an operator with a finite trace and an operator from Macaev class. This is actually the natural class of dissipative operators for which the symmetric functional model may be obtained by the procedure described above. This class can be extended via considering the corresponding relative classes with the imaginary part subordinated to the real part. It will be done elsewhere.

Based on the explicit formulae for the eigenfunctions one can prove that the characteristic function obtained above as a stationary transmission coefficient can be also interpreted in non-stationary terms.

Considering the non-perturbed shift generator in the space $D_{\text{in}} \oplus D_{\text{out}} = L_2(R_-, E) \oplus L_2(R_+, E)$:

$$\mathcal{L}_0 = i\frac{d}{dx}.$$

Then the characteristic function of the original dissipative operator as Adamjan-Arov scattering matrix, see [1] for the pair $\mathcal{L}$, $\mathcal{L}_0$ is

$$s - \lim_{t\to\infty} \mathcal{J}_0 e^{-i\mathcal{L}_0 t} P_+ e^{2i\mathcal{L}t} P_- e^{-i\mathcal{L}_0 t} \mathcal{J}_0^+.$$

Theorem 3.3. *The Adamjan-Arov scattering matrix for the pair $\mathcal{L}$, $\mathcal{L}_0$ coincides with the transmission coefficient*[4]*:*

$$\mathbf{S}^+(k - i0) = I - i \lim_{\lambda\to k-i0} \Gamma\frac{I}{L-\lambda I}\Gamma^+ =$$

$$= \lim_{\lambda\to k-i0} \frac{I - \frac{i}{2}\Gamma\frac{I}{A-\lambda I}\Gamma^+}{I + \frac{i}{2}\Gamma\frac{I}{A-\lambda I}\Gamma^+}, \tag{3.6}$$

Proof is obtained by the straightforward calculation using the fact that the spectral representation $\mathcal{J}_0$ for the non-perturbed operator is defined by Fourier transform. Hence the scattering matrix coincides with the transmission coefficient $\mathbf{S}$ in front of the exponential e^{-ikx} in the formula for the scattered wave ψ_- in the outgoing subspace. □

We construct now the symmetric functional model for the original operator L based on eigenfunctions $\psi_\pm$ of its self-adjoint dilation, see (3.2).

Theorem 3.4. *Consider the maps $\mathcal{J}_\pm$ of the spaces $L_2(E)$ into $\mathcal{E}_\pm$:*

$$\mathcal{J}_+ h_+ = \frac{1}{\sqrt{2\pi}}\int_{-\infty}^{\infty} \psi_+ (h_+(p))\, dp, \ h_+ \in L_2(E),$$

$$\mathcal{J}_- h_- = \frac{1}{\sqrt{2\pi}}\int_{-\infty}^{\infty} \psi_- (h_-(p))\, dp, \ h_- \in L_2(E),$$

and the map $\mathcal{J}$ of the column $\begin{pmatrix} h_+ \\ h_- \end{pmatrix} := \mathbf{h}$ *into $\mathcal{E}$:*

$$\mathcal{J}\mathbf{h} = \mathcal{J}_+ h_+ + \mathcal{J}_- h_-.$$

[4]The numerator and denominator of the announced representation for the Scattering matrix are commuting, so the order of them is not important.

Then the following Parseval identity is true:

$$\langle \mathcal{J}\mathbf{f},\, \mathcal{J}\mathbf{g}\rangle = \langle f_+,\, g_+\rangle + \langle \mathbf{S}^+ f_-,\, g_+\rangle + \langle \mathbf{S} f_+,\, g_-\rangle + \langle f_-,\, g_-\rangle =$$

$$\int \langle \begin{pmatrix} \mathbf{I} & \mathbf{S}^+ \\ \mathbf{S} & \mathbf{I} \end{pmatrix} \mathbf{f},\ \mathbf{g}\rangle_{E\oplus E}\, dk \tag{3.7}$$

Proof. Note that for $h_- \in H^2_-(E)$

$$\mathcal{J}\begin{pmatrix} h_- \\ 0 \end{pmatrix} = \begin{pmatrix} h_-(x) \\ 0 \\ 0 \end{pmatrix}$$

with $h_-(x) = \int_{-\infty}^{\infty} e^{-ikx} h_-(k)dk$ non equal to zero identically if $x \in R_-$. Similarly for $h_+ \in H^2_+(E)$

$$\mathcal{J}\begin{pmatrix} 0 \\ h_+ \end{pmatrix} = \begin{pmatrix} 0 \\ 0 \\ h_+(x) \end{pmatrix}$$

with $h_+(x) = \int_{-\infty}^{\infty} e^{-ikx} h_+(k)dk \neq 0$ if $x \in R_+$. Hence incoming and outgoing subspaces$\mathcal{D}_{\text{in, out}} = L_2(R_\pm)$ are mutually orthogonal. The invariant subspaces $\mathcal{E}_\pm \in \mathcal{E}$ of the dilation developed from the incoming and outgoing subspaces,

$$\mathcal{E}_\pm = \bigvee_{t=-\infty}^{t=\infty} \int_{-\infty}^{\infty} e^{ikt} \psi_\pm(h_\pm)\, dk,\ h_\pm \in H^2_\pm,$$

are represented as

$$\mathcal{E}_+ = \mathcal{J}\begin{pmatrix} 0 \\ L_2 \end{pmatrix},\ \mathcal{E}_- = \mathcal{J}\begin{pmatrix} L_2 \\ 0 \end{pmatrix}.$$

Then for $f_+ \in H^2_+$, $g_- \in H^2_-$ we obtain:

$$\langle \mathcal{J}\begin{pmatrix} 0 \\ f_+ \end{pmatrix}, \mathcal{J}\begin{pmatrix} e^{ikt} g_- \\ 0 \end{pmatrix}\rangle$$

$$= \int_{-\infty}^{0} \langle \mathbf{S} f_+, e^{ikt} g_-\rangle_{L_2} dx$$

for any finite t. Following the pattern of the previous section one may derive from it that for any $f_+, g_- \in L_2$

$$\langle \mathcal{J}\begin{pmatrix} 0 \\ f_+ \end{pmatrix}, \mathcal{J}\begin{pmatrix} g_- \\ 0 \end{pmatrix}\rangle$$

$$= \frac{1}{2\pi} \int_{-\infty}^{0} dx \int dk \int d\hat{k} e^{-ikx} e^{i\hat{k}x} \mathbf{S}(k) f_+ \bar{g}_-$$

$$+ \frac{1}{2\pi} \int_0^{\infty} dx \int dk \int d\hat{k} e^{-ikx} e^{i\hat{k}x} f_+ \overline{\mathbf{S}^+(\hat{k}) g_-}$$

$$+ \langle u_+(f_+), u_-(g_-)\rangle_K = \langle \mathbf{S} f_+,\, g_-\rangle_{L_2},$$

since $\langle u_+(f_+), u_-(g_-)\rangle_K = 0$, and remaining integrals over semi-axes should be combined to the delta-function $\delta(k - \hat{k})$. This way the announced statement is verified for special elements

$$\begin{pmatrix} 0 \\ f_+ \end{pmatrix}, \begin{pmatrix} g_- \\ 0 \end{pmatrix}.$$

The proof is accomplished based on similar arguments for various choice of special elements and linearity of the map $\mathcal{J}$. □

Consider the non-perturbed operator $\mathcal{E}_0 = \mathcal{L}_0 \oplus \mathcal{L}_0$ in $L_2(R, E) \oplus L_2(R, E)$. The corresponding evolution group $U_t : u(x) \to u(x - t)$ has unilateral invariant subspaces $L_2(R_-, E) := \mathcal{D}_{\text{in}}$, $L_2(R_+, E) := \mathcal{D}_{\text{out}}$. We denote by $P_{\text{in, out}}$ the orthogonal projections onto $\mathcal{D}_{\text{in ,out}}$ respectively. Similarly to the above reasoning in section 2 we calculate the symmetric spectral representation via Arov-Adamyan wave operators, see [1].

Theorem 3.5. *The wave-operators*

$$W_- = s - \lim_{t \to -\infty} e^{-i\mathcal{L}_0 t} P_{\text{in}} e^{i\mathcal{L} t}$$

$$W_+ = s - \lim_{t \to \infty} e^{-i\mathcal{L}_0 t} P_{\text{out}} e^{i\mathcal{L} t}$$

exist as strong limits and are isometric operators from the invariant subspaces $\mathcal{E}_{\text{in, out}} \subset \mathcal{E}$ *obtained by development of the incoming and outgoing subspaces* $L_2(R_-, E)$ *and* $L_2(R_+, E)$ *with evolution generated by* $\mathcal{L}$. *The column*

$$\begin{pmatrix} f_{\text{in}} \\ f_{\text{out}} \end{pmatrix} := \mathbf{f}$$

defines the symmetric spectral map as

$$\mathcal{J}\mathbf{f} = \mathcal{J}_- f_{\text{in}} + \mathcal{J}_+ f_{\text{out}}$$

which is calculated from the column of Cauchy data as $\begin{pmatrix} f_0 \\ f_1 \end{pmatrix} = \mathbf{f}(0)$ *as*

$$\begin{pmatrix} f_{\text{in}} \\ f_{\text{out}} \end{pmatrix} = \begin{pmatrix} \mathcal{F} W_- \mathbf{f}(0) \\ \mathcal{F} W_+ \mathbf{f}(0) \end{pmatrix}$$

where $\mathcal{F}$ *is the standard Fourier transform in* L_2*:*

$$f(x) \to \frac{1}{\sqrt{2\pi}} \int e^{ikx} f(x) dx = f(k).$$

Note that we suggested a unique recipe of construction of coordinates of the symmetric spectral representation of the dilation, but, once constructed, the column of coordinates may be a subject to change within proper limits caused by possible presence of intervals where the scattering matrix is unitary, see the discussion in [14].

Remark. Note that the eigenfunctions of the complementary component are found uniquely, up to the parametrization with the direction vectors. Their mid-components $u^<$, $u^>$ $\mathcal{E}^<$, $\mathcal{E}^>$ may serve a *canonic system of eigenfunctions* of the absolutely continuous spectrum of the original dissipative operator and adjoint operator, respectively. The corresponding spectral expansion

$$u = \frac{1}{2\pi} \int_{\sigma_a} \frac{|\mathbf{S}(k)|^2 - 1}{\mathbf{S}^+(k)} u^<(k) \langle u, u^>(k) \rangle dk, \tag{3.8}$$

is converging for elements u represented as orthogonal projections of elements of the complementary subspace $\mathcal{E}^<$ onto K. This set is dense in the absolutely continuous subspace of the operator L, see [10] and the detailed discussion of the eigenfunction expansion of the dissipative Schrödinger operator with complex potential in [14, 18]. Thus the incoming-outgoing eigenfunctions of the dilation and eigenfunctions in the complementary subspaces $\mathcal{E}^<$, $\mathcal{E}^>$ play essentially different roles in spectral problem for the dissipative operator. The above formula (3.8) shows that the problem of proper choice of the *canonic system* of eigenfunction of the absolutely-continuous spectrum for dissipative operators is naturally resolved. Note that similar question about a *canonic system* of eigenfunctions of absolutely continuous spectrum of a self-adjoint operator remains obscure. The only bridge between the General Spectral Theorem for self-adjoint operators and the *expansion theorem* is formed by classical results of I. Gelfand-A. Kostyuchenko [3] on differentiation of the spectral measure of a self-adjoint operator in properly rigged spaces. We hope to discuss the important question of the construction of the canonic system of eigenvectors of the abstract self-adjoint operator somewhere. For discussion of choice of the canonic system of eigenfunctions of the absolutely continuous spectrum in case of spectral multiplicity one for a unitary operator and a canonic system of eigenfunction of its contracting perturbation see [16].

Acknowledgement

The author is grateful to the referee for the suggestion to rewrite one of statements in more formal manner (Theorem 3.2).

References

[1] V. Adamjan, V. Arov *On unitary couplings of semi-unitary perators*, Matematicheskie Isslenovanija, **1**, 2, 1966, 3–64.

[2] N.I. Akhiezer, I.M. Glazman, *Theory of linear operators in Hilbert space.* Translated from the Russian and with a preface by Merlynd Nestell. Reprint of the 1961 and 1963 translations. Bd. 1, 2 Dover Publications, Inc., New York, 1993.

[3] I.M. Gelfand, A.G. Kostyuchenko *Expansion in eigenfunctions of differential operators* In: Dokl. AN SSSR **103** (1955), 349–352.

[4] I.M. Gelfand *Some questions of analysis and differential equations* In: Uspekhi Mat. nauk (in Russian) **14**, 3 (87), 1959, 3–19.

[5] I.M. Gelfand, G.E. Shilov *generalized functions Volume 3: Theory of differential equations*, Translated from Russian by Menhard E. Mayer, Academic Press [Harcourt Brace Iovanovich Publisher] New-York-London, 1967 [1977]

[6] S. Khruschev, N. Nikolskij A functional model and some problems of spectral theory of functions, Trudy Mat. Inst. Steklov, **176**, 1987. 97–210.

[7] P. Lax, R. Phillips *Scattering theory* Academic press, New York, 1967.

[8] S. Naboko A functional model of perturbation theory and its application to scattering theory, Proceedings of Steklov Institute, **2**, 1981, 85–116.

[9] S.N. Naboko *Non-tangential boundary values of operator-valued R-functions in a half-plane*, Leningrad Math. Journal, Vol. 1, **5**, 1990, 1255–1277.

[10] B. Sz.-Nagy, C. Foias *Analyse harmonique des operateurs de l'espace de Hilbert*, Academiai Kiado, Budapest, 1970.

[11] N. Nikolski, Operators, Functions and Systems: An Easy Reading, AMS, 2002, Vol **2**.

[12] B. Pavlov, *The continuous spectrum of resonances on a non-physical sheet* Dokl. AN USSR, **206**, 1972, 1301–1304.

[13] B. Pavlov, *On conditions of separability of the spectral componentrs of a dissipative operators*, Izv. AN USSR, **39**, 1975, 123–148.

[14] B. Pavlov, *Self-adjoint dilation of the dissipative Schroedinger operator and its resolution in terms of eigenfunctions*, Math. USSR Sbornik, v31, 4, 457–478 (1977).

[15] B. Pavlov, Spectral Analusis of a Dissipative Schrödinger operator in terms of a functional model, in the book: "Partial Differential Equations", ed. by M. Shubin in the series Encycl. Math. Sciences, **65**, Springer 1995, 87–153.

[16] B.Pavlov, M.Faddeev, *Spectral analysis of unitary perturbations of contractions*, Proc. LOMI, v115,1982. (English translation in: J. of Sov. Math. vol. 28, 768–776 (1985))

[17] V.A. Ryzhov *Construction of the functional model of non-self-adjoint operator with nonempty regular set*, Dep. VINITI 21.12.98, 389–B90, St. Petersburg (1990), 92 pages.

[18] V.A. Ryzhov Absolutely-continuous subspace of the hon-self-adjoint operator and Scattering theory, PhD Thesis, supervised by S. Naboko, St. Petersburg 1994, 147 pages.

[19] E.C. Titchmarsh *Eigenfunctions expansions associated with second-order differential equations*, Part1, Clarendon Press, Oxford(1946) 247 pages.

[20] P.Lax, R. Phillips *Scattering Theory* Academic press, New-York – London,1967.

Boris Pavlov
Department of Mathematics University of Auckland
Private Bag 92019
Auckland, New Zealand
e-mail: pavlov@math.auckland.ac.nz

Operator Theory:
Advances and Applications, Vol. 154, 179–184

A Remark on Equivalence of Weak and Strong Definitions of the Absolutely Continuous Subspace for Nonself-adjoint Operators

Roman Romanov

Abstract. We prove the equivalence of weak and strong definitions of the absolutely continuous subspace for nonself-adjoint dissipative operators.

Mathematics Subject Classification (2000). 47B44.

Keywords. absolutely continuous spectrum, dissipative operators.

In this note we analyze two known natural definitions of the absolutely continuous (a.c.) subspace for an abstract nonself-adjoint operator. So far, their equivalence has been proved [1] under certain strong assumptions about boundary behavior of the resolvent. Here we establish the equivalence without any additional assumptions in the case of dissipative operators.

Throughout the paper, L is a closed operator in a Hilbert space H such that $\sigma_{ess}(L) \subset \mathbb{R}$. In particular, the resolvent of L is defined for all $z \notin \mathbb{R}$ except for at most countably many points as a bounded operator in H. The vector Hardy classes $\mathbf{H}^2_\pm$ are the collections of analytic functions $f : \mathbb{C}_\pm \to H$ satisfying $\sup_{\varepsilon>0} \int_{\mathbb{R}} \|f(k \pm i\varepsilon)\|^2 \, dk < \infty$, respectively.

Definition 1. The subspace

$$H^w_{ac}(L) \stackrel{def}{=} \operatorname{clos} \widetilde{H^w_{ac}}(L),$$

$$\widetilde{H^w_{ac}}(L) \stackrel{def}{=} \left\{ \begin{array}{ll} u \in H : & (i)\, (L-z)^{-1} u \text{ is analytic in } \mathbb{C} \setminus \mathbb{R} \\ & (ii)\, \left\langle (L-z)^{-1} u, v \right\rangle \Big|_{\mathbb{C}_\pm} \in \mathbf{H}^2_\pm \text{ for all } v \in H \end{array} \right\},$$

is called the *weak a.c. subspace* of the operator L. Elements of the linear set $\widetilde{H^w_{ac}}(L)$ are called weak smooth vectors of L.

This work was supported in part by EPSRC Grant GR/ R20885 and by RFBR Grant 03-01-00090.

The weak a.c. subspace has first been introduced and studied by A. Tikhonov [2]. In the case of a self-adjoint L, it is well known, see, e.g., [3], that the definition above agrees with the standard one.

Another, historically first, definition of the a.c. subspace suggested by L. Sakhnovich [4] initially referred to the case when L is dissipative. A convenient equivalent formulation and an extension of it to general nonself-adjoint case was given in [5]. For clarity, we first restrict our consideration to the situation of the perturbation theory and discuss the general case afterwards. Namely, except for two remarks following Theorem 2 it is assumed throughout that

- The operator L is a completely nonself-adjoint operator[1] of the form $L = A + iV$, $A = A^*$, $V = V^*$, $\mathcal{D}(L) := \mathcal{D}(A) \subset \mathcal{D}(V)$, and V is A-bounded with a relative bound less than 1, that is, $\|Vu\|^2 \le a\|Au\|^2 + b\|u\|^2$, $a < 1$, for all $u \in \mathcal{D}(A)$.

Definition 2. The subspace

$$H_{ac}(L) \overset{def}{=} \operatorname{clos} \widetilde{H_{ac}}(L),$$

$$\widetilde{H_{ac}}(L) \overset{def}{=} \left\{ u \in H : \begin{array}{l} (i)\ (L - z)^{-1} u \text{ is analytic in } \mathbb{C} \setminus \mathbb{R} \\ (ii)\ |V|^{1/2} (L - z)^{-1} u\Big|_{\mathbb{C}_\pm} \in \mathbf{H}^2_\pm \end{array} \right\},$$

is called the (*strong*) *a.c. subspace* of the operator L. Elements of the linear set $\widetilde{H_{ac}}(L)$ are called (strong) smooth vectors of L.

In the dissipative case the subspace $H_{ac}(L)$ coincides with the invariant subspace corresponding to the canonical factorization of the characteristic function of the operator L in the sense of the Szökefalvi-Nagy-Foias functional model [5, 6].

It is easy to see that $(L - \lambda)^{-1} H^w_{ac} \subset H^w_{ac}$, $(L - \lambda)^{-1} H_{ac} \subset H_{ac}$ for all $\lambda \in \rho(L)$. An important property of strong smooth vectors is expressed by the following

Proposition [5]. *There exists a Hilbert space $\mathcal{N}$, an a.c. self-adjoint operator A_0 in $\mathcal{N}$, and a bounded operator $P \colon \mathcal{N} \to H$,* $\operatorname{Ran} P = \widetilde{H_{ac}}$, *such that for all $g \in \mathcal{N}$ and $z \notin \mathbb{R}$, $z \in \rho(L)$*

$$(L - z)^{-1} P g = P (A_0 - z)^{-1} g.$$

Corollary 1. $H^w_{ac}(L) \supset H_{ac}(L)$.

Proof. Let $g \in \mathcal{N}$ satisfy $\frac{d\mu_g}{dt} \in L^\infty(\mathbb{R})$ where the measure $d\mu_g$ is the matrix element of the spectral measure of A_0 on the vector g (an L^∞-vector of A_0), and let $u = Pg$. The spectral theorem for operator A_0 gives

$$\left\langle (L - z)^{-1} u, v \right\rangle = \left\langle (A_0 - z)^{-1} g, P^* v \right\rangle = \int \frac{1}{k - z} \rho(k) dk,$$

[1] That is, L has no reducing subspaces on which it induces a self-adjoint operator.

where $\rho \in L^2$, by the choice of g. This shows that the restrictions of the left-hand side are in $\mathbf{H}^2_\pm$ in the respective half-planes for all v, that is, $u \in \widetilde{H^w_{ac}}$. The result follows since the set of L^∞-vectors is dense in $\mathcal{N}$ by absolute continuity of A_0. □

Theorem 2. *If L is dissipative ($V \geq 0$), then $H^w_{ac}(L) = H_{ac}(L)$.*

Using a functional model for nonself-adjoint operators due to Naboko [5], Ryzhov has proved [1] the equality $H^w_{ac} = H_{ac}$ assuming that the characteristic function, $\Theta(z)$, of L has weak non-tangential boundary values, $\Theta(k \pm i0)$, a.e. on the real axis. In the dissipative case, this condition is equivalent to the requirement that the inverse characteristic function has weak boundary values from above. It appears to have not been noticed so far that in the dissipative case the inclusion $H^w_{ac}(L) \subset H_{ac}(L)$ holds unconditionally. Our proof of it is elementary and is based on the following simple property of weak smooth vectors.

Lemma 3. *Let B be a closed operator such that $\sigma_{ess}(B) \subset \mathbb{R}$. Then*

$$\sup_{z \in \mathbb{C}_\pm} |\operatorname{Im} z| \left\| (B - z)^{-1} w \right\|^2 < \infty \tag{1}$$

for any $w \in \widetilde{H^w_{ac}}(B)$.

Proof. From the Riesz integral representation and Schwartz inequality we have (signs $\pm$ refer to $z \in \mathbb{C}_\pm$, respectively)

$$\left| \left\langle (B - z)^{-1} w, v \right\rangle \right| = \left| \int \frac{1}{k - z} \mu_\pm(k) dk \right| \leq \|\mu_\pm\|_{L^2(\mathbb{R})} \frac{\pi^{1/2}}{|\operatorname{Im} z|^{1/2}},$$

where $\mu_\pm \in L^2(\mathbb{R})$ satisfy $\|\mu_\pm\|_{L^2} \leq C_w \|v\|$ in view of the uniform boundedness principle. By arbitrariness of v, (1) follows. □

Proof of the theorem. We have to show that $H^w_{ac}(L) \subset H_{ac}(L)$. Define

$$u = (L - z_0)^{-1} w, \quad w \in \widetilde{H^w_{ac}}, \quad z_0 \in \mathbb{C}_-, \quad \text{then} \quad u \in \widetilde{H^w_{ac}}.$$

Let us integrate the identity ($z = k + i\varepsilon$)

$$\left\| V^{1/2} (L - z)^{-1} u \right\|^2 = \varepsilon \left\| (L - z)^{-1} u \right\|^2 - \operatorname{Im} \left\langle (L - z)^{-1} u, u \right\rangle$$

in k from $-N$ to N,

$$\begin{aligned} \int_{-N}^{N} \left\| V^{1/2} (L - z)^{-1} u \right\|^2 dk = \varepsilon \int_{-N}^{N} \left\| (L - z)^{-1} u \right\|^2 dk \\ - \operatorname{Im} \int_{-N}^{N} \left\langle (L - z)^{-1} u, u \right\rangle dk. \end{aligned} \tag{2}$$

Consider first the case when $\varepsilon > 0$. The first term in the r.h.s., denoted by (I), is estimated as follows

$$(I) = \varepsilon \int_{-N}^{N} \left\| (L-z)^{-1} (L-z_0)^{-1} w \right\|^2 dk \le \int_{-N}^{N} \frac{1}{|z-z_0|^2} \varepsilon \left\| (L-z)^{-1} w \right\|^2 dk$$

$$+\varepsilon \left\| (L-z_0)^{-1} w \right\|^2 \int_{-N}^{N} \frac{dk}{|z-z_0|^2} \le C_w \frac{1+\varepsilon}{|\varepsilon - \operatorname{Im} z_0|} \le C_w.$$

Here we have taken (1) into account. We conclude that (I) is bounded uniformly in N and ε. In a similar way, we have

$$\int_{-N}^{N} \left\langle (L-z)^{-1} u, u \right\rangle dk = \int_{-N}^{N} \frac{1}{z-z_0} \left(\langle (L-z)^{-1} w, u \rangle - \langle u, u \rangle \right) dk$$

$$= \int_{-N}^{N} \frac{1}{z-z_0} \left\langle (L-z)^{-1} w, u \right\rangle dk - \ln \frac{N + i\varepsilon - z_0}{-N + i\varepsilon - z_0} \|u\|^2.$$

When $N \to \infty$, the first term in the right-hand side vanishes, the second tends to $-i\pi \|u\|^2$, hence the left-hand side has a finite limit, and this limit is independent of ε. Combining these, we find that the limit $N \to \infty$ of the left-hand side in (2) exists and is bounded in ε. This and a similar consideration for the case[2] $\varepsilon < 0$ yield that $u \in \widetilde{H_{ac}}$. It remains to notice that $\bigvee_{z_0 \in \mathbb{C}_-} (L-z_0)^{-1} \widetilde{H_{ac}^w} = H_{ac}^w$ because

$$i\tau (L + i\tau)^{-1} \xrightarrow{s} I, \ \tau \to +\infty. \tag{3}$$

The latter easily follows from dissipativity of L [7]. □

Remark. Let L be an arbitrary maximal dissipative completely nonself-adjoint operator. The definition of the strong a.c. subspace of L is obtained [1] by substituting the condition *(ii)* in definition 2 with the following one (see (2))

$$\sup_{\varepsilon > 0} \int_{\mathbb{R}} \left(\varepsilon \left\| (L-k-i\varepsilon)^{-1} u \right\|^2 - \operatorname{Im} \left\langle (L-k-i\varepsilon)^{-1} u, u \right\rangle \right) dk < \infty.$$

With this definition, Theorem 2 holds in the general case, with the same proof.

Remark. The proof of Theorem 2 shows that the linear sets of smooth and weak smooth vectors of a maximal dissipative operator L satisfy $(L-z_0)^{-1} \widetilde{H_{ac}^w}(L) \subset \widetilde{H_{ac}}(L)$ for all non-real $z_0 \in \rho(L)$.

In the situation of the perturbation theory, the following simple sufficient condition of triviality of the subspace $H_{ac}^w(L)$ is useful.

Proposition 4. $H_{ac}^w(L) = \{0\}$ *if for a. e.* $k \in \mathbb{R}$ *we have* $(z = k + i\varepsilon)$

$$D(z) \equiv \sqrt{\varepsilon} (L^* - z)^{-1} |V|^{1/2} \xrightarrow{s} 0 \tag{4}$$

as $\varepsilon \downarrow 0$.

---

[2]Because of the dissipativity of L, it actually suffices to consider the case $\varepsilon > 0$ only, for the function $V^{1/2} (L-z)^{-1} u$ restricted to $\mathbb{C}_-$ is in $\mathbf{H}^2_-$ for all $u \in H$ [5]. In order to keep the elementary nature of the proof, we prefer not to use this fact, since the proof of it in [5] is based on the existence of a self-adjoint dilation of L.

In the case when L is dissipative the condition (4) is also necessary, see [8]. It was used in [9] to establish triviality of the a.c. subspace for dissipative Schrödinger operators with slowly decaying imaginary part of the potential.

Proof. Given a $w \in \widetilde{H^w_{ac}}$ and $v \in \mathcal{D}(V)$, define the functions $F_\pm(z)$ to be the restrictions of the function $\left\langle (L-z)^{-1} w, |V|^{1/2} v \right\rangle$ to $\mathbb{C}_\pm$, respectively. We shall show that if (4) is satisfied, then $F_\pm$ vanish identically for all $v \in H$. Let us first derive the result from this assertion. We have, $\left\langle w, (L^*-z)^{-1} |V|^{1/2} v \right\rangle = \left\langle (L-\overline{z})^{-1} w, |V|^{1/2} v \right\rangle = 0$ for all non-real $z \in \rho(L)$ and $v \in \mathcal{D}(V)$. On the other hand, the closed linear hull of the linear sets $(L^*-z)^{-1} \operatorname{Ran} V$, $z \notin \mathbb{R}$, $z \in \rho(L)$, coincides with H, since L is a completely nonself-adjoint operator [5]. Hence, $w = 0$, as required.

Since $F_\pm \in \mathbf{H}^2_\pm$ by the choice of w, it suffices to show that the boundary values $F_+(k) = F_-(k)$ for a.e. $k \in \mathbb{R}$. Indeed, we have ($z = k + i\varepsilon$, $\varepsilon > 0$),

$$\begin{aligned} F_+(z) - F_-(\overline{z}) &= 2i\varepsilon \left\langle (L-\overline{z})^{-1} (L-z)^{-1} w, |V|^{1/2} v \right\rangle \\ &= 2i \left\langle \sqrt{\varepsilon}\, (L-z)^{-1} w, D(z) v \right\rangle \underset{\varepsilon \downarrow 0}{\longrightarrow} 0 \end{aligned}$$

for a.e. $k \in \mathbb{R}$. □

References

[1] V.A. Ryzhov, On a singular and an absolutely continuous subspace of a non-self-adjoint operator whose characteristic function has boundary values on the real axis. *Funktsional. Anal. i Prilozhen.* **32** (1998), no. 3, 83–87 (*Russian*); English transl. in: *Funct. Anal. Appl.* **32** (1999), No. 3, 208–212.

[2] A.S. Tikhonov, Absolutely continuous spectrum of a linear operator and problems of scattering theory and factorization of operator-functions. Ph. D. Thesis, Simferopol State University, Simferopol (1989) (*in Russian*).

[3] M. Reed and B. Simon, "Methods of Modern Mathematical Physics", v. 4, Academic Press, New York, 1978.

[4] L.A. Sahnovič, Dissipative operators with absolutely continuous spectrum, *Trans. Moscow Math. Soc.* **19** (1968).

[5] S.N. Naboko, A functional model of perturbation theory and its applications to scattering theory, *Trudy MIAN* **147** (1980), 86–114 (*Russian*); English transl. in: *Proc. Steklov Inst. Math.* (1981), No. 2, 85–116.

[6] B. Sz.-Nagy and C. Foias, "Analyse Harmonique des Opérateurs de l'Espase de Hilbert", Masson et Cie/ Academiai Kiado, 1967.

[7] E. Hille and R. Phillips, "Functional analysis and semi-groups", AMS Colloquium Publications, Vol. XXXI, AMS, Providence, R.I., 1974.

[8] B.S. Pavlov, Self-adjoint dilation of the dissipative Schrödinger operator and its resolution in terms of eigenfunctions, *Math. USSR Sbornik* **31** (1977), No. 4, 457–478.

[9] R. Romanov, On instability of the absolutely continuous spectrum of dissipative Schrödinger operators and Jacobi matrices, submitted.

Roman Romanov
School of Computer Science
Cardiff University, Cardiff
Queen's Buildings, PO Box 916
Newport Road, Cardiff CF24 3XF
UK
and
Laboratory of Quantum Networks
Institute for Physics
Saint Petersburg State University
198904 St. Petersburg
Russia
e-mail: `roma@rvr.stud.pu.ru`, `R.V.Romanov@cs.cf.ac.uk`

Operator Theory:
Advances and Applications, Vol. 154, 185–201

On a Transformation of the Sturm-Liouville Equation with Slowly Decaying Potentials and the Titchmarsh-Weyl m-function

Alexei Rybkin

Dedicated to the memory of F. Atkinson

Abstract. We put forward a new transformation of the half-line Sturm-Liouville equation with non-smooth potentials from L_p with $p \geq 2$. This transformation yields existence of the Weyl solution with higher order WKB-type asymptotic behavior (spatial and spectral parameter). We apply our approach to the study of high-energy asymptotics for the Titchmarsh-Weyl m-function, improving on some relevant results of others.

Mathematics Subject Classification (2000). 34E10, 34E20, 34B20, 34L40.

Keywords. Sturm-Liouville problem, Schrödinger operator, long-range potentials, WKB-type asymptotics, Titchmarsh-Weyl m-function.

1. Introduction

We will be concerned with the Sturm-Liouville equation

$$-u'' + q(x)\,u = \lambda^2 u \text{ for } x \in \mathbb{R}_+ := (0,\infty) \tag{1.1}$$

with real potentials q essentially bounded on $\mathbb{R}_+$. With applications to the spectral analysis in mind we will be especially interested in the behavior of the solutions to (1.1) at infinity. By variation of parameters (1.1) formally transforms to the integral equation

$$y(x,\lambda) = 1 + \int_x^\infty K(x,s,\lambda)\, y(s,\lambda)\, ds, \tag{1.2}$$

where $y := e^{-i\lambda x} u$ and

$$K(x,s,\lambda) := \frac{e^{-2i\lambda(s-x)} - 1}{2i\lambda}\, q(s). \tag{1.3}$$

If $q \in L_1(\mathbb{R}_+)$ then (1.2) can be solved by iteration and hence the original equation (1.1) has the unique solution $u(x,\lambda)$ such that

$$u(x,\lambda) \sim e^{i\lambda x}, x \to \infty, \tag{1.4}$$

for all $\lambda \in \mathbb{R}$. Such a solution is referred to as Jost and its existence is the main feature of the short-range scattering. Unfortunately if we go beyond the condition $q \in L_1(\mathbb{R}_+)$ the transformation from (1.1) to (1.2) becomes singular and in general one has to look for some other transformations. Eq. (1.1) with non-smooth potentials from $L_p(\mathbb{R}_+)$ with some $p < 2$ was the main object of the recent research due to Christ-Kiselev (see, e.g., [7]). They found a sequence of transformations for (1.1) that, combined with some subtle results on almost everywhere convergence of certain integral, yields a series solution $u(x,\lambda)$ to (1.1), which is absolutely convergent for a.e. real λ and has the WKB-type asymptotic behavior

$$u(x,\lambda) \sim \exp\left\{ i\lambda x + \frac{1}{2i\lambda}\int_0^x q(s)\,ds \right\}, x \to \infty, \tag{1.5}$$

for a.e. $\lambda \in \mathbb{R}$. By the Gilbert-Pearson subordinacy theory [12] the existence of asymptotics (1.5) implies that the absolutely continuous spectrum of the Schrödinger operators defined by (1.1) is $\mathbb{R}_+$. Note that if q is merely L_2 then it is still an open question if (1.5) holds for a.e. $\lambda \in \mathbb{R}$. In the general case of $q \in L_p(\mathbb{R}_+), p > 2$, we can only claim that $\mathbb{R}_+$ is the essential spectrum and hence solutions with behavior (1.5) for real λ need not exist.

Of course, if we assume some smoothness of q then the situation may considerably improve. For instance, the Green-Liouville transformation of (1.1) (see, e.g., [25])

$$y'' + y + \left[\frac{1}{4}\frac{q''(x)}{\{\lambda^2 - q(x)\}^2} + \frac{5}{16}\frac{q'^2(x)}{\{\lambda^2 - q(x)\}^3} \right] y = 0, \tag{1.6}$$

where $y = \{\lambda^2 - q(x)\}^{1/4} u$, may come into play. Transformation (1.1) $\Rightarrow$ (1.6) is a crucial ingredient in the WKB-analysis and serves a very wide range of potentials (even growing at infinity) but requires the existence of q' and q'' which appears to be too much in the setting of slowly decaying non-smooth and random potentials. Even for smooth potentials like $q(x) = x^{-\alpha}\sin x^{\beta}, 0 < \alpha \leq \beta \leq 1$, the transformation is not of much help since q' and q'' unboundedly oscillate at ∞. Note, that, as it was shown by Buslaev-Matveev [5], the Green-Liouville transformation (1.6) works for slowly decaying potentials subject to

$$\left| q^{(n)}(x) \right| \leq C x^{-\alpha - n}, \alpha > 0, n = 0,1,2. \tag{1.7}$$

In the present paper we are going to deal with the case of potentials $q \in L_p(\mathbb{R}_+)$, $p \geq 2$, without any smoothness assumptions. We put forward a transformation of the original equation (1.1) that yields solutions with higher-order WKB-type asymptotic for every complex λ. We then use it to study the Titchmarsh-Weyl m-function. We leave the background information and a discussion of the intensive literature for Section 4.

Notation. $L_p, 1 \leq p \leq \infty$, as usual, stands for the Lebesgue class of functions f with the finite norm[1]

$$\|f\|_p^p := \int_{\mathbb{R}} |f(x)|^p \, dx, p < \infty; \|f\|_\infty := \text{ess-sup}\,\{|f(x)|, x \in \mathbb{R}\}.$$

$C^n, n \in \mathbb{N}$, is a set of functions on $\mathbb{R}$ such that $\left\|f^{(j)}\right\|_\infty < \infty, j = 0, 1, \ldots, n$. We write $f \in X_0$ if $f \in X$ and $\lim_{x \to \pm\infty} f(x) = 0$. The class $l_p(L_1), 1 \leq p \leq \infty$, with the norm

$$\|f\|_{l_p(L_1)}^p := \sum_{n=-\infty}^{\infty} \left(\int_n^{n+1} |f(x)| \, dx \right)^p, \ p < \infty;$$

$$\|f\|_{l_\infty(L_1)} := \sup \left\{ \int_n^{n+1} |f(x)| \, dx, n \in \mathbb{R} \right\}.$$

We write $f \in c_0(L_1)$ if $f \in l_\infty(L_1)$ and $\lim_{n \to \pm\infty} \int_n^{n+1} |f(x)| \, dx = 0$. It is clear that $l_1(L_1) = L_1, L_p + L_1 \subset l_p(L_1)$ but $L_p + L_1 \neq l_p(L_1)$. With compactness of our exposition in mind and whenever it leads to no confusion, we write

$$\int_a^b f := \int_a^b f(x) \, dx.$$

2. A Fourier-type transform and a Riemann-Lebesgue-type lemma

Let f be a measurable function on $\mathbb{R}_+$ which class will be specified later. Fixed $\lambda \in \mathbb{C}_+$ consider the following Fourier-type transform

$$\widetilde{f}(x, \lambda) := \int_0^\infty e^{i\lambda s} f(s + x) \, ds. \tag{2.1}$$

For every $\lambda \in \mathbb{C}_+$, the decaying exponential under the integral sign in (2.1) makes the "transformed function" $\widetilde{f}(x, \lambda)$ well defined for a broad class of functions.

Lemma 1. *If* $f \in l_p(L_1), 1 \leq p \leq \infty$. *Then*

$$\left\| \widetilde{f}(\cdot, \lambda) \right\|_p \leq 2 \left(1 + \frac{1}{\operatorname{Im} \lambda} \right) \|f\|_{l_p(L_1)}. \tag{2.2}$$

Proof. Assume $p \in [1, \infty)$. By Jensen's inequality

$$\begin{aligned} \left| \widetilde{f}(x, \lambda) \right|^p &= \left| \sum_{n \geq 0} \int_n^{n+1} e^{i\lambda s} f(s + x) \, ds \right|^p \\ &\leq \left\{ \sum_{n \geq 0} e^{-\operatorname{Im} \lambda n} \left(\int_n^{n+1} |f(s + x)| \, ds \right) \right\}^p \end{aligned} \tag{2.3}$$

[1] If a function f is initially defined only on a subset of $\mathbb{R}$ we continue it to the whole $\mathbb{R}$ as zero.

$$\leq \left(\sum_{n\geq 0} e^{-\operatorname{Im}\lambda n}\right)^{p-1} \sum_{n\geq 0} e^{-\operatorname{Im}\lambda n}\left(\int_n^{n+1} |f(s+x)|\, ds\right)^p$$
$$= \left(\frac{1}{1-e^{-\operatorname{Im}\lambda}}\right)^{p-1} \sum_{n\geq 0} e^{-\operatorname{Im}\lambda n}\left(\int_n^{n+1} |f(s+x)|\, ds\right)^p .$$

Integrating (2.3) with respect to x, one has

$$\left\|\widetilde{f}(\cdot,\lambda)\right\|_p^p \leq \left(1+\frac{1}{\operatorname{Im}\lambda}\right)^{p-1} \sum_{n\geq 0} e^{-\operatorname{Im}\lambda n} \int_{\mathbb{R}}\left(\int_n^{n+1} |f(s+x)|\, ds\right)^p dx. \tag{2.4}$$

Observe now that

$$\int_{\mathbb{R}}\left(\int_n^{n+1} |f(s+x)|\, ds\right)^p dx = \sum_{m\in\mathbb{Z}} \int_m^{m+1}\left(\int_x^{x+1} |f|\right)^p dx$$
$$\leq \sum_{m\in\mathbb{Z}}\left(\int_m^{m+2} |f|\right)^p \leq 2^{p-1}\sum_{m\in\mathbb{Z}}\left\{\left(\int_m^{m+1}|f|\right)^p+\left(\int_{m+1}^{m+2}|f|\right)^p\right\}$$
$$= 2^p \sum_{m\in\mathbb{Z}}\left(\int_m^{m+1}|f|\right)^p = 2^p \left\|f\right\|_{l_p(L_1)}^p . \tag{2.5}$$

Plugging (2.5) into (2.4), one has

$$\left\|\widetilde{f}(\cdot,\lambda)\right\|_p^p \leq \left(1+\frac{1}{\operatorname{Im}\lambda}\right)^{p-1} \sum_{n\geq 0} e^{-\operatorname{Im}\lambda n}\cdot 2^p \left\|f\right\|_{l_p(L_1)}^p$$
$$\leq 2^p\left(1+\frac{1}{\operatorname{Im}\lambda}\right)^{p} \left\|f\right\|_{l_p(L_1)}^p$$

that yields (2.2). Similarly one proves (2.2) for $p=\infty$. □

Given a function $\varphi(x)\geq 0, x\in\mathbb{R}_+$, set $\Lambda(\varphi) := \{\lambda : \operatorname{Im}\lambda \geq \varphi(|\lambda|)\}$. The following statement will play an important role in our consideration.

Proposition 1. *Let $f\in C^1 + c_0(L_1)$ then*

$$\left\|\widetilde{f}(\cdot,\lambda)\right\|_\infty \to 0, |\lambda|\to\infty, \lambda\in\Lambda(\varphi). \tag{2.6}$$

where φ is monotonically decreasing function (depending on f) such that

$$\lim_{x\to\infty}\varphi(x) = 0.$$

The function φ will be specified in the proof.

Proof. By Lemma 2 every function $f\in C^1 + c_0(L_1)$ can be represented as

$$f = f_1 + f_2, \text{where } f_1\in C^1 \text{ and } \left\|f_2\right\|_{l_\infty(L_1)} < \varepsilon. \tag{2.7}$$

Consider $\widetilde{f_1}$ and $\widetilde{f_2}$ separately. Integrating (2.1) by parts and taking into account that $f_1 \in C^1$ yield

$$\widetilde{f_1}(x,\lambda) = -\frac{1}{i\lambda} f_1(x) - \frac{1}{i\lambda}\int_0^\infty e^{i\lambda s} f_1'(s+x)\,ds \tag{2.8}$$

and hence

$$\begin{aligned}\left\|\widetilde{f_1}(\cdot,\lambda)\right\|_\infty &\le \frac{\|f_1\|_\infty}{|\lambda|} + \frac{\|f_1'\|_\infty}{|\lambda|}\int_0^\infty e^{-\operatorname{Im}\lambda s} ds \\ &= \frac{\|f_1\|_\infty}{|\lambda|} + \frac{\|f_1'\|_\infty}{|\lambda|\operatorname{Im}\lambda} \le \left(\frac{1}{|\lambda|} + \frac{1}{|\lambda|\operatorname{Im}\lambda}\right)\|f_1\|_{C^1}.\end{aligned} \tag{2.9}$$

For $\widetilde{f_2}$ we have:

$$\left|\widetilde{f_2}(x,\lambda)\right| \le \int_0^\infty e^{-\operatorname{Im}\lambda s}\,|f_2(s+x)|\,ds$$

$$= \sum_{n\ge 0}\int_n^{n+1} e^{-\operatorname{Im}\lambda s}\,|f_2(s+x)|\,ds \le \sum_{n\ge 0} e^{-\operatorname{Im}\lambda n}\int_n^{n+1} |f_2(s+x)|\,ds \tag{2.10}$$

$$\le \frac{1}{1-e^{-\operatorname{Im}\lambda}}\,\|f_2\|_{l_\infty(L_1)} < \left(1+\frac{1}{\operatorname{Im}\lambda}\right)\varepsilon.$$

Combining (2.9) and (2.10), one has

$$\left\|\widetilde{f}(\cdot,\lambda)\right\|_\infty \le \left(\frac{1}{|\lambda|} + \frac{1}{|\lambda|\operatorname{Im}\lambda}\right)\|f_1\|_{C^1} + \left(1+\frac{1}{\operatorname{Im}\lambda}\right)\varepsilon. \tag{2.11}$$

Setting $\Lambda_1 := \{\lambda : \operatorname{Im}\lambda \ge 1\}$ and letting in (2.11) $|\lambda| \to \infty, \lambda \in \Lambda_1$, we obtain that for any $\varepsilon > 0$

$$\lim \left\|\widetilde{f}(\cdot,\lambda)\right\|_\infty \le 2\varepsilon.$$

That is

$$\lim \left\|\widetilde{f}(\cdot,\lambda)\right\|_\infty = 0, |\lambda| \to \infty, \lambda \in \Lambda_1. \tag{2.12}$$

Consider now the case when $\operatorname{Im}\lambda \to 0$. Assuming $\operatorname{Im}\lambda < 1$, (2.11) implies

$$\begin{aligned}\left\|\widetilde{f}(\cdot,\lambda)\right\|_\infty &\le \frac{1}{\operatorname{Im}\lambda}\left\{\frac{\operatorname{Im}\lambda+1}{|\lambda|}\|f_1\|_{C^1} + (1+\operatorname{Im}\lambda)\,\varepsilon\right\} \\ &\le \frac{2}{\operatorname{Im}\lambda}\left(\frac{1}{|\lambda|}\|f_1\|_{C^1} + \varepsilon\right)\end{aligned} \tag{2.13}$$

Define now a function $g(t), t > 0$, as follows

$$g(t) := \inf_{\varepsilon>0}\inf\left(t^{-1}\|f_1\|_{C^1} + \varepsilon\right),$$

where the first inf is taken over all representations (2.7) with fixed $\varepsilon > 0$. It is a monotonically decreasing function and $g(t) \to 0, t \to \infty$. Take now any continuous non-negative function φ such that $\varphi(t) \to 0, t \to \infty$, and

$$g(t) = o(\varphi(t)), t \to \infty,$$

(e.g., $\varphi = \sqrt{g}$) and set $\Lambda_2(\varphi) := \{\lambda : \varphi(|\lambda|) \le \operatorname{Im}\lambda < 1\}$. It follows then from (2.13) that

$$\left\|\widetilde{f}(\cdot,\lambda)\right\|_\infty \le \frac{2}{\operatorname{Im}\lambda} g(|\lambda|) \le \frac{o(\varphi(|\lambda|))}{\varphi(|\lambda|)} = O(\varphi(|\lambda|))$$

and hence

$$\lim \left\|\widetilde{f}(\cdot,\lambda)\right\|_\infty = 0, |\lambda| \to \infty, \lambda \in \Lambda_2(\varphi). \tag{2.14}$$

Combining (2.12) and (2.14) proves the proposition with $\Lambda(\varphi) = \Lambda_1 \cup \Lambda_2(\varphi)$. □

Remark 1. *The standard Riemann-Lebesgue Lemma follows from the proof of Proposition* 1. *Indeed, every* $f \in L_1$ *admits representation* (2.7) *with* $f_1 \in C_0^\infty$ *and* f_2 *such that* $\|f_2\|_1 < \varepsilon$. *Estimate* (2.11) *for real* λ *can be improved to read*

$$\left|\widetilde{f}(x,\lambda)\right| \le |\lambda|^{-1}\left(|f_1(x)| + \|f_1\|_1\right) + \varepsilon, \forall\lambda, \operatorname{Im}\lambda = 0. \tag{2.15}$$

It follows from (2.1) *that* ($\operatorname{Im}\lambda = 0$)

$$\lim_{x \to -\infty} \left|\widetilde{f}(x,\lambda)\right| = \left|\widetilde{f}(\lambda)\right|,$$

where $\widetilde{f}(\lambda)$ *is the usual Fourier transform and* (2.15) *implies the Riemann-Lebesgue Lemma.*

Remark 2. *If we suppose that* $f \in l_p(L_1), p > 1$, *then the conclusion of Proposition* 1 *can be improved to read*

$$\left\|\widetilde{f}(\cdot,\lambda)\right\|_p \to 0, |\lambda| \to \infty, \lambda \in \Lambda(\varphi_p), \tag{2.16}$$

with some monotonic function $\varphi_p(x) \to 0, x \to \infty$. *Moreover, if* $p_1 < p_2$ *then* $\varphi_{p_1} \le \varphi_{p_2}$ *and* $\lim_{x\to\infty} \frac{\varphi_{p_1}(x)}{\varphi_{p_2}(x)} = 0$, *in particular,* $\varphi_p \le \varphi_\infty \le \varphi$ *where* φ *is as in Proposition* 1.

Remark 3. *If* $f \in C^n$ *then similarly to* (2.9) *one has*

$$\|F(\cdot,\lambda)\|_\infty \le \left(|\lambda|^n \operatorname{Im}\lambda\right)^{-1} \|f\|_{C^n}$$

and (2.6) *holds with* $\varphi(t) = o\left(1/t^{n-1}\right), t \to \infty$.

3. A transformation of the original equation

Given potential $q(x)$ and fixed $\lambda \in \mathbb{C}_+$ consider the following chain of transformations

$$q_1(x,\lambda) = -\int_0^\infty e^{2i\lambda s} q(s+x)\,ds = -\widetilde{q}(x, 2\lambda). \tag{3.1}$$

$$q_{n+1}(x,\lambda) = \int_0^\infty \Theta_n^2(x,s,\lambda)\, q_n^2(s+x,\lambda)\, ds,\ n \in \mathbb{N} \tag{3.2}$$

$$\Theta_n(x,s,\lambda) := \exp\left\{ i\lambda s + \int_x^{s+x} \sum_{m=1}^{n} q_m(t,\lambda)\, dt \right\},$$

$$\Theta_n(s,\lambda) := \Theta_n(0,s,\lambda).$$

This way we get the sequence of functions $\{q_n(\cdot,\lambda)\}_{n\geq 1}$ which can be viewed as "momentum dependent" transformations of the original potential q. Note that for $n \geq 2$ these transformations are highly nonlinear and were previously considered by many authors (see, e.g., [15], [17], [18]). The main feature of the transformation $q_n(\cdot,\lambda) \to q_{n+1}(\cdot,\lambda)$ is that in a way it improves the rate of decay at infinity.

The following theorem is our main result.

Theorem 1. *Let $q \in l_p(L_1), p = 2^n$ with some $n \in \mathbb{N}$, then for any λ from*

$$\Lambda := \left\{ \lambda \in \mathbb{C}_+ : \left\| \int_0^\infty e^{2i\lambda s} q(s+x)\, ds \right\|_\infty < \frac{1}{4} \operatorname{Im} \lambda \right\}$$

the equation

$$-u'' + q(x)\, u = \lambda^2 u, x \in \mathbb{R}_+, \tag{3.3}$$

can be transformed into the Volterra type integral equation:

$$y(x,\lambda) = 1 + \int_x^\infty K_n(x,s,\lambda)\, y(s,\lambda)\, ds, \tag{3.4}$$

$$u(x,\lambda) = \Theta_n(x,\lambda)\, y(x,\lambda),$$

with the kernel

$$K_n(x,s,\lambda) := (q_n\Theta_n)^2(s,\lambda) \int_x^s \Theta_n^{-2}(t,\lambda)\, dt. \tag{3.5}$$

The kernel $K_n(x,s,\lambda)$ satisfies the bound ($\lambda \in \Lambda$)

$$\int_x^\infty |K_n(x,s,\lambda)|\, ds \leq \left(\frac{1}{\operatorname{Im}\lambda}\right)^{p-1} \left(2 + \frac{1}{\operatorname{Im}\lambda}\right)^p \|q\chi_x\|_{l_p(L_1)}^p, \tag{3.6}$$

where χ_x is the characteristic function of (x,∞), and therefore equation (3.4) is solvable by iteration.

Proof. Observe that for any $n \in \mathbb{N}$

$$\Theta_n(x,\lambda) = \exp\left\{ i\lambda x + \int_0^x \sum_{m=1}^{n} q_m(s,\lambda)\, ds \right\}$$

is a solution to

$$-u'' + \left(q(x) + q_n^2(x,\lambda)\right) u = \lambda^2 u \tag{3.7}$$

for any $x \geq 0$ and $\lambda \in \mathbb{C}_+$. Indeed, assume that (3.7) holds for $n-1, n \geq 2$. Then

$$\Theta_n = \Theta_{n-1} \exp \int_0^x q_n. \tag{3.8}$$

Note that q_n is differentiable and

$$q_n' = -q_{n-1}^2 - 2\Lambda_{n-1} q_n, \tag{3.9}$$

where

$$\Lambda_n(x,\lambda) := i\lambda x + \int_0^x \sum_{m=1}^n q_m(s,\lambda)\,ds.$$

Differentiating (3.8) twice and using (3.9) one has

$$\begin{aligned}\Theta_n'' &= \left\{\Theta_{n-1}'' + 2\Theta_{n-1}' q_n + \left(q_n^2 + q_n'\right)\Theta_{n-1}\right\}\exp\int_0^x q_n \\ &= \left(q_{n-1}^2 + q - \lambda^2 + 2\Lambda_{n-1}q_n + q_n^2 - q_{n-1}^2 - 2\Lambda_{n-1}q_n\right)\Theta_n \\ &= \left(q + q_n^2 - \lambda^2\right)\Theta_n,\end{aligned}$$

which is (3.7) for $n \geq 2$. By a straightforward computation one verifies (3.7) for $n = 1$ and by induction (3.7) holds for any $n \in \mathbb{N}$.

If $u_1(x,\lambda) = \Theta_n(x,\lambda)$ is a solution to (3.7) then as the other solution we choose

$$u_2(x,\lambda) = \Theta_n(x,\lambda)\int_0^x \Theta_n^{-2}(s,\lambda)\,ds.$$

Rewriting the original equation (3.3) as

$$-u'' + \left(q(x) + q_n^2(x,\lambda)\right)u - \lambda^2 u = q_n^2(x,\lambda)\,u \tag{3.10}$$

and by variation of parameters one easily verifies (formally) that (3.10) implies (3.4). To justify this formal computation it is enough to show that $K_n(x,s,\lambda) \in L_1(\mathbb{R}_+, ds)$ for every $x \in \mathbb{R}_+, \lambda \in \Lambda$ and, by choosing x large enough, we can achieve $\|K_n(x,\cdot,\lambda)\|_1 < 1$. Let us prove first that for any $p \geq 1$ and natural m

$$q_m \in L_{2p} \Longrightarrow q_{m+1} \in L_p. \tag{3.11}$$

Since $l_p(L_1) \subset c_0(L_1)$, by Proposition 1 we can achieve

$$\delta := \frac{1}{\operatorname{Im}\lambda}\|q_1(\cdot,\lambda)\|_\infty = \frac{1}{\operatorname{Im}\lambda}\|\widetilde{q}(\cdot,2\lambda)\|_\infty < \frac{1}{4}$$

and by Lemma 3 of Appendix

$$\frac{1}{\operatorname{Im}\lambda}\sum_{j=1}^m \|q_j(\cdot,\lambda)\|_\infty < \sum_{j=1}^m \delta^j < 2\delta < \frac{1}{2}$$

and hence it follows from (3.2) that

$$\begin{aligned}|\Theta_{m+1}(x,s,\lambda)|^2 &\leq \exp\left\{-2\operatorname{Im}\lambda s\left(1 - \frac{1}{\operatorname{Im}\lambda}\sum_{j=1}^m \|q_j(\cdot,\lambda)\|_\infty\right)\right\} \\ &\leq \exp\{-2(1-2\delta)\operatorname{Im}\lambda s\} < \exp(-\operatorname{Im}\lambda s).\end{aligned} \tag{3.12}$$

By Jensen's inequality then

$$|q_{m+1}(x,\lambda)|^p \tag{3.13}$$

$$\le \int_0^\infty |\Theta_{m+1}(x,s,\lambda)|^2 |q_m(s+x,\lambda)|^{2p}\ ds \cdot \left(\int_0^\infty |\Theta_{m+1}(x,s,\lambda)|^2 ds\right)^{p-1}$$

$$\le (\operatorname{Im}\lambda)^{-p+1} \int_0^\infty \exp(-\operatorname{Im}\lambda s)\, |q_m(s+x,\lambda)|^{2p}\ ds.$$

Integrating (3.13) with respect to x, one has

$$\|q_{m+1}(\cdot,\lambda)\|_p^p \le \left(\frac{1}{\operatorname{Im}\lambda}\right)^{p-1} \int_0^\infty \exp(-\operatorname{Im}\lambda s)\left(\int_s^\infty |q_m(x,\lambda)|^{2p}\ dx\right) ds$$

$$\le \left(\frac{1}{\operatorname{Im}\lambda}\right)^{p} \|q_m(\cdot,\lambda)\|_{2p}^{2p}$$

and finally

$$\|q_{m+1}(\cdot,\lambda)\|_p \le \frac{1}{\operatorname{Im}\lambda} \|q_m(\cdot,\lambda)\|_{2p}^2, p\ge 1, n\in\mathbb{N}, \tag{3.14}$$

that proves (3.11). By induction, (3.14) yields

$$\|q_n(\cdot,\lambda)\|_2^2 \le \left(\frac{1}{\operatorname{Im}\lambda}\right)^{p-2} \|q_1(\cdot,\lambda)\|_p^p,\ p=2^n,$$

and by Lemma 1:

$$\|q_n(\cdot,\lambda)\|_2^2 \le \left(\frac{1}{\operatorname{Im}\lambda}\right)^{p-2} 2^p\left(1+\frac{1}{2\operatorname{Im}\lambda}\right)^p \|q\|_{l_p(L_1)}^p$$

$$= \left(\frac{1}{\operatorname{Im}\lambda}\right)^{p-2} \left(2+\frac{1}{\operatorname{Im}\lambda}\right)^p \|q\|_{l_p(L_1)}^p,$$

that is

$$\|q_n^2(\cdot,\lambda)\|_1 \le \left(\frac{1}{\operatorname{Im}\lambda}\right)^{p-2} \left(2+\frac{1}{\operatorname{Im}\lambda}\right)^p \|q\|_{l_p(L_1)}^p,\ p=2^n. \tag{3.15}$$

By (3.15), $q_n^2(\cdot,\lambda)\in L_1$ and it is only left to demonstrate that $\Theta_n^2(s,\lambda)\int_x^s\Theta_n^{-2}(t,\lambda)dt$ is bounded. Indeed,

$$\left|\Theta_n^2(s,\lambda)\int_x^s \Theta_n^{-2}(t,\lambda)\,dt\right| = \left|\int_x^s \exp^2\left\{i\lambda(s-t)+\int_t^s\sum_{m=1}^n q_m(\cdot,\lambda)\right\}dt\right|$$

$$\le \left|\int_0^{s-x} \exp^2\left\{-\operatorname{Im}\lambda t\left(1-\frac{1}{\operatorname{Im}\lambda}\sum_{m=1}^n \|q_m(\cdot,\lambda)\|_\infty\right)\right\}\right|$$

$$\le \int_0^\infty \exp(-\operatorname{Im}\lambda t)\,dt = \frac{1}{\operatorname{Im}\lambda}.$$

Estimate now $\int_x^\infty |K_n(x,s,\lambda)|\,ds$. It immediately follows from (3.1) and (3.2) that

$$q_1(x,\lambda) = -\widetilde{q\chi_x}(x,2\lambda),\ (q\chi_x)_n(x,\lambda) = q_n(x,\lambda),$$

and hence

$$\int_x^\infty |K_n(x,s,\lambda)|\,ds \le \frac{1}{\operatorname{Im}\lambda}\int_x^\infty |q_n^2(s,\lambda)|\,ds = \frac{1}{\operatorname{Im}\lambda}\int_x^\infty \left|(q\chi_s)_n^2(s,\lambda)\right|ds$$
$$\le \left(\frac{1}{\operatorname{Im}\lambda}\right)^{p-1}\left(2+\frac{1}{\operatorname{Im}\lambda}\right)^p \|q\chi_x\|_{l_p(L_1)}^p .$$

that proves (3.6). The theorem is proven. □

Remark 4. *The domain $\Lambda \subset \mathbb{C}_+$ but it follows from Proposition 1 that*

$$\lim_{k\to\infty} dist\,(\Lambda, k) = 0,$$

that is Λ approaches the real line.

Remark 5. *The conditions of Theorem 1 are clearly not optimal. Using arguments of Proposition 1 one can achieve the contractive property of the integral operator in (3.4) for any $\operatorname{Im}\lambda > 0$ and $x > 0$ by choosing $|\lambda|$ large enough.*

Remark 6. *We do not assume potentials q in Theorem 1 real.*

Corollary 1. *Under conditions of Theorem 1, equation (3.3) has the Weyl solution Ψ (that is $\Psi(x,\lambda) \in L_2(\mathbb{R}_+), \lambda \in \mathbb{C}_+$) satisfying the following asymptotics*

$$\Psi(x,\lambda) \sim \exp\left\{ i\lambda x + \int_0^x \sum_{m=1}^n q_m(s,\lambda)\,ds \right\}, x \to \infty, \tag{3.16}$$

for every $\lambda \in \mathbb{C}_+$. This immediately follows from Theorem 1 if $\lambda \in \Lambda$. However with some additional effort (see Remark 5) (3.16) can be proven as stated.

Remark 7. *The structure of $\{q_n\}$ is very messy. One can prove though that if $\lambda \to \infty$ in some sector in the upper half-plane then*

$$q_1(x,\lambda) \sim (2i\lambda)^{-1} q(x),\, q_n(x,\lambda) \sim -(2i\lambda)^{-p+1} q^{p/2}(x),\, p = 2^n.$$

If we call $\exp\left\{ i\lambda x + \frac{1}{2i\lambda}\int_0^x q(s)\,ds \right\}$ the WKB-type asymptotics then (3.16) can be referred as to higher-order WKB-type asymptotics.

Remark 8. *It is natural to ask when (3.16) holds for real λ. As we have already mentioned in Introduction, in general there is no answer to this question even for $q \in L_2$. The best-known result here belongs to Christ-Kiselev [7] (see also the extensive literature cited therein) saying that if $q \in l_p(L_1), p < 2$, then equation (3.3) has a bounded solution $u(x,\lambda)$ for almost all $\lambda \in \mathbb{R}$ with the WKB-type asymptotics*

$$u(x,\lambda) \sim \exp\left\{ i\lambda x + \frac{1}{2i\lambda}\int_0^x q(s)\,ds \right\}, x \to \infty. \tag{3.17}$$

On the other hand it is proven that under the only condition $q \in L_p, p > 2$, the spectrum of $H = -d^2/dx^2 + q(x), u(0) = 0$,[2] may be purely singular. In such

[2] Or any other self-adjoint boundary condition

situations (3.16) *holds for real* λ *only on a set of Lebesgue measure zero. As it was recently proven by the author* [22] *that* $q \in l_2(L_1)$ *implies that the absolutely continuous spectrum is* $\mathbb{R}_+$. *However, the existence of the asymptotics* (3.17) *is still an open problem related to some difficult issues of harmonic analysis (see, e.g.,* [21]*)*

In the conclusion of this section we note that in the setting of Schrödinger operators with long-range potentials solutions of type (3.16) play a role of the Jost solution in the short-range scattering. It can be proven (see also [22] in this context) that if $q \in l_2(L_1)$ then the perturbation 2-determinant Δ_2 of the pair (H, H_0), (where $H_0 = H$ with $q = 0$) defined by

$$\Delta_2(z) := \det{}_2 \left\{ I + (H_0 - z)^{-1/2} q (H_0 - z)^{-1/2} \right\}$$

exists and admits the representation:

$$\Delta_2(\lambda^2) = \Psi(0, \lambda), \lambda \in \mathbb{C}_+, \tag{3.18}$$

where Ψ is the Weyl solution to (3.3) subject to

$$\Psi(x, \lambda) \sim \exp\left\{ i\lambda x + \int_0^x q_1(s, \lambda)\, ds \right\}, x \to \infty. \tag{3.19}$$

This will be discussed in detail elsewhere. Note that (3.18) was first proven by Koplienko [20] under conditions (1.7).

Theorem 1 type assertions are important tools in the study of perturbation p-determinants $\Delta_p(\lambda^2)$ with $p > 2$ for which similar to (3.18) relations hold. Theorem 1 for $p = 2$ was a crucial argument in [22] (see Remark 8). Note however, that (3.18) does not hold for $\Delta_p(\lambda^2)$ with $p > 2$ – one should consider some different (but quite similar) from (3.1)–(3.2) transformations resulting in a different expression for the phase in (3.19). We hope to return to these important issues elsewhere.

4. Applications to the Weyl-Titchmarsh m-function

Let us consider

$$-u'' + q(x) u = \lambda^2 u \tag{4.1}$$

on the half-line $\mathbb{R}_+$ with $q \in L_{1,\mathrm{loc}}(\mathbb{R}_+)$. Assume that q is real and in the limit point case at ∞ (see, e.g., [25]). Then the Weyl solution $\Psi(x, \lambda)$ of (4.1) is unique up to a multiplicative constant and one can define the (Dirichlet) Titchmarsh-Weyl m-function

$$m(\lambda^2) = \frac{\Psi'(0, \lambda)}{\Psi(0, \lambda)}, \lambda^2 \in \mathbb{C}_+. \tag{4.2}$$

It is well known that the Titchmarsh-Weyl m-function is analytic from $\mathbb{C}_+$ to $\mathbb{C}_+$ and plays a crucial role in spectral analysis of the one-dimensional Schrödinger operator. Since there is no explicit formula for m in terms of q, it becomes important to study m for large λ. The research in this direction was originated by Everitt [11] in 1972 who proved that

$$m\left(\lambda^2\right) = i\lambda + o\left(1\right), \lambda \to \infty, \lambda \in S_\varepsilon := \left\{\lambda : \varepsilon \leq \arg \lambda \leq \pi/2 - \varepsilon, \varepsilon > 0\right\}. \quad (4.3)$$

In 1981 Atkinson [1] improved (4.3) to read (a is any positive)

$$m\left(\lambda^2\right) = i\lambda + \int_0^a e^{2i\lambda x} q\left(x\right) dx + O\left(\left|\lambda\right|^{-1}\right), \quad (4.4)$$

as $\lambda \to \infty$ in S_ε (or some parabolic domains allowing $\operatorname{Im} \lambda \to 0$). Atkinson's representation (4.4) received much of attention and has been improved and generalized in many directions by Bennewitz [4], Brown-Knowles-Weikard [6], Clark-Gesztesy [8], [9], Gesztesy-Simon [13], Harris [14], [15], [16], Hinton-Klaus-Shaw [17], Kaper-Kwong [18], [19], Rybkin [23], [24] and many others (see the extensive literature cited therein). Note that one of the last papers by Atkinson [2] was also devoted to generalization of (4.4) for potentials with strong singularities at the origin.

The result below is not new and in different forms has been discussed by many authors but our statement appears to be optimal and its proof comes as a simple corollary of Theorem 1 and the following proposition.

Proposition 2. *Assume q to be real and from $L_{1,\mathrm{loc}}\left(\mathbb{R}_+\right), \widetilde{q} = q\chi_{[0,a]}, a > 0$. Let $m, \widetilde{m}$ be the Titchmarsh-Weyl m-functions corresponding to $q, \widetilde{q}$ respectively. If $\operatorname{Im} \lambda^2 > 0$ and $\operatorname{Im} \lambda > \max\left\{4 \int_0^a |q|, a^{-1} \ln 6\right\}$, then*

$$\left|m\left(\lambda^2\right) - \widetilde{m}\left(\lambda^2\right)\right| \leq \frac{864}{5} \left|\lambda\right| \exp\left(-2a \operatorname{Im} \lambda\right). \quad (4.5)$$

In an implicit form, Proposition 2 is due to Atkinson [1]. In the present form, it was recently found by Simon and studied in great detail in Gesztesy-Simon [13]. Note that in [13] Gesztesy-Simon established that the converse for Proposition 2 is also true generalizing the Borg-Marchenko uniqueness result. (See also [3] for a simpler proof.)

Theorem 2. *Assume $q \in L_{1,\mathrm{loc}}\left(\mathbb{R}_+\right)$, real and in the limit point case at ∞. Let $\left(q_a\right)_n\left(\lambda\right) = \left(q_a\right)_n\left(0, \lambda\right)$ be as before defined for $q_a = q\chi_{[0,a]}, a > 0$. Then for $\lambda^2 \in \mathbb{C}_+$*

$$m\left(\lambda^2\right) = i\lambda + \sum_{n \geq 1} \left(q_a\right)_n\left(\lambda\right) + r\left(\lambda\right), \quad (4.6)$$

where the remainder $r\left(\lambda\right)$ admits the estimate

$$\left|r\left(\lambda\right)\right| \leq \frac{864}{5} \left|\lambda\right| e^{-2a \operatorname{Im} \lambda}$$

holding uniformly in $\Lambda = \left\{\lambda^2 \in \mathbb{C}_+ : \operatorname{Im} \lambda \geq \max\left\{4 \int_0^a |q|, a^{-1} \ln 6\right\}\right\}$.

Proof. Let m_a be the Titchmarsh-Weyl m-function corresponding to q_a. Since $q_a \in L_1$, Theorem 1 applies with any $n \in \mathbb{N}$ and equation (4.1) can be solved by iteration. By Proposition 1, for any fixed $\operatorname{Im}\lambda > 0$ by choosing $|\lambda|$ large enough one can make $(\operatorname{Im}\lambda)^{-1}\|\widetilde{q_a}(\cdot,\lambda)\|_\infty < 1/2$ and by Lemma 3 then

$$\left|\sum_{n\geq 1}(q_a)_n(\lambda)\right| \leq \sum_{n\geq 1}\|(q_a)_n(\cdot,\lambda)\|_\infty \leq \frac{\|\widetilde{q_a}(\cdot,\lambda)\|_\infty}{1-(\operatorname{Im}\lambda)^{-1}\|\widetilde{q_a}(\cdot,\lambda)\|_\infty}.$$

Corollary 1 applies with any n and one has

$$\widetilde{m}\left(\lambda^2\right) = i\lambda + \sum_{n\geq 1}(q_a)_n(\lambda), \tag{4.7}$$

where the series on the right converges at least in the domain $\|\widetilde{q_a}(\cdot,\lambda)(\cdot,\lambda)\|_\infty < 1/2\operatorname{Im}\lambda$. Applying Proposition 2 completes the proof. □

Remark 9. *Our representation* (4.6) *is not asymptotic: the series in* (4.6) *is absolutely convergent in* $\mathbb{C}_+$. *In the previous literature cited above the series in* (4.6) *is usually considered as asymptotic in sectorial domains*

$$S_\varepsilon = \{\lambda : \varepsilon \leq \arg\lambda \leq \pi/2-\varepsilon, \varepsilon > 0\}.$$

The next assertion says when the error term in (4.6) can be dropped and appears to be new.

Theorem 3. *If* q *is real and from* $C^1 + c_0(L_1)$ *then*

$$m\left(\lambda^2\right) = i\lambda + \sum_{n\geq 1} q_n(\lambda), \tag{4.8}$$

for any λ *subject to the condition*

$$\frac{2}{\operatorname{Im}\lambda}\left\|\int_0^\infty e^{2i\lambda s} q(s+x)\,ds\right\|_\infty < 1. \tag{4.9}$$

Proof. By Proposition 1, condition (4.9) is satisfied for some domain Λ asymptotically approaching the real line. Consider (4.7). By Lemma 2, $\sum_{n\geq 1} q_n(0,\lambda)$ converges absolutely for any $\lambda \in \Lambda$ and it is clear from the proof of the same lemma that $(q_a)_n \to q_n, a \to \infty$, and the right-hand side of (4.7) converges to that of (4.8). Since by our condition on q guarantees the limit point case at ∞, one concludes [25] that $m_a\left(\lambda^2\right) \to m\left(\lambda^2\right), a \to \infty$, on any compact set in $\mathbb{C}_+$ and a passage to the limit in (4.7) yields (4.8). □

Remark 10. *The question when representation of type* (4.8) *holds was a focus of* [15] *where representation* (4.8) *was proven for* λ *from sectorial domains* S_ε *under the condition*

$$\left|\int_0^\infty e^{2i\lambda s} q(s+x)\,ds\right| \leq a(x)\,b(\lambda), \tag{4.10}$$

with $a(x) \in L_1(\mathbb{R}_+)$ *and* $b(\lambda) \to 0, |\lambda| \to \infty, \lambda \in S_\varepsilon$. *Condition* (4.10), *as apposed to ours, assumes a strong decay of* q *at infinity (for instance,* $\left(\int_x^\infty |q|^p\right)^{1/p} \in L_1(\mathbb{R}_+), p > 1$*). Note in this connection, that the conclusion of Theorem 3 holds even for periodic potentials* q *integrable on the periods.*

Remark 11. *Since the spectrum of the operator* $-d^2/dx^2 + q(x), u(0) = 0$, *coincides with the complement of the set* $\{t \in \mathbb{R} : \operatorname{Im} m(t + i0) = 0\}$, *representations of type* (4.8) *can be used in spectral analysis of Schrödinger operators on the line. However, to make substantial assertions about the spectrum one needs to improve condition* (4.9) *allowing* λ^2 *in* (4.8) *to approach the real line (or at least be parallel to it). In view of Remark* 3, *one of the trivial ways to achieve it is to require uniform boundedness of* q, q', *and* q''. *We have a strong reason to believe that a more suibtle analysis of* q_n *and a proper refinement of Proposition* 1 *should yield in Theorem* 3 *a much stronger result on the domain of convergence of the series in* (4.8). *We hope to address this issue elsewhere.*

5. Appendix. Some auxiliary facts

We give here two known facts which we have used in the previous sections. For the reader convenience we provide their simple proofs.

Lemma 2. $c_0(L_1)$ *is a Banach space with respect to the norm*

$$\|f\| := \left\| \int_x^{x+1} |f| \right\|_\infty .$$

Proof. We are going to show that for any $\varepsilon > 0$ a function $f \in c_0(L_1)$ can be represented as

$$f = f_1 + f_2,\ f_1 \in C_0^\infty, \|f_2\| < \varepsilon.$$

Let $\chi_{[a,b]}$ be the characteristic function of a compact interval $[a,b]$. We have

$$f = f\chi_{[a,b]} + f\chi_{\mathbb{R}\setminus[a,b]}. \tag{5.1}$$

Choose now $a, b \in \mathbb{R}$ so that

$$\left\| f\chi_{\mathbb{R}\setminus[a,b]} \right\| = \sup\left\{ \int_x^{x+1} |f| : x+1 < a, x > b \right\} = \frac{\varepsilon}{2}. \tag{5.2}$$

Since clearly $f\chi_{[a,b]} \in L_1(a-1, b+1)$ and $f\chi_{[a,b]} = 0, x \notin [a,b]$, the function $f\chi_{[a,b]}$ can be represented as

$$f\chi_{[a,b]} = f_1 + g,$$

where $f_1 \in C_0^\infty[a-1, b+1]$ and $f_1 = 0$ on $[a-1, b+1] \setminus (a,b)$, and g is such that $\int_a^b |g| = \frac{\varepsilon}{2}$. Set now

$$f_2 = f\chi_{\mathbb{R}\setminus[a,b]} + g.$$

By (5.1) and (5.2), we have

$$\|f_2\| \le \left\| f\chi_{\mathbb{R}\setminus[a,b]} \right\| + \|g\| \le \frac{\varepsilon}{2} + \int_a^b |g| = \varepsilon$$

and the lemma is proven. □

Lemma 3. *Let q_n be defined by (3.2) and*

$$\delta_n := \frac{1}{\operatorname{Im}\lambda} \|q_n(\cdot,\lambda)\|_\infty, n \in \mathbb{N}. \tag{5.3}$$

If $\delta_1 = \delta < 1/2$ then

$$\delta_n < \delta^n. \tag{5.4}$$

Proof. We conduct our proof by induction. It follows from (3.2) with $n = 1$ that ($\alpha := \operatorname{Im}\lambda$)

$$\begin{aligned}\delta_2 &= \frac{1}{\alpha}\|q_2(\cdot,\lambda)\|_\infty \le \frac{1}{\alpha}\sup_{x\in\mathbb{R}}\int_x^\infty \left|\exp\left\{i\lambda(s-x) + \int_x^s q_1(\cdot,\lambda)\right\}\right|^2 |q_n(s,\lambda)|^2 ds \\ &\le \frac{1}{\alpha}\|q_1(\cdot,\lambda)\|_\infty^2 \cdot \int_x^\infty \exp^2\{(-\alpha + \|q_1(\cdot,\lambda)\|_\infty)(s-x)\} ds \\ &= \alpha\delta_1^2 \cdot \int_0^\infty \exp\{-2\alpha(1-\delta_1)s\} ds = \frac{\delta_1^2}{2(1-\delta_1)} < \delta^2. \end{aligned}\tag{5.5}$$

Assume that (5.4) holds for some n, then

$$\sum_{m=1}^n \delta_m < \sum_{m=1}^n \delta^m < \sum_{m\ge 1}^n \left(\frac{1}{2}\right)^m = 1 - \left(\frac{1}{2}\right)^n$$

and similarly to (5.5) one has

$$\begin{aligned}\delta_{n+1} &= \frac{1}{\alpha}\|q_{n+1}(\cdot,\lambda)\|_\infty \\ &\le \frac{1}{\alpha}\|q_n(\cdot,\lambda)\|_\infty^2 \cdot \int_0^\infty \exp\left\{-2\alpha s\left(1 - \frac{1}{\alpha}\sum_{m=1}^n \|q_m(\cdot,\lambda)\|_\infty\right)\right\} ds \\ &= \frac{1}{\alpha}\|q_n(\cdot,\lambda)\|_\infty^2 \left\{2\alpha\left(1 - \frac{1}{\alpha}\sum_{m=1}^n \|q_m(\cdot,\lambda)\|_\infty\right)\right\}^{-1} \\ &= \frac{1}{2}\delta_n^2\left(1 - \sum_{m=1}^n \delta_m\right)^{-1} < \frac{2^n}{2}\delta^{2n} = 2^{n-1}\left(\frac{1}{2}\right)^{n-1}\delta^{n+1} < \delta^{n+1}. \end{aligned}$$

□

References

[1] F.V. Atkinson, *On the location of the Weyl circles*, Proc. Roy. Soc. Edinburgh Sect. A 88 (1981), no. 3-4, 345–356.

[2] F.V. Atkinson and C. T. Fulton, *Asymptotics of the Titchmarsh-Weyl m-coefficient for non-integrable potentials*, Proc. Roy. Soc. Edinburgh Sect. A 129 (1999), no. 4, 663–683

[3] C. Bennewitz, *A proof of the local Borg-Marchenko theorem*, Comm. Math. Phys. 218 (2001), no. 1, 131–132.

[4] C. Bennewitz, *Spectral asymptotics for Sturm-Liouville equations*, Proc. London Math. Soc. (3) 59 (1989), no. 2, 294–338.

[5] V.S. Buslaev and V.B. Matveev, *Wave operators for the Schrödinger equation with slowly decreasing potential*, Theoret. and Math. Phys. 2 (1970), no. 3, 266–274.

[6] B.M. Brown, I. Knowles, and R. Weikard, *On the inverse resonace problem*, J. London Math. Soc. (2) 68 (2003), no. 2, 383–401.

[7] M. Christ and A. Kiselev, *Scattering and wave operators for one-dimensional Schrödinger operators with slowly decaying nonsmooth potentials.* Geom. Funct. Anal. 12 (2002), no. 6, 1174–1234.

[8] S. Clark and F. Gesztesy, *Weyl-Titchmarsh M-function asymptotics, local uniqueness results, trace formulas, and Borg-type theorems for Dirac operators*, Trans. Amer. Math. Soc. 354 (2002), no. 9, 3475–3534

[9] S. Clark and F. Gesztesy, *Weyl-Titchmarsh M-function asymptotics for matrix-valued Schrödinger operators*, Proc. London Math. Soc. (3) 82 (2001), no. 3, 701–724.

[10] P. Deift and R. Killip, *On the absolutely continuous spectrum of one-dimensional Schrödinger operators with square summable potentials*, Commun. Math. Phys. 203 (1999), 341–347.

[11] W.N. Everitt, *On a property of the m -coefficient of a second-order linear differential equation,* J. London Math. Soc. (2) 4 (1971/72), 443–457.

[12] D.J. Gilbert and D.B. Pearson, *On subordinacy and analysis of the spectrum of one-dimensional Schrödinger operators*, J. Math. Anal. Appl. 128 (1987), 30–56.

[13] F. Gesztesy and B. Simon, *A new approach to inverse spectral theory. II. General real potentials and the connection to the spectral measure,* Ann. of Math. (2) 152 (2000), no. 2, 593–643.

[14] B.J. Harris, *The asymptotic form of the Titchmarsh-Weyl m-function associated with a non-definite, linear, second order differential equation,* Mathematika 43 (1996), no. 1, 209–222.

[15] B.J. Harris, *An exact method for the calculation of certain Titchmarsh-Weyl m-functions.* Proc. Roy. Soc. Edinburgh Sect. A 106 (1987), no. 1–2, 137–142.

[16] B.J. Harris, *The asymptotic form of the Titchmarsh-Weyl m-function associated with a second-order differential equation with locally integrable coefficient.* Proc. Roy. Soc. Edinburgh Sect. A 102 (1986), no. 3–4, 243–251.

[17] D.B. Hinton, M. Klaus, and J.K. Shaw, *Series representation and asymptotics for Titchmarsh-Weyl m*-functions., Differential Integral Equations 2 (1989), no. 4, 419–429.

[18] H.G. Kaper and M.K. Kwong, *Asymptotics of the Titchmarsh-Weyl m-coefficient for integrable potentials. II.* Differential equations and mathematical physics (Birmingham, Ala., 1986), 222–229, Lecture Notes in Math., 1285, Springer, Berlin, 1987.

[19] H.G. Kaper and M.K. Kwong, *Asymptotics of the Titchmarsh-Weyl m-coefficient for integrable potentials*, Proc. Roy. Soc. Edinburgh Sect. A 103 (1986), no. 3–4, 347–358.

[20] L.S. Koplienko, *Regularized function of spectral shift for a one-dimensional Schrödinger operator with slowly decreasing potential*, Siberian Math. J. 26 (1985), no. 3, 365–369.

[21] C. Muscalu, T. Tao and C. Thiele, *A counterexample to a multilinear endpoint question of Christ and Kiselev.* Math. Res. Lett. 10 (2003), no. 2–3, 237–246.

[22] A. Rybkin, *On the absolutely continuous spectrum and negative discrete spectra of Schrödinger operators on the line with locally integrable globally square summable potentials,* to appear in JMP.

[23] A. Rybkin, *Some new and old asymptotic representations of the Jost solution and the Weyl m-function for Schrödinger operators on the line.* Bull. London Math. Soc. 34 (2002), no. 1, 61–72.

[24] A. Rybkin, *On a complete analysis of high-energy scattering matrix asymptotics for one-dimensional Schrödinger operators with integrable potentials*, Proc. Amer. Math. Soc. 130 (2002), no. 1, 59–67.

[25] E. C. Titchmarsh, *Eigenfunction expansions associated with second-order differential equations. Part I,* Second Edition, Clarendon Press, Oxford 1962, 203 pp.

Alexei Rybkin
Department of Mathematical Sciences
University of Alaska Fairbanks
PO Box 756660
Fairbanks, AK 99775, USA
e-mail: `ffavr@uaf.edu`

Operator Theory:
Advances and Applications, Vol. 154, 203–218

Uniform Levinson Type Theorems for Discrete Linear Systems

Luis O. Silva

Abstract. For the solutions of parametric discrete linear systems we obtain an asymptotic formula whose remainder is estimated uniformly with respect to the parameter. This result is an extension of the discrete Levinson theorem and allows us, in particular, to study the parametric discrete linear systems that arise when dealing with the generalized eigenvectors of Jacobi matrices.

Mathematics Subject Classification (2000). Primary 39A11;Secondary 47B36.

Keywords. Asymptotic behavior, Levinson theorem, Jacobi Matrices.

1. Introduction

The Levinson theorem, in its discrete version, describes the asymptotic behavior as $n \to \infty$ of the sequence of d-dimensional vectors $\{\vec{x}_n\}_{n\geq n_0}$ being a solution of the linear system

$$\vec{x}_{n+1} = A_n\vec{x}_n \quad \text{for } n \geq n_0 \,, \tag{1}$$

where the sequence of $d \times d$ matrices $\{A_n\}_{n\geq n_0}$ satisfies certain conditions beginning at some $n_1 \geq n_0$. If, for instance, (1) is a diagonal system, that is, $A_n = \operatorname{diag}\{\nu_n^{(k)}\}_{k=1}^d$ for all $n \geq n_0$, then a fundamental system of solutions of (1) is given by the sequences $\{\vec{x}_n^{(k)}\}_{n\geq n_0}$ $(k = 1, \dots, d)$, such that $\vec{x}_n^{(k)} = \prod_{i=n_0}^{n-1} \nu_i^{(k)} \vec{e}_k$, where $\{\vec{e}_n\}_{n=1}^d$ is the canonical basis in $\mathbb{C}^d$ ($\vec{e}_k$ is the vector with all components zero except the k-th, which is one). Therefore, in this case, a basis in the space of solutions of (1) behaves asymptotically, as $n \to \infty$, as $\prod_{i=n_0}^{n-1} \nu_i^{(k)} \vec{e}_k$. The discrete Levinson theorem asserts that a similar asymptotic behavior takes place when $\{A_n\}_{n\geq n_0}$ is perturbed by a sequence of matrices whose norms form a summable series, and even when A_n is not diagonal, but diagonalizable in a specific sense (see below). This assertion is the discrete analogue of the well-known Levinson theorem on the asymptotics of solutions of a system of ordinary differential equations [2].

The asymptotic analysis of solutions of discrete linear systems has already a long history. Important results on the asymptotics of the solutions of (1), including the discrete Levinson theorem, are in [1] and [3]. A modern approach to the asymptotic analysis of solutions of difference equations is in [7], where various discrete Levinson type theorems are proved.

In 1992 the theory of subordinacy, developed some years earlier as a tool in the spectral analysis of ordinary differential operators by Gilbert and Pearson [5] [4], was successfully carried over into the discrete domain [11]. Since then, various asymptotic methods have been used to study the spectral properties of difference operators, in particular, Jacobi matrices. One of these techniques has been the discrete Levinson theorem (cf. [8], [7] and [13]).

It is well known that the three term recurrence equation for determining the generalized eigenvectors of a Jacobi matrix is equivalent to a discrete linear system similar to (1), which depends on the parameter $\lambda \in \mathbb{R}$, i.e.,

$$\vec{x}_{n+1}(\lambda) = A_n(\lambda)\vec{x}_n(\lambda) \qquad \text{for} \quad n \geq n_0 \,, \tag{2}$$

where, $\forall\lambda \in \mathbb{R}$, $\{A_n(\lambda)\}_{n\geq n_0}$ is a sequence of 2×2 matrices and $\{\vec{x}_n(\lambda)\}_{n\geq n_0}$ is a sequence of two-dimensional vectors (see [9], [10] and [7]). Since we are most interested in applications to general difference operators, we shall consider systems of the form (2), but in a broader setting, namely, we regard $\{A_n(\lambda)\}_{n\geq n_0}$ to be a sequence of $d \times d$ complex matrices for every λ in some interval $\mathfrak{I} \subset \mathbb{R}$. Besides, in order to avoid trivial cases, we always assume that

$$\det A_n(\lambda) \neq 0 \qquad \forall n \geq n_0, \quad \forall\, \lambda \in \mathfrak{I}\,.$$

Notice that, under this assumption, the solutions of the system (2) form a d-dimensional space and any solution $\{\vec{x}_n(\lambda)\}_{n\geq n_0}$ is uniquely determined by

$$\vec{x}_n(\lambda) = \prod_{i=n_0}^{n-1} A_i(\lambda)\vec{x}_{n_0}(\lambda)\,, \qquad \vec{x}_{n_0} \in \mathbb{C}^d\,. \tag{3}$$

Here, and later on in this paper, the product of matrices is considered in "chronological" order, i.e., $\prod_{i=n_0}^{n-1} A_i(\lambda) = A_{n-1} \ldots A_{n_0}$. The simplicity of (3) is delusive, since it generally does not give direct information about the asymptotic behavior of the solutions.

Traditionally, equation (2) has been studied pointwise with respect to λ. However, in many cases, there is a uniform behavior of the system with respect to this parameter ([7] and [13]). Making use of this uniformity to obtain a uniform estimate of the remainder in the asymptotic expansion of the solutions as $n \to \infty$ turns out to be important in the spectral analysis of Jacobi matrices.

Recent developments [13] have shown that, when studying Jacobi matrices, a complete spectral analysis requires an extension of the discrete Levinson theorem to deal with the parametric recurrence systems that arise from the equation for determining the generalized eigenvectors of operators associated with Jacobi matrices. The point is that the spectral analysis given by the theory of subordinacy is not complete in the sense that it left unanswered the question of whether pure point parts of the spectrum have points of accumulation on any finite interval. The method used to answer this question requires, among other things, a uniform estimate of the asymptotic remainder of the solutions of (2) with respect to λ. Peculiarities of this method are its wide applicability and that it is based on the asymptotic behavior of solutions, i.e., it is in the spirit of subordinacy theory. In this paper we develop the first "ingredient" of this method, the discrete uniform Levinson theorem. The other parts are planned to be the material of other papers. It was decided to present the uniform Levinson theorem separately because the author thinks that it may have other applications besides the one described in this introduction.

2. Preliminaries

For the definitions below, we consider a sequence $V = \{V_n(\lambda)\}_{n\geq n_0}$, such that, for every λ in some interval $\mathfrak{I} \subset \mathbb{R}$, $V_n(\lambda)$ is a complex $d \times d$ matrix with the eigenvalues $\{\nu_n^{(k)}(\lambda)\}_{k=1}^d$.

Definition 2.1. We shall say that the sequence $V = \{V_n(\lambda)\}_{n\geq n_0}$ satisfies the Levinson condition for k (**L.c.(k)**) when there exist an $N \geq n_0$ and a constant number $M > 1$ such that, k being fixed, each j ($1 \leq j \leq d$) falls into one and only one of the two classes I_1 or I_2, where

(a) $j \in I_1$ if $\forall \lambda \in \mathfrak{I}$

$$\frac{|\prod_{i=N}^{n-1} \nu_i^{(k)}(\lambda)|}{|\prod_{i=N}^{n-1} \nu_i^{(j)}(\lambda)|} \to \infty \qquad \text{as } n \to \infty\,, \qquad \text{and}$$

$$\frac{|\prod_{i=n_1}^{n_2-1} \nu_i^{(k)}(\lambda)|}{|\prod_{i=n_1}^{n_2-1} \nu_i^{(j)}(\lambda)|} > \frac{1}{M}\,, \qquad \forall n_2\,, n_1 \text{ such that } n_2 > n_1 \geq N\,.$$

(b) $j \in I_2$ if $\forall \lambda \in \mathfrak{I}$

$$\frac{|\prod_{i=n_1}^{n_2-1} \nu_i^{(k)}(\lambda)|}{|\prod_{i=n_1}^{n_2-1} \nu_i^{(j)}(\lambda)|} < M\,, \qquad \forall n_2\,, n_1 \text{ such that } n_2 > n_1 \geq N\,.$$

Remark 1. It is easy to show that when the sequence $V = \{V_n(\lambda)\}_{n\geq n_0}$ is such that

$$V_n(\lambda) \overset{\lambda\in\mathfrak{I}}{\rightrightarrows} V_\infty,$$

where V_∞ is a matrix whose eigenvalues are different from zero and have pairwise distinct absolute values, the sequence V satisfies the (**L.c.(k)**) for every $k = 1, \ldots, d$.

Definition 2.2. Assuming that $\nu_n^{(k)}(\lambda) \neq 0$ $\forall\, k, n, \lambda$, we define for each k the normed space $X_k(n_0)$ that consists of all sequences $\vec{\varphi} = \{\vec{\varphi}_n(\lambda)\}_{n=n_0}^{\infty}$ of functions defined on some interval $\mathfrak{J} \subset \mathbb{R}$ and with range in $\mathbb{C}^d$, such that

$$\sup_{n \geq n_0} \sup_{\lambda \in \mathfrak{J}} \{\|\vec{\varphi}_n(\lambda)\|_{\mathbb{C}^d} \frac{1}{|\prod_{i=n_0}^{n-1} \nu_i^{(k)}(\lambda)|}\} < \infty$$

and where the norm is defined by

$$\|\vec{\varphi}\|_{X_k(n_0)} = \sup_{n \geq n_0} \sup_{\lambda \in \mathfrak{J}} \{\|\vec{\varphi}_n(\lambda)\|_{\mathbb{C}^d} \frac{1}{|\prod_{i=n_0}^{n-1} \nu_i^{(k)}(\lambda)|}\}. \tag{4}$$

Clearly, this space is complete.

In the space $X_k(n_0)$ we shall also consider the subspace[1] $X_k^0(n_0)$ which contains all functions of $X_k(n_0)$ such that

$$\sup_{\lambda \in \mathfrak{J}} \{\|\vec{\varphi}_n(\lambda)\|_{\mathbb{C}^d} \frac{1}{|\prod_{i=n_0}^{n-1} \nu_i^{(k)}(\lambda)|}\} \to 0 \text{ as } n \to \infty \tag{5}$$

Consider now the linear space of sequences $V = \{V_n(\lambda)\}_{n=n_0}^{\infty}$. In this space we define the operator Δ by

$$(\Delta V)_n = V_{n+1}(\lambda) - V_n(\lambda). \tag{6}$$

Definition 2.3. The sequence $V = \{V_n\}_{n=n_0}^{\infty}$ is said to be in the class $\widetilde{\mathcal{D}}^1$ iff

$$\{\sup_{\lambda \in \mathfrak{J}} \|(\Delta V)_n\|\}_{n=n_0}^{\infty} \in l^1.$$

This class is just the uniform analogue of one of the classes defined in [14]

In addition to these notations and concepts, we also shall need the following projectors acting in the linear space $\mathbb{C}^d$. Consider again the canonical orthonormal basis in $\mathbb{C}^d$, $\{\vec{e}_k\}_{k=1}^d$. Let P_k be the projector to the one-dimensional space generated by $\vec{e}_k$, i.e., P_k is the $d \times d$ diagonal matrix whose diagonal has the coordinates of $\vec{e}_k$ as its elements:

$$P_k = \operatorname{diag}\{(\vec{e}_k)_l\}_{l=1}^d \qquad k = 1, \ldots, d. \tag{7}$$

Using these projectors and the classes I_1, I_2 from the definition of (**L.c.(k)**), we define

$$P^{(i)} = \sum_{j \in I_i} P_j \qquad i = 1, 2. \tag{8}$$

3. Auxiliary results

In the following lemma we introduce, with the help of two sequences of matrices satisfying certain conditions, an operator in $X_k(n_0)$ and establish some of its properties.

---

[1]That is a closed subset with norm induced from $X_k(n_0)$.

Lemma 3.1. *Consider the sequences* $\{\Lambda_n(\lambda)\}_{n=n_0}^{\infty}$ *and* $\{R_n(\lambda)\}_{n=n_0}^{\infty}$*, such that, for every* λ *in some interval* $\mathfrak{I} \subset \mathbb{R}$*,* $\Lambda_n(\lambda)$ *and* $R_n(\lambda)$ *are* $d \times d$ *complex matrices and* $\Lambda_n(\lambda)$ *has the eigenvalues* $\{\nu_n^{(k)}(\lambda)\}_{k=1}^{d}$*, such that* $\nu_n^{(k)}(\lambda) \neq 0 \ \forall\, k, n, \lambda$*. If the following conditions hold*

i. *the sequence* $\{\Lambda_n(\lambda)\}_{n=n_0}^{\infty}$ *satisfies* (**L.c.(k)**) *for* $k = 1, \ldots, d$*.*
ii. *there exists an* $N_1 \geq n_0$ *such that for all* $n > N_1$*, and all* $\lambda \in \mathfrak{I}$*, the matrices* $\Lambda_n(\lambda)$ *are diagonal matrices.*
iii. *there exists a* $C > 0$ *such that* $\sup_{\lambda\in\mathfrak{I}} \sum_{n=n_0}^{\infty} \frac{\|R_n(\lambda)\|}{|\nu_n^{(k)}(\lambda)|} < C$
iv. *for any* $\epsilon > 0$ *there exists an* N_ϵ *(which depends only on* ϵ*) such that* $\forall \lambda \in \mathfrak{I}$ *we have*

$$\sum_{n=N_\epsilon}^{\infty} \frac{\|R_n(\lambda)\|}{|\nu_n^{(k)}(\lambda)|} < \epsilon,$$

then there exists a natural number $N_0 \geq N_1$ *such that the operator* T_k *defined for every* $\vec{\varphi} = \{\vec{\varphi}_n(\lambda)\}_{n=N_0}^{\infty}$ *in* $X_k(N_0)$ $(k = 1, \ldots, d)$ *by the following expression*

$$\begin{aligned}(T_k\vec{\varphi})_n(\lambda) = &\sum_{m=N_0}^{n-1} P^{(1)}\Psi_n(\lambda)\Psi_{m+1}^{-1}(\lambda)R_m(\lambda)\vec{\varphi}_m(\lambda) \\ &- \sum_{m=n}^{\infty} P^{(2)}\Psi_n(\lambda)\Psi_{m+1}^{-1}(\lambda)R_m(\lambda)\vec{\varphi}_m(\lambda)\,,\end{aligned} \tag{9}$$

where $\Psi_n(\lambda) = \prod_{i=N_0}^{n-1} \Lambda_i(\lambda)$*, has the following properties:*

1. T_k *is a correctly defined operator on* $X_k(N_0)$*, that is, the series in* (9) *converges, provided that* $\vec{\varphi}$ *is in* $X_k(N_0)$*.*
2. $\|T_k\| < 1$
3. $T_k X_k(N_0) \subset X_k^0(N_0)$

Proof. We begin by choosing a natural number $N_0 \geq \max\{N_1, N\}$, where N_1 is from condition ii and N is from i (see definition of **L.c.(k)**) and such that

$$M \sup_{\lambda\in\mathfrak{I}} \sum_{n=N_0}^{\infty} \frac{\|R_n(\lambda)\|}{|\nu_n^{(k)}(\lambda)|} < 1\,. \tag{10}$$

Firstly we show that T_k is correctly defined on the space $X_k(N_0)$. Let us begin by estimating the series in (9).

$$P^{(2)}\Psi_n(\lambda)\Psi_{m+1}^{-1}(\lambda) = \operatorname{diag}\{h_{n,m}^{(l)}(\lambda)\}_{l=1}^{d}\,,$$

where

$$h_{n,m}^{(l)}(\lambda) = \begin{cases} \dfrac{\prod_{i=N_0}^{n-1} \nu_i^{(l)}(\lambda)}{\prod_{i=N_0}^{m} \nu_i^{(l)}(\lambda)} & l \in I_2, \\ 0, & \text{else}\,. \end{cases}$$

Therefore, taking into account that $m \geq n$,

$$|h_{n,m}^{(l)}| = |\frac{\prod_{i=N_0}^{n-1} \nu_i^{(l)}(\lambda)}{\prod_{i=N_0}^{m} \nu_i^{(l)}(\lambda)}| = \frac{1}{|\nu_n^{(l)}\nu_{n+1}^{(l)}\cdots\nu_m^{(l)}|} = \frac{1}{|\nu_n^{(k)}\nu_{n+1}^{(k)}\cdots\nu_m^{(k)}|}\frac{|\nu_n^{(k)}\nu_{n+1}^{(k)}\cdots\nu_m^{(k)}|}{|\nu_n^{(l)}\nu_{n+1}^{(l)}\cdots\nu_m^{(l)}|}$$

$$\leq M\frac{1}{|\nu_n^{(k)}\nu_{n+1}^{(k)}\cdots\nu_m^{(k)}|} = M|\frac{\prod_{i=N_0}^{n-1} \nu_i^{(k)}(\lambda)}{\prod_{i=N_0}^{m} \nu_i^{(k)}(\lambda)}|.$$

Here, we have used (b). Thus, from (4) and condition iii, it is thus clear that for any sequence $\vec{\varphi} = \{\vec{\varphi}_n(\lambda)\}_{n=N_0}^{\infty}$ in $X_k(N_0)$ we have convergence of the series in (9).

Let us now study the first term of the equation

$$P^{(1)}\Psi_n(\lambda)\Psi_{m+1}^{-1}(\lambda) = \operatorname{diag}\{f_{n,m}^{(l)}(\lambda)\}_{l=1}^{d},$$

where

$$f_{n,m}^{(l)}(\lambda) = \begin{cases} \frac{\prod_{i=N_0}^{n-1} \nu_i^{(l)}(\lambda)}{\prod_{i=N_0}^{m} \nu_i^{(l)}(\lambda)} & l \in I_1, \\ 0, & \text{else.} \end{cases}$$

Hence, taking into account that in this case $m \leq n$ and using (a), we have

$$|f_{n,m}^{(l)}| = |\frac{\prod_{i=N_0}^{n-1} \nu_i^{(l)}(\lambda)}{\prod_{i=N_0}^{m} \nu_i^{(l)}(\lambda)}| = |\nu_{m+1}^{(l)}\nu_{m+2}^{(l)}\cdots\nu_{n-1}^{(l)}|$$

$$= |\nu_{m+1}^{(k)}\nu_{m+2}^{(k)}\cdots\nu_{n-1}^{(k)}|\frac{|\nu_{m+1}^{(l)}\nu_{m+2}^{(l)}\cdots\nu_{n-1}^{(l)}|}{|\nu_{m+1}^{(k)}\nu_{m+2}^{(k)}\cdots\nu_{n-1}^{(k)}|}$$

$$\leq M|\nu_{m+1}^{(k)}\nu_{m+2}^{(k)}\cdots\nu_{n-1}^{(k)}| = M|\frac{\prod_{i=N_0}^{n-1} \nu_i^{(k)}(\lambda)}{\prod_{i=N_0}^{m} \nu_i^{(k)}(\lambda)}|.$$

We now show that operator T_k is a contraction.

$$\|T_k\vec{\varphi}\|_{X_k(N_0)} = \sup_{n\geq N_0}\sup_{\lambda\in\mathfrak{I}}\{\frac{1}{|\prod_{i=N_0}^{n-1} \nu_i^{(k)}(\lambda)|}$$

$$\times \sum_{m=N_0}^{n-1} P^{(1)}\Psi_n\Psi_{m+1}^{-1}R_m(\lambda)\vec{\varphi}_m(\lambda) - \sum_{m=n}^{\infty} P^{(2)}\Psi_n\Psi_{m+1}^{-1}R_m(\lambda)\vec{\varphi}_m(\lambda)\|\}$$

$$\leq \sup_{n\geq N_0}\sup_{\lambda\in\mathfrak{I}}\{\frac{M}{|\prod_{i=N_0}^{n-1} \nu_i^{(k)}(\lambda)|}(\sum_{m=N_0}^{n-1} + \sum_{m=n}^{\infty})\|R_m(\lambda)\|\frac{|\prod_{i=N_0}^{n-1} \nu_i^{(k)}(\lambda)|}{|\prod_{i=N_0}^{m} \nu_i^{(k)}(\lambda)|}\|\vec{\varphi}_m(\lambda)\|\}$$

$$\leq \sup_{\lambda\in\mathfrak{I}}\{M\sum_{m=N_0}^{\infty}\frac{\|R_m(\lambda)\|}{|\nu_m^{(k)}(\lambda)|}\frac{\|\vec{\varphi}_m(\lambda)\|}{|\prod_{i=N_0}^{m-1} \nu_i^{(k)}(\lambda)|}\} \leq \|\vec{\varphi}\|_{X_k(N_0)}.$$

Note that in the second line we have written Ψ_n instead of $\Psi_n(\lambda)$ to simplify the writing of the formula.

It remains to prove property 3. We must show that if $\vec{\varphi} \in X_k(N_0)$ then (5) holds for $\{(T_k\vec{\varphi})_n(\lambda)\}_{n\geq N_0}$. Using the definition of T_k (see (9)) we obtain

$$\sup_{\lambda\in\mathfrak{J}}\{\|(T_k\vec{\varphi})_n(\lambda)\|_{\mathbb{C}^d}\frac{1}{|\prod_{i=N_0}^{n-1}\nu_i^{(k)}(\lambda)|}\}$$

$$\leq \sup_{\lambda\in\mathfrak{J}}\Big\{\frac{1}{|\prod_{i=N_0}^{n-1}\nu_i^{(k)}(\lambda)|}\|\sum_{m=N_0}^{n-1}P^{(1)}\Psi_n(\lambda)\Psi_{m+1}^{-1}(\lambda)R_m(\lambda)\vec{\varphi}_m(\lambda)\|\Big\} \quad (11)$$

$$+\sup_{\lambda\in\mathfrak{J}}\Big\{\frac{1}{|\prod_{i=N_0}^{n-1}\nu_i^{(k)}(\lambda)|}\|\sum_{m=n}^{\infty}P^{(2)}\Psi_n(\lambda)\Psi_{m+1}^{-1}(\lambda)R_m(\lambda)\vec{\varphi}_m(\lambda)\|\Big\}.$$

Since $\vec{\varphi}$ is in $X_k(N_0)$ it follows at once from (b) that the second term on the right-hand side of the last expression can be done as little as we want if we take n sufficiently big.

Consider now the first term on the right-hand side of (11).

$$\sup_{\lambda\in\mathfrak{J}}\Big\{\frac{1}{|\prod_{i=N_0}^{n-1}\nu_i^{(k)}(\lambda)|}\|\sum_{m=N_0}^{n-1}P^{(1)}\Psi_n(\lambda)\Psi_{m+1}^{-1}(\lambda)R_m(\lambda)\vec{\varphi}_m(\lambda)\|\Big\}$$

$$\leq \sup_{\lambda\in\mathfrak{J}}\Big\{\frac{1}{|\prod_{i=N_0}^{n-1}\nu_i^{(k)}(\lambda)|}\sum_{m=N_0}^{s-1}\|P^{(1)}\Psi_n(\lambda)\Psi_{m+1}^{-1}(\lambda)R_m(\lambda)\vec{\varphi}_m(\lambda)\|\Big\}$$

$$+\sup_{\lambda\in\mathfrak{J}}\Big\{\frac{1}{|\prod_{i=N_0}^{n-1}\nu_i^{(k)}(\lambda)|}\sum_{m=s}^{\infty}\|P^{(1)}\Psi_n(\lambda)\Psi_{m+1}^{-1}(\lambda)R_m(\lambda)\vec{\varphi}_m(\lambda)\|\Big\}.$$

Provided that $s > N_0$ this inequality holds for every $n \geq N_0$ no matter how big it is. Now, since

$$\sup_{\lambda\in\mathfrak{J}}\Big\{\frac{1}{|\prod_{i=N_0}^{n-1}\nu_i^{(k)}(\lambda)|}\sum_{m=s}^{\infty}\|P^{(1)}\Psi_n(\lambda)\Psi_{m+1}^{-1}(\lambda)R_m(\lambda)\vec{\varphi}_m(\lambda)\|\Big\}$$

$$\leq M\|\vec{\varphi}\|_{X_k(s)}\sup_{\lambda\in\mathfrak{J}}\sum_{m=s}^{\infty}\frac{\|R_m(\lambda)\|}{|\nu_n^{(k)}(\lambda)|},$$

and taking into account that $\{\vec{\varphi}_n(\lambda)\}_{n\geq N_0} \in X_k(N_0)$ implies $\{\vec{\varphi}_n(\lambda)\}_{n\geq s} \in X_k(s)$, there exists an $s > N_0$, such that for every $\epsilon_1 > 0$

$$\sup_{\lambda\in\mathfrak{J}}\Big\{\frac{1}{|\prod_{i=N_0}^{n-1}\nu_i^{(k)}(\lambda)|}\sum_{m=s}^{\infty}\|P^{(1)}\Psi_n(\lambda)\Psi_{m+1}^{-1}(\lambda)R_m(\lambda)\vec{\varphi}_m(\lambda)\|\Big\} \leq \epsilon_1.$$

Thus, we have shown that the first term on the right-hand side of (11) is estimated as follows

$$\sup_{\lambda\in\mathfrak{J}}\Big\{\frac{1}{|\prod_{i=N_0}^{n-1}\nu_i^{(k)}(\lambda)|}\|\sum_{m=N_0}^{n-1}P^{(1)}\Psi_n(\lambda)\Psi_{m+1}^{-1}(\lambda)R_m(\lambda)\vec{\varphi}_m(\lambda)\|\Big\}$$
$$\leq \sup_{\lambda\in\mathfrak{J}}\Big\{\frac{1}{|\prod_{i=N_0}^{n-1}\nu_i^{(k)}(\lambda)|}\sum_{m=N_0}^{s-1}\|P^{(1)}\Psi_n(\lambda)\Psi_{m+1}^{-1}(\lambda)R_m(\lambda)\vec{\varphi}_m(\lambda)\|\Big\}+\epsilon_1 .$$

But

$$\sup_{\lambda\in\mathfrak{J}}\Big\{\frac{1}{|\prod_{i=N_0}^{n-1}\nu_i^{(k)}(\lambda)|}\sum_{m=N_0}^{s-1}\|P^{(1)}\Psi_n(\lambda)\Psi_{m+1}^{-1}(\lambda)R_m(\lambda)\vec{\varphi}_m(\lambda)\|\Big\}$$
$$\leq C\sup_{\lambda\in\mathfrak{J}}\Big\{\frac{\|P^{(1)}\Psi_n(\lambda)\|}{|\prod_{i=N_0}^{n-1}\nu_i^{(k)}(\lambda)|}\Big\},$$

where C is a positive constant. Now, as a consequence of (a) we know that

$$\sup_{\lambda\in\mathfrak{J}}\Big\{\frac{\|P^{(1)}\Psi_n(\lambda)\|}{|\prod_{i=N_0}^{n-1}\nu_i^{(k)}(\lambda)|}\Big\}=\sup_{\lambda\in\mathfrak{J}}\Big\{\frac{|\prod_{i=N_0}^{n-1}\nu_i^{(l)}(\lambda)|}{|\prod_{i=N_0}^{n-1}\nu_i^{(k)}(\lambda)|}\Big\}\to 0 \qquad \text{as } n\to\infty .$$

This completes the proof of the lemma. □

Now we show that, using the operator introduced in the previous lemma, we can write an equation which turns out to be equivalent to a recurrence linear system of the form (2) defined by the sequences $\{\Lambda_n(\lambda)\}_{n\geq n_0}$ and $\{R_n(\lambda)\}_{n\geq n_0}$.

Lemma 3.2. *Let the sequences of function matrices* $\{\Lambda_n(\lambda)\}_{n\geq n_0}$ *and* $\{R_n(\lambda)\}_{n\geq n_0}$ *satisfy the conditions of the previous lemma. Let* $N_0 \geq n_0$ *be such that* T_k *given by (9) is correctly defined*[2]*. The sequence of function vectors* $\vec{\varphi}=\{\vec{\varphi}_n^{(k)}(\lambda)\}_{n\geq N_0}$ *in* $X_k(N_0)$ *is a solution of the recurrence equation*

$$\vec{\varphi}_{n+1}(\lambda)=(\Lambda_n(\lambda)+R_n(\lambda))\vec{\varphi}_n(\lambda) \qquad n\geq N_0 \tag{12}$$

iff $\vec{\varphi}$ *satisfies equation*

$$\vec{\varphi}=\vec{\psi}^{(k)}+T_k\vec{\varphi} \tag{13}$$

where the sequence $\vec{\psi}^{(k)}=\{\vec{\psi}_n^{(k)}(\lambda)\}_{n\geq N_0}$ *is defined by*

$$\vec{\psi}_n^{(k)}=\Psi_n(\lambda)\vec{e}_k=\prod_{i=N_0}^{n-1}\nu_i^{(k)}(\lambda)\vec{e}_k$$

Proof. The proof is straightforward. Using the definition of T_k in (13) and substituting this into (12) one obtains an identity (it has to be taken into account that $\Lambda_n(\lambda)\Psi_n(\lambda)=\Psi_{n+1}(\lambda)$ and that $P^{(1)}+P^{(2)}=I$, where I is the $d\times d$-identity matrix). □

---

[2]The existence of N_0 is guaranteed by Lemma 3.1.

Remark 2. A straightforward consequence of T_k's property 3, stated in Lemma 3.1, is that if $\vec{\varphi} \in X_k(N_0)$ satisfies equation (13) then

$$\vec{\varphi} - \vec{\psi}^{(k)} \in X_k^0(N_0)\,. \tag{14}$$

It is of practical importance to deal with systems such as (12), but when the sequence $\{\Lambda_n(\lambda)\}_{n\geq n_0}$ is not necessarily diagonal. The following lemma gives sufficient conditions for a sequence of matrices to be diagonalizable in a specific sense.

Lemma 3.3. *Let $\{V_n(\lambda)\}_{n\geq n_0}$ be a sequence of function matrices defined on every λ in some interval $\mathfrak{I} \subset \mathbb{R}$, such that*

i. $\{V_n(\lambda)\}_{n\geq n_0}$ *is in* $\widetilde{\mathcal{D}}^1$
ii. $V_n(\lambda) \overset{\lambda\in\mathfrak{I}}{\rightrightarrows} V_\infty$ *as* $n \to \infty$, *where* V_∞ *is a* $d \times d$ *matrix with pairwise distinct eigenvalues* $\{\nu^{(k)}\}_{k=1}^d$

Then we can find an $m \geq n_0$ such that there exists a sequence of diagonal matrices $\{\Lambda_n(\lambda)\}_{n\geq m}$ and a sequence of invertible matrices $\{G_n(\lambda)\}_{n\geq m}$ such that

1. $\{\Lambda_n(\lambda)\}_{n\geq m} \in \widetilde{\mathcal{D}}^1$
2. $\{G_n(\lambda)\}_{n\geq m} \in \widetilde{\mathcal{D}}^1$
3. $V_n(\lambda) = G_n(\lambda)\Lambda_n(\lambda)G_n^{-1}(\lambda)$ *for* $n \geq m$ *for all* $\lambda \in \mathfrak{I}$
4. $\Lambda_n(\lambda) \overset{\lambda\in\mathfrak{I}}{\rightrightarrows} \Lambda$ *as* $n \to \infty$, *where* $\Lambda = \operatorname{diag}\{\nu^{(k)}\}_{k=1}^d$
5. $G_n(\lambda) \overset{\lambda\in\mathfrak{I}}{\rightrightarrows} G$ *as* $n \to \infty$, *where* G *is invertible and* $V_\infty = G\Lambda G^{-1}$

Proof. The proof is very similar to the one in [6] for the non-parametric analogous assertion.

As we have done before, let us denote by $\{\nu_n^{(k)}(\lambda)\}_{k=1}^d$ the set of eigenvalues of $V_n(\lambda)$. Since the matrix V_∞ has pairwise distinct eigenvalues, by choosing $\epsilon > 0$ small enough, the operators

$$\mathcal{P}^{(k)} = \oint\limits_{|z-\nu^{(k)}|=\epsilon} (V_\infty - zI)^{-1} dz \quad k = 1, \ldots, d \tag{15}$$

are projectors onto the eigenspaces of V_∞ corresponding to $\nu^{(k)}$ $(k = 1, \ldots, d)$. Now, taking ϵ even smaller if necessary, it is clear that there exists some $N \geq n_0$, such that the operators, defined for all $\lambda \in \mathfrak{I}$ and each $n \geq N$ by

$$\mathcal{P}_n^{(k)}(\lambda) = \oint\limits_{|z-\nu_n^{(k)}(\lambda)|=\epsilon} (V_n(\lambda) - zI)^{-1} dz \quad k = 1, \ldots, d \tag{16}$$

are projectors onto the eigenspaces of $V_n(\lambda)$ corresponding to the eigenvalue $\nu_n^{(k)}(\lambda)$ $(k = 1, \ldots, d)$. From the conditions of the lemma it is clear that $\nu_n^{(k)}(\lambda) \overset{\lambda\in\mathfrak{I}}{\rightrightarrows} \nu^{(k)}$ and, therefore, $k \neq j$ implies

$$\nu_n^{(k)}(\lambda) \neq \nu_n^{(j)}(\lambda) \quad \forall\, n \geq N, \text{ and } \forall\, \lambda \in \mathfrak{I}\,. \tag{17}$$

We define the matrices

$$\Lambda_n(\lambda) = \operatorname{diag}\{\nu_n^{(k)}(\lambda)\}_{k=1}^d$$

From (17) we have property 4. Furthermore, using (17) and ii, from (15), (16) we have

$$\mathcal{P}_n^{(k)}(\lambda) \overset{\lambda\in\mathfrak{I}}{\rightrightarrows} \mathcal{P}^{(k)} \quad \text{and} \quad \{\mathcal{P}_n^{(k)}(\lambda)\}_{n\geq N} \in \widetilde{\mathcal{D}}^1$$

Now, for $k = 1, \ldots, d$ we choose an arbitrary non-zero vector $\vec{v}_k$ in $\operatorname{Ran}(P_k)$ and construct the function matrix $G_n(\lambda)$ with columns

$$\mathcal{P}_n^{(1)}(\lambda)\vec{v}_1, \ldots, \mathcal{P}_n^{(d)}(\lambda)\vec{v}_d\,.$$

Therefore

$$V_n(\lambda)G_n(\lambda) = G_n(\lambda)\Lambda_n(\lambda) \quad \text{for } n \geq N, \quad \forall\lambda \in \mathfrak{I}.$$

Now, since $\mathcal{P}_n^{(k)}(\lambda)\vec{v}_k \overset{\lambda\in\mathfrak{I}}{\rightrightarrows} \mathcal{P}^{(k)}\vec{v}_k = \vec{v}_k \neq 0$ as $n \to \infty$, we can find an $m \geq N$ such that $G_n(\lambda)$ are invertible for $n \geq m$. Thus property 3 holds. Moreover, $\{\mathcal{P}_n^{(k)}(\lambda)\}_{n\geq m} \in \widetilde{\mathcal{D}}^1$ implies 2. Now, let us consider the matrix G with columns given by the vectors $\vec{v}_1, \ldots, \vec{v}_d$. Clearly

$$G_n(\lambda) \overset{\lambda\in\mathfrak{I}}{\rightrightarrows} G \tag{18}$$

and G is invertible. Thus 5 holds. So this, together with 2, implies

$$\{G_n(\lambda)^{-1}\}_{n\geq m} \in \widetilde{\mathcal{D}}^1\,.$$

Finally property 1 takes place since $\widetilde{\mathcal{D}}^1$ is an algebra. □

Remark 3. It is not difficult to understand that if the sequence $\{V_n(\lambda)\}_{n\geq m}$ satisfies the Levinson condition (**L.c.(k)**) for a certain k, then the diagonalized sequence $\{\Lambda_n(\lambda)\}_{n\geq m}$ also satisfies this condition for the same k.

4. The asymptotic behavior of solutions

In this section we obtain Levinson type assertions that give a uniform estimate of the remainder in the asymptotic expansion.

Theorem 4.1. *Let the sequences $\{\Lambda_n(\lambda)\}_{n\geq n_0}$ and $\{R_n(\lambda)\}_{n\geq n_0}$ satisfy the conditions of Lemma* 3.1. *Then we can find an $N_0 \in \mathbb{N}$ such that there exists a basis $\{\vec{\varphi}_n^{(k)}(\lambda)\}_{k=1}^d$ in the space of solutions of* (12) *satisfying*

$$\sup_{\lambda\in\mathfrak{I}}\left\|\frac{\vec{\varphi}_n^{(k)}(\lambda)}{\prod_{i=m}^{n-1}\nu_i^{(k)}(\lambda)} - \vec{e}_k\right\| \to 0, \quad \textit{as } n \to \infty\,, \quad \textit{for } k = 1, \ldots, d\,. \tag{19}$$

Proof. The statement of the theorem easily follows from Lemmas 3.1 and 3.2. Note that property 2 implies that (13) yields linearly independent solutions for $k = 1, \ldots, d$. Equation (19) follows directly from (14). □

Now we use Lemma 3.3 to replace the restricting condition of diagonality that in Theorem 4.1 is imposed on $\{\Lambda_n(\lambda)\}_{n\geq n_0}$ by a weaker assumption, viz., that a sequence is diagonalizable in the sense specified in Lemma 3.3.

Theorem 4.2. *Let V_∞ be a $d\times d$ matrix whose non-zero eigenvalues $\nu^{(k)}$ are pairwise distinct. Consider the sequences $\{V_n(\lambda)\}_{n=n_0}^{\infty}$ and $\{R_n(\lambda)\}_{n=n_0}^{\infty}$, such that, for every λ in some interval $\mathfrak{I}\subset\mathbb{R}$, $V_n(\lambda)$ and $R_n(\lambda)$ are $d\times d$ complex matrices and $V_n(\lambda)$ has the eigenvalues $\{\nu_n^{(k)}(\lambda)\}_{k=1}^{d}$, such that $\nu_n^{(k)}(\lambda)\neq 0$ $\forall\, k,n,\lambda$. If the following conditions hold*

i. $V_n(\lambda) \overset{\lambda\in\mathfrak{I}}{\rightrightarrows} V_\infty$ *as* $n\to\infty$.
ii. *the sequence* $\{V_n(\lambda)\}_{n=n_0}^{\infty}$ *satisfies* (**L.c.(k)**) *for* $k=1,\dots,d$.
iii. $\{V_n(\lambda)\}_{n=n_0}^{\infty}\in\widetilde{\mathcal{D}}^1$
iv. *there exists a* $C>0$ *such that* $\sup_{\lambda\in\mathfrak{I}}\sum_{n=n_0}^{\infty}\|R_n(\lambda)\|<C$.
v. *for any* $\epsilon>0$ *there exists a* N_ϵ *(which depends only on* ϵ*) such that* $\forall\lambda\in\mathfrak{I}$ *we have*

$$\sum_{n=N_\epsilon}^{\infty}\|R_n(\lambda)\|<\epsilon,$$

then we can find an $m\in\mathbb{N}$ such that there exists a basis $\{\vec{x}_n^{(k)}(\lambda)\}_{k=1}^{d}$ in the space of solutions of the recurrence relation

$$\vec{x}_{n+1}^{(k)}(\lambda)=(V_n(\lambda)+R_n(\lambda))\vec{x}_n^{(k)}(\lambda)\quad \text{for } n\geq m \tag{20}$$

such that

$$\sup_{\lambda\in\mathfrak{I}}\Big\|\frac{\vec{x}_n^{(k)}(\lambda)}{\prod_{i=m}^{n-1}\nu_i^{(k)}(\lambda)}-\vec{p}_k\Big\|\to 0,\qquad \text{as } n\to\infty,$$

where $\vec{p}_k$ is the eigenvector of V_∞ corresponding to $\nu^{(k)}$.

Proof. Since $\{V_n(\lambda)\}_{n=n_0}^{\infty}$ satisfies here the conditions of Lemma 3.3 we can find an $m_1\geq n_0$ such that $\forall\lambda\in\mathfrak{I}$

$$V_n(\lambda)=G_n^{-1}(\lambda)\Lambda_n(\lambda)G_n(\lambda)\quad \text{for } n\geq m_1\,.$$

Therefore, letting $\vec{\varphi}_n^{(k)}(\lambda)=G_n(\lambda)\vec{x}_n^{(k)}(\lambda)$ in (20), considered for $n\geq m_1$, we have

$$\vec{\varphi}_{n+1}^{(k)}(\lambda)=G_{n+1}(\lambda)\Big(V_n(\lambda)+R_n(\lambda)\Big)G_n^{-1}(\lambda)\vec{\varphi}_n^{(k)}(\lambda)\quad \text{for } n\geq m_1$$

This recurrence system can be rewritten as follows

$$\vec{\varphi}_{n+1}^{(k)}=\Big(\Lambda_n(\lambda)+\widetilde{R}_n(\lambda)\Big)\vec{\varphi}_n^{(k)}, \tag{21}$$

where $\widetilde{R}_n(\lambda)=(\Delta G)_n(\lambda)V_n(\lambda)G_n^{-1}(\lambda)+G_{n+1}(\lambda)R_n(\lambda)G_n^{-1}(\lambda)$ (recall the operator Δ introduced in Sec. 2). Clearly the sequence $\{\Lambda_n(\lambda)\}_{n=m_1}^{\infty}$, by Lemma 3.3, satisfies the conditions of Lemma 3.1, in particular (**L.c.(k)**) for $k=1,\dots,d$. Next we show that the sequence $\{\widetilde{R}_n(\lambda)\}_{n\geq m_1}$ satisfies the conditions iii and iv of Lemma 3.1. Indeed, considering condition i and property 5 of Lemma 3.3, it is not difficult to show that we can find an $m_2 \geq m_1$, such that there exist $C_1, C_2, C_3 > 0$ for which the following estimates hold

$$\begin{aligned} &\|\widetilde{R}_n(\lambda)\| \leq C_1\|(\Delta G)_n(\lambda)\| + C_2\|R_n(\lambda)\| \quad \forall\lambda \in \mathfrak{I}\,,\ \forall n \geq m_2 \\ &|\nu_n^{(k)}(\lambda)| \geq C_3 \quad \forall\lambda \in \mathfrak{I}\,,\ \forall n \geq m_2 \quad (k = 1, \ldots, d)\,. \end{aligned} \tag{22}$$

Now, since $\{G_n(\lambda)\}_{n=m_1}^{\infty}$ is in $\widetilde{\mathcal{D}}^1$, there exist a $C > 0$ such that

$$\sum_{n=m_1}^{\infty} \sup_{\lambda\in\mathfrak{I}}\|(\Delta G)_n(\lambda)\| < C\,.$$

This last estimate, together with (22) and conditions iv and v, yields the boundedness of

$$\sup_{\lambda\in\mathfrak{I}} \sum_{n=m_2}^{\infty} \frac{\|\widetilde{R}_n(\lambda)\|}{|\nu_n^{(k)}(\lambda)|}$$

and the uniform tail's estimate

$$\sup_{\lambda\in\mathfrak{I}} \sum_{n=l}^{\infty} \frac{\|\widetilde{R}_n(\lambda)\|}{|\nu_n^{(k)}(\lambda)|} \to 0 \quad \text{as } l \to \infty$$

Hence the conditions of Lemmas 3.1 and 3.2 are satisfied. This means that we can apply Theorem 4.1 to the system (21). Therefore for an $m \geq m_2$ there exists a basis $\{\vec{\varphi}^{(k)}(\lambda)\}_{k=1}^{d}$ in the space of solutions of (21) such that

$$\sup_{\lambda\in\mathfrak{I}}\|\vec{\varphi}_n^{(k)}(\lambda)/\prod_{i=m}^{n-1}\nu_i^{(k)}(\lambda) - \vec{e}_k\| \to 0, \quad \text{as } n \to \infty\,,$$

Thus, (20) has the linearly independent solutions $\vec{x}_n^{(k)}(\lambda) = G_n^{-1}(\lambda)\vec{\varphi}_n^{(k)}(\lambda)$ for $k = 1, \ldots, d$. It is also not difficult to show that

$$\sup_{\lambda\in\mathfrak{I}}\|\vec{x}_n^{(k)}(\lambda)/\prod_{i=m}^{n-1}\nu_i^{(k)}(\lambda) - \vec{p}_k\| \to 0, \quad \text{as } n \to \infty\,,$$

where $\vec{p}_k = G^{-1}\vec{e}_k$ is the eigenvector of V_∞ corresponding to $\nu^{(k)}$. □

Remark 4. If in Theorem 4.2 we consider the sequences $\{V_n(\lambda)\}_{n\geq n_0}$ and V_∞ defined as in Remark 1, then V_∞ satisfies the conditions of the theorem and i and ii are fulfilled. In this case we have the "classical" discrete Levinson theorem [7], but in its uniform variant.

5. Asymptotics of the generalized eigenvectors of a Jacobi matrix with rapidly increasing weights

Let us consider an operator J in $l_2(\mathbb{N})$ whose matrix representation with respect to the canonical basis in $l_2(\mathbb{N})$ is a Jacobi matrix of the form

$$\begin{pmatrix} 0 & b_1 & 0 & 0 & \cdots \\ b_1 & 0 & b_2 & 0 & \cdots \\ 0 & b_2 & 0 & b_3 & \cdots \\ \vdots & \vdots & \vdots & & \end{pmatrix}, \tag{23}$$

where $b = \{b_n\}_{n\in\mathbb{N}}$ is a sequence of positive numbers. By means of the subordinacy theory, the spectral characterization of J can be done by analyzing the asymptotic behavior, as $n \to \infty$, of the solutions of the following equation

$$b_{n-1}u_{n-1} + b_n u_{n+1} = \lambda u_n \qquad n > 1\,, \quad \lambda \in \mathbb{R}\,. \tag{24}$$

The solutions $u = \{u_n(\lambda)\}_{n=1}^{\infty}$ of (24) are called the generalized eigenvectors of J corresponding to λ. Equation (24) can be written as follows

$$\vec{u}_{n+1} = B_n(\lambda)\vec{u}_n\,, \qquad n > 1\,, \quad \lambda \in \mathbb{R}\,. \tag{25}$$

where $\vec{u}_n(\lambda) = \begin{pmatrix} u_{n-1}(\lambda) \\ u_n(\lambda) \end{pmatrix}$ and $B_n(\lambda) = \begin{pmatrix} 0 & 1 \\ -\frac{b_{n-1}}{b_n} & \frac{\lambda}{b_n} \end{pmatrix}$ is the so-called transfer matrix. It turns out that it is possible to group the transfer matrices $B_n(\lambda)$ in such a way that the resulting system satisfies the conditions of Theorem 4.2 or, at least Theorem 4.1. Let us illustrate this for the case when the sequence $b = \{b_n\}_{n=1}^{\infty}$ is defined by

$$b_n = n^{\alpha}\bigl(1 + \frac{c_n}{n}\bigr), \qquad n \in \mathbb{N}\,, \tag{26}$$

where $1 < \alpha < 2$ and $\{c_n\}_{n\in\mathbb{N}}$ is a positive periodic sequence of even period, i.e., $c_{n+T} = c_n$ for all $n \in \mathbb{N}$ and a fixed even natural number $T = 2L$. This class of Jacobi matrices with rapidly growing weights is based on an example suggested by A.G. Kostyuchenko and K.A. Mirzoev [12][3]. The asymptotics of the generalized eigenvectors for this class, has been studied in [7] and [13] with different approaches. We use here mainly the same reasoning as in [7], but applying our uniform version of the Levinson theorem.

First we define how we should group the transfer matrices in order to apply the main result of this paper. Consider the following recurrence equation for the sequences of two-dimensional vectors $\vec{x}(\lambda) = \{\vec{x}_n(\lambda)\}_{n=1}^{\infty}$

$$\vec{x}_{n+1}(\lambda) = A_n(\lambda)\vec{x}_n(\lambda)\,, \qquad n \in \mathbb{N}\,, \quad \lambda \in \mathbb{R}\,. \tag{27}$$

where $A_n(\lambda) = \prod_{k=0}^{T-1} B_{nT+k}(\lambda)$. It is clear that for any solution $\vec{x}(\lambda)$ of (27) there exists a unique sequence u being a solution of (24) whose elements are determined by the second component of the vector $\vec{u}_{nT}(\lambda) = \vec{x}_n(\lambda)$. Moreover, there is an

[3] Actually, this class is not restricted to the case $\alpha < 2$. This restriction is not important for this example, but simplifies some formulas.

isomorphism $\vec{x} \mapsto u$ of the space of solutions of (27) to the space of solutions of (24) and we have for every $n \in \mathbb{N}$ and the integers $s = 1, \ldots, T$

$$u_{nT+s}(\lambda) = \left(\prod_{k=0}^{s-1} B_{nT+k} \vec{x}_n(\lambda), \ \vec{e}_2 \right)_{\mathbb{C}^2}, \tag{28}$$

where $\vec{e}_2 = \binom{0}{1}$ ($\vec{e}_1 = \binom{1}{0}$) and $(\cdot,\cdot)_{\mathbb{C}^2}$ is the inner product in $\mathbb{C}^2$, which is used to single out the second component of the vector.

Let us calculate $A_n(\lambda)$ for our particular matrix. Simple algebraic computations give us

$$B_{nT+k}(\lambda) = \mathcal{E} + \frac{1}{n}\mathcal{M}_k + \frac{1}{n^\alpha}\mathcal{N}_k(n,\lambda) + \frac{1}{n^2}\mathcal{Q}_k(n), \tag{29}$$

where

$$\mathcal{E} = \begin{pmatrix} 0 & 1 \\ -1 & 0 \end{pmatrix}, \quad \mathcal{M}_k = \begin{pmatrix} 0 & 0 \\ \frac{\alpha + c_k - c_{k-1}}{T} & 0 \end{pmatrix}, \quad \mathcal{N}_k(n,\lambda) = \begin{pmatrix} 0 & 0 \\ 0 & \frac{\lambda}{T^\alpha(1+O(\frac{1}{n}))} \end{pmatrix},$$

and $\mathcal{Q}_k(n)$ is a matrix which does not depend on λ and tends to a constant matrix when $n \to \infty$. Thus,

$$A_n(\lambda) = (-1)^L I + \frac{1}{n}\sum_{k=0}^{T-1} \mathcal{E}^{T-1-k}\mathcal{M}_k\mathcal{E}^k + \frac{1}{n^\alpha}\mathcal{R}_n(\lambda), \tag{30}$$

where I is the 2×2-identity matrix and where the sequence of matrices $\{\mathcal{R}_n(\lambda)\}_{n\in\mathbb{N}}$ tends, uniformly with respect to λ in any finite interval $\mathfrak{I}$ of $\mathbb{R}$, to a constant matrix. Indeed, from (29) it straightforwardly follows that each element of $\mathcal{R}_n(\lambda)$ is a product of a polynomial of λ and a function of n which does not depend on λ and tends to a constant as $n \to \infty$. As regards the second term of the right-hand side of (30) we have

$$\frac{1}{n}\sum_{k=0}^{T-1} \mathcal{E}^{T-1-k}\mathcal{M}_k\mathcal{E}^k = \frac{(-1)^{L-1}}{n}\begin{pmatrix} \frac{\alpha}{2} + \frac{\sum_{j=1}^{2L}(-1)^j c_j}{T} & 0 \\ 0 & \frac{\alpha}{2} - \frac{\sum_{j=1}^{2L}(-1)^j c_j}{T} \end{pmatrix}. \tag{31}$$

Now, assuming that

$$V_n(\lambda) := (-1)^L\left(I - \frac{1}{n}\begin{pmatrix} \frac{\alpha}{2} + \frac{\sum_{j=1}^{2L}(-1)^j c_j}{T} & 0 \\ 0 & \frac{\alpha}{2} - \frac{\sum_{j=1}^{2L}(-1)^j c_j}{T} \end{pmatrix}\right)$$

and

$$R_n(\lambda) := \frac{1}{n^\alpha}\mathcal{R}_n(\lambda)$$

it is easy to show that the sequences $\{V_n(\lambda)\}_{n\in\mathbb{N}}$ and $\{R_n(\lambda)\}_{n\in\mathbb{N}}$ satisfy the conditions of Theorem 4.1 for the system (27). Thus, there exists an $m \in \mathbb{N}$ such that, as $n \to \infty$,

$$\sup_{\lambda\in\mathfrak{I}}\|x_n^{(k)}(\lambda) - (-1)^{nL}\prod_{l=m}^{n-1}\left(1 - l^{-1}\left(\frac{\alpha}{2} + (-1)^{k+1}\frac{\sum_{j=1}^{2L}(-1)^j c_j}{T}\right)\right)\vec{e}_k\| \to 0\,,$$

where $k = 1, 2$.

Now, reasoning as in [7], one can obtain a basis $u^{(k)} = \{u_n^{(k)}\}_{n\geq m}$, $k = 1, 2$, in the space of solutions of (24) such that, for $s = 1, \dots, T$,

$$u_{nT+s}^{(k)}(\lambda) = (-1)^{nL}\prod_{l=m}^{n-1}\left(1 - l^{-1}\left(\frac{\alpha}{2} + (-1)^{k+1}\frac{\sum_{j=1}^{2L}(-1)^j c_j}{T}\right)\right)w_{nT+s}^{(k)}(\lambda)\,,$$

where

$$\sup_{\lambda\in\mathfrak{I}}\|w_n^{(k)}(\lambda) - (\mathcal{E}^n\vec{e}_k, \vec{e}_2)_{\mathbb{C}^2}\| \to 0\,, \qquad \text{as } n \to \infty\,, \quad k = 1, 2\,.$$

Hence we have obtained uniform estimates of the asymptotic remainder term for the generalized eigenvectors of J.

References

[1] Z. Benzaid and D. A. Lutz. Asymptotic representation of solutions of perturbed systems of linear difference equations. *Studies Appl. Math.*, **77**(1987) 195–221.

[2] E. Coddington and N. Levinson. *Theory of Ordinary Differential Equations.* McGraw-Hill, New York, 1955.

[3] S. N. Elaydi. *An Introduction to Difference Equations.* Springer-Verlag, New York, 1999.

[4] D. Gilbert. On subordinacy and analysis of the spectrum of Schrödinger operators with two singular end points. *Proc. Roy. Soc. Edinb.*, **112A**(1989) 213–229.

[5] D. Gilbert and D. Pearson. On subordinacy and analysis of the spectrum of one-dimensional Schrödinger operators. *J. Math. Anal. Appl.*, **128**(1987) 30–56.

[6] J. Janas and M. Moszyński. Alternative approaches to the absolute continuity of Jacobi matrices with monotonic weights. *Integr. Equ. Oper. Theory*, **43**(2002) 397–416.

[7] J. Janas and M. Moszyński. Spectral properties of Jacobi matrices by asymptotic analysis. *J. Approx. Theory*, **120**(2003) 309–336.

[8] J. Janas and S. Naboko. Multithreshold spectral phase transitions for a class of Jacobi matrices. In *Recent advances in operator theory (Groningen, 1998)*, volume 124 of *Oper. Theory Adv. Appl.* Birkhäuser, Basel, 2001, 267–285.

[9] J. Janas and S.N. Naboko. Jacobi matrices with absolutely continuous spectrum. *Proc. Amer. Math. Soc.*, **127**(1999) 791–800.

[10] J. Janas and S.N. Naboko. Jacobi matrices with power-like weights, grouping in blocks approach. *J. Functional Analysis*, **166**(1999) 218–243.

[11] S. Khan and D. Pearson. Subordinacy and spectral theory for infinite matrices. *Helv. Phys. Acta*, **65**(1992) 505–527.

[12] A.G. Kostyuchenko and K.A. Mirzoev. Generalized Jacobi matrices and deficiency indices of differential operators with polynomial coefficients. *Funct. Anal. Appl.*, **33**(1999) 30–45.

[13] L.O. Silva. Spectral properties of Jacobi matrices with rapidly growing power-like weights. In *Operator methods in ordinary and partial differential equations (Stockholm, 2000)*, volume 132 of *Oper. Theory Adv. Appl.* Birkhäuser, Basel, 2002, 387–394.

[14] G. Stolz. Spectral theory of slowly oscillating potentials I. Jacobi matrices. *Manuscripta Math.*, **84**(1994) 245–260.

Luis O. Silva
Departamento de Métodos Matemáticos y Numéricos
IIMAS-UNAM
Apdo. Postal 20-726
Mexico, D.F. 01000
e-mail: silva@leibniz.iimas.unam.mx

Operator Theory:
Advances and Applications, Vol. 154, 219–231

Free Functional Model Related to Simply-connected Domains

Alexey Tikhonov

Abstract. The aim of this paper is to extend the S.-Nagy-Foiaş functional model to the case of functions for contractive operators. This model meets general requirements and can be used for studying trace class perturbations of normal operators with spectrum on a curve.

Mathematics Subject Classification (2000). Primary 47A45; Secondary 47A48, 47A55.

Keywords. functional model, operator, Hilbert space.

Introduction

The functional model of S.-Nagy-Foiaş [1] of contractive (dissipative) operators is well known and has many important and useful applications. It was generalized many times and in various directions. Here we single out three of them: 1) models of unitary nodes [2]; 2) models of operators those are analogous of contractive operators [3, 4]; 3) models of operators those are close to unitary (self-adjoint) operators but are not contractive (dissipative) [5, 6].

On the other hand, trace class perturbations of normal operators with spectrum on a curve is an important object to study. In [7] the author revealed that such operators can be represented in the form $\varphi(T_0) + N_0 \varkappa M_0$, where φ is a conformal map of the unit disk, T_0, N_0, M_0 are entries of an unitary operator-matrix $\mathfrak{A}_0$, and $\varkappa$ is a bounded operator. This representation enables us to extend scheme from [5, 6] to the trace class perturbations of normal operators provided that we have a suitable functional model for the triple of operators $(\varphi(T_0), M_0, N_0)$. Problems, which are studied in [7], impose some requirements on this model. For instance, the definition of spectral components yields that the Hardy-Smirnov spaces $E^2(G_\pm)$ [8] and the corresponding (non-orthogonal) projections $P_\pm$ ($\operatorname{Ran} P_\pm = E^2(G_\pm)$, $\operatorname{Ker} P_\pm = E^2(G_\mp)$) should be important ingredients of the model (here $G_+ = \operatorname{int} C$, $G_- = \operatorname{ext} C$ are the interior and exterior domains for a closed curve C, respectively). Besides, it is desirable that we be able to work with a "dilation" space and with a projection of it onto the model subspace. Both of them should be consistent with Hardy-Smirnov spaces. Note that models from [3, 4] do not meet all our requirements on a model.

In the present paper we construct linearly similar functional model for triples of operators

$$(T, M, N) = (\varphi(T_0),\, M_0\chi_+(T_0),\, \chi_-(T_0)N_0) \qquad (*)$$

where $\varphi \in CM(\mathbb{D}, G_+)$, the operators T_0, N_0, M_0 are entries of an unitary node

$$\mathfrak{A}_0 = \begin{pmatrix} T_0 & N_0 \\ M_0 & L_0 \end{pmatrix} \in [H \oplus \mathfrak{N}]\,, \quad \mathfrak{A}_0^*\mathfrak{A}_0 = I, \quad \mathfrak{A}_0\mathfrak{A}_0^* = I\,,$$

and $\chi_+ = \sqrt{\varphi'}/(\psi_+ \circ \varphi)$, $\chi_- = \sqrt{\varphi'}\,(\psi_- \circ \varphi)$. We assume that the curve C is simple closed $C^{1+\varepsilon}$-smooth, the unitary node $\mathfrak{A}_0$ is simple (that is, the contraction T_0 is completely nonunitary), and $\psi_\pm,\ 1/\psi_\pm \in H^\infty(G_+)$. (Note that $C^{1+\varepsilon}$–smoothness of the curve C implies $\varphi',\ 1/\varphi' \in H^\infty(G_+)$ [9]). In the paper the class of all conformal maps of G_1 onto G_2 is denoted by $CM(G_1, G_2)$, the set of all bounded operators acting in a Hilbert space H is denoted by $[H]$. A linearly similar model of triple (T, M, N) is understood to be a triple of operators $(\widehat{T}, \widehat{M}, \widehat{N})$ such that there exists bounded invertible operator W and

$$\widehat{T}W = WT\,, \qquad \widehat{M}W = M\,, \qquad \widehat{N} = WN\,.$$

Basic requirements for our variant of a functional model are: 1) possibility to work with the "dilation" space and the projection onto the model subspace; 2) possibility to work with the dual model; 3) possibility to treat the absolutely continuous subspace; 4) existence of simple explicit transformations between a model of an operator and a model of a function of it; 5) possibility to construct a model for a basic operator T and channel operators M, N simultaneously. Note that, although these requirements arose in the process of solving of the duality problem for spectral components [7], they are not specific for this problem only. At the same time they are quite general and natural. Possibility to construct such model and, of course, the construction itself is a matter of this paper.

The paper is organized as follows. In Section 1 we review the known construction of the S.-Nagy-Foiaş functional model [1] for which we use the coordinate-free approach from [10]. In so doing, we consider features of the model that are essential for our generalization. In Section 2 the construction of the model is extended to the case of a closed smooth curve. We employ certain "change of variable" from the model for a curve to the model for the unit circle. Next, we consider the model subspace $\mathcal{K}_\Theta$ and the projection P_Θ onto this subspace and study their relationships to the dual model and the "change of variable". In Section 3 we introduce model operators $\widehat{T}, \widehat{M}, \widehat{N}$ and find the transformations of them when we do the "change of variable". Using these transformations we establish the main result of the paper: triple $(\widehat{T}, \widehat{M}, \widehat{N})$ is a linearly similar model for certain triple (T, M, N)

of the form (*) and, conversely, for any triple (T, M, N) of the form (*) there exists a linearly similar model $(\widehat{T}, \widehat{M}, \widehat{N})$. The most part of presenting results was announced without proofs in [7].

1. Model for the unit circle

Let $\mathfrak{N}$, $\mathcal{H}$ be separable Hilbert spaces and let $\pi_{0\pm} \in [L^2(\mathbb{T}, \mathfrak{N}), \mathcal{H}]$ be mappings satisfying the conditions:

$$\begin{aligned} &(\mathrm{i})^0 \quad \pi_{0\pm}^* \pi_{0\pm} = I\,;\\ &(\mathrm{ii})_1^0 \quad (\pi_{0-}^* \pi_{0+}) z = z(\pi_{0-}^* \pi_{0+})\,; \qquad (\mathrm{ii})_2^0 \quad P_- \pi_{0-}^* \pi_{0+} P_+ = 0\,;\\ &(\mathrm{iii})^0 \quad \operatorname{Ran} \pi_{0+} \vee \operatorname{Ran} \pi_{0-} = \mathcal{H}\,, \end{aligned}$$

where P_+ and P_- are orthogonal projections on the Hardy spaces $H^2(\mathfrak{N})$ and $H^2_-(\mathfrak{N})$, respectively. We set $\Theta_0^\pm = \pi_{0\mp}^* \pi_{0\pm}$ and $\Delta_0^\pm = (I - \Theta_0^\pm \Theta_0^\mp)^{1/2}$. It is easy to show that $\Theta_0^\pm, \Delta_0^\pm \in L^\infty(\mathbb{T}, [\mathfrak{N}])$. (Usually, we employ one and the same symbol for an operator-valued function and for the operator of multiplication by it.) Moreover, operator-valued functions $\Theta_0^\pm$ admit analytic continuation to $\mathbb{D}_\pm$ and have contractive values there. This follows from $(\mathrm{ii})^0$ and the self-adjointness of projections $P_\pm$. Isometries $\tau_{0\pm} : \operatorname{clos} \operatorname{Ran} \Delta_0^\pm \mapsto \mathcal{H}$ are uniquely determined by the relations $\tau_{0\pm} \Delta_0^\pm = (I - \pi_{0\pm} \pi_{0\pm}^*) \pi_{0\mp}$. We extend them to $(\operatorname{Ran} \Delta_0^\pm)^\perp$ by 0. For them, we have $\tau_{0\pm} = ((\Delta_0^\pm)^{-1} (\pi_{0\mp}^* - \Theta_0^\pm \pi_{0\pm}^*))^* \in [L^2(\mathbb{T}, \mathfrak{N}), \mathcal{H}]$ and the following identities hold:

1) $$\begin{pmatrix} \pi_{0\pm}^* \\ \tau_{0\pm}^* \end{pmatrix} (\pi_{0\pm}, \tau_{0\pm}) = \begin{pmatrix} I & 0 \\ 0 & P_{\operatorname{clos} \operatorname{Ran} \Delta_0^\pm} \end{pmatrix};$$

2) $$\begin{pmatrix} \pi_{0\pm}^* \\ \tau_{0\pm}^* \end{pmatrix} (\pi_{0\mp}, \tau_{0\mp}) = \begin{pmatrix} \Theta_0^\mp & \Delta_0^\mp \\ \Delta_0^\pm & -\Theta_0^\pm P_{\operatorname{clos} \operatorname{Ran} \Delta_0^\mp} \end{pmatrix};$$

3) $$\pi_{0\pm} \pi_{0\pm}^* + \tau_{0\pm} \tau_{0\pm}^* = I\,.$$

We define also the dual mappings $\pi_{*0\pm} = -i\pi_{0\mp} C_{1/z}$, where $(C_{1/z} f(\cdot))(z) = (i/z) f(1/z)$. It is easy to check that the maps $\pi_{*0\pm}$ satisfy conditions $(\mathrm{i})^0$, $(\mathrm{ii})^0$, $(\mathrm{iii})^0$ and we have also $\Theta_{*0}^\pm = (\Theta_0^\pm)^\sim$, $\Delta_{*0}^\pm = (\Delta_0^\mp)^\sim$, where $A^\sim(z) = A(\bar{z})^*$. The following duality relations hold

$$(f, g)_{\mathcal{H}} = < \pi_{0\pm}^* f, \pi_{*0\mp}^* g >_{\mathbb{T}} + < \tau_{0\pm}^* f, \tau_{*0\mp}^* g >_{\mathbb{T}}, \quad f, g \in \mathcal{H}\,,$$

where

$$< u, v >_{\mathbb{T}} = \frac{1}{2\pi i} \int_{\mathbb{T}} (u(z), v(\bar{z}))_{\mathfrak{N}}\, dz, \;\; u \in L^2(\mathbb{T}, \mathfrak{N}),\; v \in L^2(\mathbb{T}, \mathfrak{N})\,.$$

Further, from the identity $\| \pi_{0+} z u_+ + \pi_{0-} z u_- \| = \| \pi_{0+} u_+ + \pi_{0-} u_- \|$ and the condition $(\mathrm{iii})^0$ it follows that there exists a unique unitary operator $U_0 \in [\mathcal{H}]$ with absolutely continuous spectrum such that $U_0 \pi_{0\pm} = \pi_{0\pm} z$. It is easy to check the following identities $U_0 \tau_{0\pm} = \tau_{0\pm} z$, $\pi_{0\pm}^* U_0 = z \pi_{0\pm}^*$, $\tau_{0\pm}^* U_0 = z \tau_{0\pm}^*$.

Define the orthogonal projections $Q_{0\pm} = \pi_{0\pm} P_\pm \pi_{0\pm}^*$ and subspaces $D_{0\pm} = \operatorname{Ran} Q_{0\pm}$. It is easy to check that $Q_{0\pm} Q_{0\mp} = 0$ and that the subspaces $D_{0\pm}$ are invariant under operators $U_0^{\pm 1}$. Then we arrive at the Lax-Phillips-Adamyan-Arov scheme [11]:

$$D_{0+} \perp D_{0-}, \quad \bigcap_{n\in\mathbb{N}} U_0^{\pm n} D_{0\pm} = \{0\}, \quad \bigcup_{n\in\mathbb{Z}} U_0^n D_{0+} \bigvee \bigcup_{n\in\mathbb{Z}} U_0^n D_{0-} = \mathcal{H}.$$

It is clear that, conversely, one can easily construct operators $\pi_{0\pm}$ that satisfy conditions (i)^0, (ii)^0, and (iii)^0 whenever the corresponding outgoing and incoming subspaces $D_{0\pm}$ are given.

As is well known [11], the generalized Lax-Phillips scheme is equivalent to the functional model of S.-Nagy-Foiaş [1]. For this model, the original Nikolski-Vasyunin [10] properties of operators $\pi_{0\pm}$ are:

$$\pi_{0\pm}^* \pi_{0\pm} = I\,; \qquad \pi_{0\pm} z = U_0 \pi_{0\pm}\,;$$
$$\pi_{0\pm} H^2(\mathbb{D}_\pm) = D_{0\pm}\,; \qquad \operatorname{Ran} \pi_{0+} \vee \operatorname{Ran} \pi_{0-} = \mathcal{H}\,,$$

where $U_0 \in [\mathcal{H}]$ is a minimal unitary dilation of some completely nonunitary contraction $T_0 \in [H]$, $D_{0+} = (\vee_{n\geq 0} U_0^n H) \ominus H$, $D_{0-} = \mathcal{H} \ominus (\vee_{n\geq 0} U_0^n H)$.

Therefore our set of axioms (i)^0, (ii)^0, (iii)^0 and Nikolski-Vasyunin's axioms are equivalent. But in our settings we use neither a unitary dilation nor orthogonal complements. This reformulation enables us to extend the functional model to the case of a closed curve. We present this generalization in the next section.

Now we define model operators $\widehat{T}_0 \in [\mathcal{K}_0]$, $\widehat{M}_0 \in [\mathcal{K}_0, \mathfrak{N}]$, $\widehat{N}_0 \in [\mathfrak{N}, \mathcal{K}_0]$:

$$\widehat{T}_0 f = U_0 f - \pi_{0+} \widehat{M}_0 f, \quad \widehat{M}_0 f = (\pi_{0+}^* U_0 f)(\infty), \quad \widehat{N}_0 n = (I - \pi_{0+} P_+ \pi_{0+}^*) \pi_{0-} n,$$

where $f \in \mathcal{K}_0$, $n \in \mathfrak{N}$, $\mathcal{K}_0 = \operatorname{Ran} P_0$, $P_0 = I - Q_{0+} - Q_{0-}$. Note that operators $\widehat{M}_0$ and $\widehat{N}_0$ are (up to unitary factor) the defect operators [1, 2] of the contraction $\widehat{T}_0$, the operator P_0 is an orthogonal projection and

$$\mathfrak{A}_0 = \begin{pmatrix} \widehat{T}_0 & \widehat{N}_0 \\ \widehat{M}_0 & \Theta_0^+(0)^* \end{pmatrix}$$

is a simple unitary node.

The inverse is also true. If $\mathfrak{A}_0 \in [H \oplus \mathfrak{N}]$ is a simple unitary node, then there exists a pair $\Pi_0 = (\pi_{0+}, \pi_{0-})$ satisfying conditions (i)^0, (ii)^0, (iii)^0, and a unitary operator $W_0 \in [H, \mathcal{K}_0]$ such that

$$\widehat{T}_0 W_0 = W_0 T_0, \quad \widehat{M}_0 W_0 = M_0, \quad \widehat{N}_0 = W_0 N_0, \quad \Theta_0^+(0)^* = L_0\,.$$

Note also that for dual objects we have the following relations

$$P_{*0} = P_0, \;\; \mathcal{K}_{*0} = \mathcal{K}_0, \;\; \widehat{T}_{*0} = \widehat{T}_0^*, \;\; \widehat{M}_{*0} = \widehat{N}_0^*, \;\; \widehat{N}_{*0} = \widehat{M}_0^*\,.$$

2. Model for a curve (geometry)

Since the curve C is $C^{1+\varepsilon}$-smooth, there exist (non-orthogonal) projections $P_\pm \in [L^2(C,\mathfrak{N})]$ such that $\operatorname{Ran} P_\pm = E^2(G_\pm,\mathfrak{N})$, $\operatorname{Ker} P_\pm = E^2(G_\mp,\mathfrak{N})$. Let mappings $\pi_\pm \in [L^2(C,\mathfrak{N}),\mathcal{H}]$ satisfy conditions:

$$\begin{array}{llll} (\mathrm{i})_1 & \forall \psi \in L^\infty(C,[\mathfrak{N}])\ (\pi_\pm^*\pi_\pm)\psi = \psi(\pi_\pm^*\pi_\pm); & (\mathrm{i})_2 & (\pi_\pm^*\pi_\pm)^{-1} \in [L^2(C,\mathfrak{N})]; \\ (\mathrm{ii})_1 & (\pi_-^\dagger\pi_+)z = z(\pi_-^\dagger\pi_+); & (\mathrm{ii})_2 & P_-(\pi_-^\dagger\pi_+)P_+ = 0; \\ (\mathrm{iii}) & \operatorname{Ran}\pi_+ \vee \operatorname{Ran}\pi_- = \mathcal{H}. & & \end{array}$$

Here $\pi_-^\dagger$ is the Moore-Penrose inverse operator for π_-: $\pi_-^\dagger f = (\pi_-|(\operatorname{Ker}\pi_-)^\perp)^{-1} f$, $f \in \operatorname{Ran}\pi_-$; $\pi_-^\dagger f = 0$, $f \perp \operatorname{Ran}\pi_-$.

Conditions (i), (ii), and (iii) are generalization of $(\mathrm{i})^0$, $(\mathrm{ii})^0$, and $(\mathrm{iii})^0$ to the case of a curve. Indeed, we have $\pi_{0-}^\dagger = \pi_{0-}^*$. On the other hand, there exists a transformation ("change of variable") between $\Pi = (\pi_+,\pi_-)$ and $\Pi_0 = (\pi_{0+},\pi_{0-})$, which we are going to describe.

First, note that from conditions $(\mathrm{i})_1$ and $(\mathrm{i})_2$ it follows that $\pi_\pm^*\pi_\pm = \delta_\pm I$, where $\delta_\pm$, $1/\delta_\pm \in L^\infty(C)$. Then there exist [1, 12] outer analytic scalar functions $\psi_\pm \in H^\infty(G_+)$ such that for their boundary values $|\psi_\pm|^2 = \delta_\pm$. Clearly, $1/\psi_\pm \in H^\infty(G_+)$. Note also that, since $\delta_\pm$ do not vanish on $\mathbb{T}$, we have $\psi_\pm \in C^{1+\varepsilon}(\operatorname{clos} G_+) \Leftrightarrow \delta_\pm \in C^{1+\varepsilon}(C)$ (see, e.g., [13]). Define the unitary operator $C_{\varphi_{21}} \in [L^2(C_2,\mathfrak{N}), L^2(C_1,\mathfrak{N})]$ by formula

$$(C_{\varphi_{21}} f(\cdot))(z_1) = \sqrt{\varphi_{21}'(z_1)}\, f(\varphi_{21}(z_1)), \quad z_1 \in C_1, \quad f \in L^2(C_2,\mathfrak{N}),$$

where $\varphi_{21} \in CM(G_{1+},G_{2+})$. Let $\varphi \in CM(\mathbb{D},G_+)$. Put $\pi_{0\pm} = \pi_\pm 1/\psi_\pm C_{\varphi^{-1}}$. We will check that $\Pi_0 = (\pi_{0+},\pi_{0-})$ satisfy conditions $(\mathrm{i})^0$, $(\mathrm{ii})^0$, and $(\mathrm{iii})^0$.

It is easy to show that operators $\pi_\pm 1/\psi_\pm$ are isometries. Hence, $\pi_{0\pm}$ satisfy the condition $(\mathrm{i})^0$. Obviously, $\pi_\pm = \pi_{0\pm} C_\varphi \psi_\pm$. Since $\operatorname{Ker}\pi_\pm = \operatorname{Ker}\pi_{0\pm} = \{0\}$, we have $\pi_\pm^\dagger = 1/\psi_\pm C_\varphi^{-1}\pi_{0\pm}^\dagger = 1/\psi_\pm C_{\varphi^{-1}}\pi_{0\pm}^* \in [\mathcal{H}, L^2(C,\mathfrak{N})]$. Whence, $\pi_{0\pm}^* = C_\varphi \psi_\pm \pi_\pm^\dagger$ and $\pi_{0-}^*\pi_{0+} = C_\varphi \psi_- \pi_-^\dagger \pi_+ 1/\psi_+ C_{\varphi^{-1}}$. Taking into account the conditions $(\mathrm{ii})_1$ and $(\mathrm{ii})_2$ we can regard the operator $\pi_-^\dagger\pi_+$ as an operator of multiplication by boundary values of bounded analytic operator-valued function $\Theta^+(z)$, $z \in G_+$. Since $\psi_\pm$, $1/\psi_\pm \in H^\infty(G_+)$, the operator-valued function $\Theta_0^+ = (\psi_-/\psi_+\Theta^\pm)\circ\varphi$ is bounded and analytic in the unit disk. Then the conditions $(\mathrm{ii})_1^0$ and $(\mathrm{ii})_2^0$ are fulfilled for the operator $\pi_{0-}^*\pi_{0+}$ of multiplication by boundary values of $\Theta_0^+(z_0)$, $z_0 \in \mathbb{D}$. The condition $(\mathrm{iii})^0$ follows from the identity $\operatorname{Ran}\pi_{0\pm} = \operatorname{Ran}\pi_\pm$.

Note that one can consider slightly more general "change of variable" $\pi_{2\pm} = \pi_{1\pm} C_{\varphi_{21}} \eta_\pm$, where $\varphi_{21} \in CM(G_{1+},G_{2+})$, $\eta_\pm$, $1/\eta_\pm \in H^\infty(G_{2+})$. Then the pair $\Pi_2 = (\pi_{2+},\pi_{2-})$ satisfy conditions (i), (ii), (iii) $\Leftrightarrow$ the pair $\Pi_1 = (\pi_{1+},\pi_{1-})$ satisfy the same conditions.

We set $\Theta^\pm = \pi_\mp^\dagger\pi_\pm$, $\Delta^\pm = (I - \Theta^\pm\Theta^\mp)^{1/2}$. It is easy to show that $\Theta^\pm$, $\Delta^\pm \in L^\infty(C,[\mathfrak{N}])$. As we already know, $\Theta^+ \in H^\infty(G_+,[\mathfrak{N}])$. However, now we cannot assert that the operator-valued function $\Theta^- = \pi_+^\dagger \pi_-$ admits analytic continuation to G_-.

Further, put $\tau_\pm = ((\Delta^\pm)^{-1}(\pi_\mp^\dagger - \Theta^\pm \pi_\pm^\dagger))^\dagger$. It is easy to check that $\tau_\pm = \tau_{0\pm} C_\varphi \psi_\mp$, $\tau_\pm^\dagger = 1/\psi_\mp C_{\varphi^{-1}} \tau_{0\pm}^\dagger$. Combining this with the corresponding relations for $\pi_{0\pm}$, $\tau_{0\pm}$ from Section 1, we obtain

Proposition 1. *One has*

1) $\begin{pmatrix} \pi_\pm^\dagger \\ \tau_\pm^\dagger \end{pmatrix} (\pi_\pm, \tau_\pm) = \begin{pmatrix} I & 0 \\ 0 & P_{\operatorname{clos Ran} \Delta^\pm} \end{pmatrix}$;

2) $\begin{pmatrix} \pi_\pm^\dagger \\ \tau_\pm^\dagger \end{pmatrix} (\pi_\mp, \tau_\mp) = \begin{pmatrix} \Theta^\mp & \Delta^\mp \\ \Delta^\pm & -\Theta^\pm P_{\operatorname{clos Ran} \Delta^\mp} \end{pmatrix}$;

3) $\pi_\pm \pi_\pm^\dagger + \tau_\pm \tau_\pm^\dagger = I$.

We set $\pi_{*\pm} = \pi_{*0\pm} C_{\varphi^\sim} 1/\psi_\mp^\sim$. It is clear that $\Pi_* = (\pi_{*+}, \pi_{*-})$ satisfy conditions (i), (ii), (iii) and $\Theta_*^\pm = (\Theta^\pm)^\sim$, $\Delta_*^\pm = (\Delta^\mp)^\sim$. Here again $A^\sim(z) = A(\bar z)^*$.

Proposition 2. *One has*

$$(f,g)_{\mathcal{H}} = \langle \pi_\pm^\dagger f, \pi_{*\mp}^\dagger g\rangle_C + \langle \tau_\pm^\dagger f, \tau_{*\mp}^\dagger g\rangle_C, \quad f,g \in \mathcal{H},$$

where

$$\langle u,v\rangle_C = \frac{1}{2\pi i}\int_C (u(z), v(\bar z))_{\mathfrak{N}}\, dz, \quad u \in L^2(C,\mathfrak{N}),\ v \in L^2(\bar C, \mathfrak{N}).$$

Proof. Using obvious properties of the pairing, we obtain

$$\langle \pi_{0\pm}^* f, \pi_{*0\mp}^* g\rangle_{\mathbb{T}} = \langle C_\varphi \psi_\pm \pi_\pm^\dagger f, C_{\varphi^\sim} 1/\psi_\mp^\sim \pi_{*\mp}^\dagger g\rangle_{\mathbb{T}}$$
$$= \langle \psi_\pm \pi_\pm^\dagger f, 1/\psi_\mp^\sim \pi_{*\mp}^\dagger g\rangle_C = \langle (1/\psi_\pm)\psi_\pm \pi_\pm^\dagger f, \pi_{*\mp}^\dagger g\rangle_C = \langle \pi_\pm^\dagger f, \pi_{*\mp}^\dagger g\rangle_C .$$

Similarly, we get the corresponding identity for $\tau_\pm$, $\tau_{*\mp}$. It remains to make use of the relations of duality for the unit circle. □

Proposition 3. *Conditions $U\pi_\pm = \pi_\pm z$ uniquely determine the normal operator U. The spectrum of U is absolutely continuous and lies on the curve C.*

Proof. Since $\pi_\pm z = \pi_{0\pm} C_\varphi \psi_\pm z = \pi_{0\pm} \varphi(z_0) C_\varphi \psi_\pm = \varphi(U_0)\pi_{0\pm} C_\varphi \psi_\pm = \varphi(U_0)\pi_\pm$, we have $U = \varphi(U_0)$. □

In the same way, we get $U\tau_\pm = \tau_\pm z$, $\pi_\pm^\dagger U = z\pi_\pm^\dagger$, $\tau_\pm^\dagger U = z\tau_\pm^\dagger$. From the duality relations we obtain $U_* = U^*$.

We pass to the model subspace and the projection onto it. To this end we consider auxiliary projections $q_\pm = \pi_\pm P_\pm \pi_\pm^\dagger$. These operators have the following properties.

Lemma 4. 1) $q_\pm^2 = q_\pm$; 2) $\operatorname{Ker} q_\pm = \operatorname{Ker} P_\pm \pi_\pm^\dagger$; 3) $\operatorname{Ran} q_\pm = \pi_\pm E^2(G_\pm) = \operatorname{Ker} P_\mp \pi_\pm^\dagger \cap \operatorname{Ker} \tau_\pm^\dagger$; 4) $q_{*\pm} = q_\mp^*$; 5) $q_{2-} q_{1+} = q_{2-}(I - q_{1-}) = (I - q_{2+}) q_{1+} = 0$, *where* $\pi_{2\pm} = \pi_{1\pm} C_{\varphi_{21}} \eta_\pm$, $\varphi_{21} \in CM(G_{1+}, G_{2+})$, $\eta_\pm$, $1/\eta_\pm \in H^\infty(G_{2+})$.

Proof. This technical lemma bases mainly on the condition $(\mathrm{ii})_2$. We have

1) $q_\pm^2 = \pi_\pm P_\pm \pi_\pm^\dagger \pi_\pm P_\pm \pi_\pm^\dagger = \pi_\pm P_\pm P_\pm \pi_\pm^\dagger = \pi_\pm P_\pm \pi_\pm^\dagger = q_\pm$.

2) This follows from identities $q_\pm = \pi_\pm(P_\pm \pi_\pm^\dagger)$, $P_\pm \pi_\pm^\dagger = \pi_\pm^\dagger q_\pm$.

3) Let $f \in \operatorname{Ran} q_\pm$. Then $f = q_\pm g = \pi_\pm u_\pm$, $u_\pm = P_\pm \pi_\pm^\dagger g$. Conversely, let $P_\mp u_\pm = 0$. Then $\pi_\pm u_\pm = \pi_\pm(P_\pm + P_\mp)u_\pm = \pi_\pm P_\pm u_\pm = (\pi_\pm P_\pm \pi_\pm^\dagger)\pi_\pm u_\pm$. Whence, $\operatorname{Ran} q_\pm = \pi_\pm E^2(G_\pm)$.

Next, since $P_\mp \pi_\pm^\dagger \pi_\pm = P_\mp$, $\tau_\pm^\dagger \pi_\pm = 0$, we have $\pi_\pm E^2(G_\pm) \subset \operatorname{Ker} P_\mp \pi_\pm^\dagger \cap \operatorname{Ker} \tau_\pm^\dagger$. Conversely, let $f \in \operatorname{Ker} P_\mp \pi_\pm^\dagger \cap \operatorname{Ker} \tau_\pm^\dagger$. Then $f = (\pi_\pm \pi_\pm^\dagger + \tau_\pm \tau_\pm^\dagger) f = \pi_\pm \pi_\pm^\dagger f = \pi_\pm(P_\pm + P_\mp)\pi_\pm^\dagger f = \pi_\pm u_\pm$, $u_\pm = P_\pm \pi_\pm^\dagger f \in E^2(G_\pm)$. Whence, $\pi_\pm E^2(G_\pm) = \operatorname{Ker} P_\mp \pi_\pm^\dagger \cap \operatorname{Ker} \tau_\pm^\dagger$.

4) Let $f \in \operatorname{Ran} q_\pm$, $g \in \operatorname{Ker} q_{*\mp}$. Then, by Proposition 2, we have $(f,g) = < \pi_\pm^\dagger f, \pi_{*\mp}^\dagger g >_C$, where $\pi_\pm^\dagger f \in E^2(G_\pm)$, $\pi_{*\mp}^\dagger g \in E^2(G_{*\pm})$. Taking into account that $E^2(G_\pm)^{<\perp>} = E^2(G_{*\pm})$, we get $(f,g) = 0$.

Conversely, let $g \perp \operatorname{Ran} q_\pm$. If we take $f = \pi_\pm u_\pm$, $u_\pm \in E^2(G_\pm)$, then $0 = (f,g) = < \pi_\pm^\dagger f, \pi_{*\mp}^\dagger g >_C = < u_\pm, \pi_{*\mp}^\dagger g >_C$. Whence, $\pi_{*\mp}^\dagger g \in E^2(G_{*\pm})$, i.e., $g \in \operatorname{Ker} q_{*\mp}$. Therefore, $(\operatorname{Ran} q_\pm)^\perp = \operatorname{Ker} q_{*\mp}$. Similarly, $(\operatorname{Ker} q_\pm)^\perp = \operatorname{Ran} q_{*\mp}$.
On the other hand, for any linear operator $(\operatorname{Ker} q_\pm)^\perp = \operatorname{Ran} q_\pm^*$ and $(\operatorname{Ran} q_\pm)^\perp = \operatorname{Ker} q_\pm^*$. Since $q_{*\mp}$ and $q_\pm^*$ are projections with the same ranges and kernels, we obtain $q_{*\mp} = q_\pm^*$.

5) Let $A_\pm = 1/\eta_\pm C_{\varphi_{21}}^{-1}$. Then we have $\pi_{1\pm} = \pi_{2\pm} A_\pm$ and $P_{2-} A_\pm P_{1+} = 0$. Using the latter observation and the obvious identity $q_{1-} q_{1+} = 0$, we get

$$\begin{aligned}
q_{2-}(I - q_{1-}) &= \pi_{2-} P_{2-} \pi_{2-}^\dagger (I - \pi_{1-} P_{1-} \pi_{1-}^\dagger) \\
&= \pi_{2-} P_{2-} A_- \pi_{1-}^\dagger (I - \pi_{1-} P_{1-} \pi_{1-}^\dagger) \\
&= \pi_{2-} P_{2-} A_- (I - P_{1-}) \pi_{1-}^\dagger = \pi_{2-} (P_{2-} A_- P_{1+}) \pi_{1-}^\dagger = 0, \\
(I - q_{2+}) q_{1+} &= (I - \pi_{2+} P_{2+} \pi_{2+}^\dagger) \pi_{1+} P_{1+} \pi_{1+}^\dagger \\
&= (I - \pi_{1+} A_+^{-1} P_{2+} A_+ \pi_{1+}^\dagger) \pi_{1+} P_{1+} \pi_{1+}^\dagger \\
&= \pi_{1+} (I - A_+^{-1} P_{2+} A_+) P_{1+} \pi_{1+}^\dagger \\
&= \pi_{1+} A_+^{-1} (P_{2-} A_+ P_{1+}) \pi_{1+}^\dagger = 0, \\
q_{2-} q_{1+} &= q_{2-} q_{1-} q_{1+} + q_{2-}(I - q_{1-}) q_{1+} = 0 + 0 = 0. \qquad \square
\end{aligned}$$

We set $Q_+ = q_+(I - q_-)$, $Q_- = q_-$ and $D_\pm = \operatorname{Ran} Q_\pm$. Since $Q_+ q_+ = q_+$, we have $\operatorname{Ran} Q_\pm = \operatorname{Ran} q_\pm$. It is easy to check that $Q_\pm^2 = Q_\pm$, $Q_\pm Q_\mp = 0$. Now we can define the projection P_Θ onto the model subspace $\mathcal{K}_\Theta = \operatorname{Ran} P_\Theta$:

$$P_\Theta = I - Q_+ - Q_- = (1 - Q_+)(1 - Q_-) = (1 - q_+)(1 - q_-).$$

This projection plays a central role in the paper and have the following properties:

Theorem 5. 1) $P_\Theta^2 = P_\Theta$; 2) $P_{*\Theta} = P_\Theta^*$; 3) $P_{3\Theta}P_{2\Theta}P_{1\Theta} = P_{3\Theta}P_{1\Theta}$, *where* $\pi_{3\pm} = \pi_{2\pm}C_{\varphi_{32}}\eta_{3\pm}$, $\pi_{2\pm} = \pi_{1\pm}C_{\varphi_{21}}\eta_{2\pm}$.

Proof. 1) This follows from the properties of $Q_\pm$.

2) Using Lemma 4(4), we get

$$P_{*\Theta} = (1 - q_{*+})(1 - q_{*-}) = (1 - q_-^*)(1 - q_+^*) = ((1 - q_+)(1 - q_-))^* = P_\Theta^*.$$

3) Using Lemma 4(5), we get

$$\begin{aligned}(1 - q_{2-})(1 - q_{1+})(1 - q_{1-}) &= (1 - q_{2-} - q_{1+})(1 - q_{1-}) \\ &= ((1 - q_{1+}) - q_{2-})(1 - q_{1-}) = (1 - q_{1+})(1 - q_{1-}), \\ (1 - q_{3+})(1 - q_{3-})(1 - q_{2+}) &= (1 - q_{3+})(1 - q_{3-} - q_{2+}) \\ &= (1 - q_{3+})((1 - q_{3-}) - q_{2+}) = (1 - q_{3+})(1 - q_{3-}). \quad \square\end{aligned}$$

Remark. *We can rewrite statement* 3) *in the form* $Z_{31} = Z_{32}Z_{21}$, *where* $Z_{ij} = P_{i\Theta}|\mathcal{K}_{j\Theta} \in [\mathcal{K}_{j\Theta}, \mathcal{K}_{i\Theta}]$. *In particular, we have* $Z_{21}^{-1} = Z_{12}$. *Note that the model subspace* $\mathcal{K}_\Theta$ *varies depending on "change of variable". In general,* $\mathcal{K}_{1\Theta} \neq \mathcal{K}_{2\Theta}$.

Remark. *In contrast to the case of the unit circle, we have* $P_\Theta^* \neq P_\Theta$, $Q_+ \neq q_+$, *and* $P_\Theta = (1 - q_+)(1 - q_-) \neq (1 - q_-)(1 - q_+) = (1 - q_+ - q_-)$.

Remark. *Besides, we have* $\forall\,\psi \in H^\infty(G_\pm)$: $(I - Q_\pm)\psi(U))Q_\pm = 0$. *That means, the subspaces* $D_\pm$ *are invariant under operators* $\psi(U)$. *But, on the other hand, we have only* $\forall\,\psi \in H^\infty(G_+)$: $Q_-\psi(U)(I - Q_-) = 0$. *Thus we have some asymmetry here as well as in the previous remark.*

Remark. *Nevertheless, the model subspace* $\mathcal{K}_\Theta$ *admits the following symmetric description:*

$$\mathcal{K}_\Theta = \{\, f \in \mathcal{H} : \pi_+^\dagger f \in E^2(G_-, \mathfrak{N}),\ \pi_-^\dagger f \in E^2(G_+, \mathfrak{N}) \,\} = \operatorname{Ker} P_+\pi_+^\dagger \cap \operatorname{Ker} P_-\pi_-^\dagger.$$

Indeed, let $\pi_+^\dagger f \in E^2(G_-, \mathfrak{N})$, $\pi_-^\dagger f \in E^2(G_+, \mathfrak{N})$. Then we get $P_\Theta f = (I - \pi_+P_+\pi_+^\dagger)(I - \pi_-P_-\pi_-^\dagger)f = (I - \pi_+P_+\pi_+^\dagger)f = f$ and $f \in \mathcal{K}_\Theta$. Conversely, we have $P_+\pi_+^\dagger P_\Theta = P_+\pi_+^\dagger(I - \pi_+P_+\pi_+^\dagger)(I - \pi_-P_-\pi_-^\dagger) = P_+P_-\pi_+^\dagger(I - \pi_-P_-\pi_-^\dagger) = 0$ and $P_-\pi_-^\dagger P_\Theta = P_-\pi_-^\dagger(I - \pi_+P_+\pi_+^\dagger)(I - \pi_-P_-\pi_-^\dagger) = P_-\pi_-^\dagger(I - \pi_-P_-\pi_-^\dagger) = P_-P_+\pi_-^\dagger = 0$.

Example. *Let* $\Theta \in H^\infty(G_+, [\mathfrak{N}])$, $\sup_{z \in C} \|\Theta(z)^{-1}\| < \infty$. *Then, for the space* $\mathcal{H} = L^2(C, \mathfrak{N})$ *and the mappings* $\pi_+ = I$, $\pi_- = \Theta^{-1}$, *we have*

$$\begin{aligned}\mathcal{K}_\Theta &= \{\, f \in L^2(C, \mathfrak{N}) : f \in E^2(G_-, \mathfrak{N}),\ \Theta f \in E^2(G_+, \mathfrak{N}) \,\} \\ &= \{\, f \in E^2(G_-, \mathfrak{N}) : \Theta f \in E^2(G_+, \mathfrak{N}) \,\} = \operatorname{Ker} P_+ \cap \operatorname{Ker} P_-\Theta.\end{aligned}$$

Thus we arrive at Yakubovich's model [4] for the class of C_{00} operators. In [4] Yakubovich considers a more general case where the domain G_+ is multiply connected and the curve C is less smooth. Note that the most part of presenting above results can be extended to this case. But, in contrast to [4], we do not restrict ourself to the case of C_{00} operators and allow them to have absolutely continuous spectrum. Another difference is the presence of the projection P_Θ in our model. This projection enables us to lift various objects (e.g., spectral components, commutant, etc.) from the model space $\mathcal{K}_\Theta$ up to the level of the dilation space $\mathcal{H}$.

Our triples of operators correspond to Yakubovich's 3-systems. In [4] Yakubovich studies the problem: how to construct a model for a given 3-system. He has solved this problem, but under the assumption that a characteristic function is known (more precisely, it must belong to some family of operator-valued functions corresponding to the transfer function of a 3-system). In the next section we also study this problem for the case of simple connected domains. We do not assume that the characteristic function Θ^+ is given. But we suppose instead that the functions $\chi_\pm$ are known. Such a statement of the problem justifies in applications of this result to the study trace class perturbations of normal operators [7].

3. Model for a curve (operators)

First we introduce model operators:

$$\widehat{T}f = Uf - \pi_+\widehat{M}f, \quad \widehat{M}f = \frac{1}{2\pi i}\int\limits_C (\pi_+^\dagger f)(z)\,dz = (\pi_+^\dagger Uf)(\infty),$$

$$\widehat{N}n = P_\Theta \pi_- n = (I - \pi_+ P_+ \pi_+^\dagger)\pi_- n, \quad \text{where } f \in \mathcal{K}_\Theta,\ n \in \mathfrak{N}.$$

Theorem 6. *The mapping $\Phi : \psi \mapsto P_\Theta \psi(U)|\mathcal{K}_\Theta$, $\psi \in H^\infty(G_+)$ is a homomorphism from the algebra $H^\infty(G_+)$ into the algebra $[\mathcal{K}_\Theta]$. This homomorphism has the following properties:* 1) $\Phi(1) = I$, $\Phi(z) = \widehat{T}$; 2) $||\Phi|| < \infty$; 3) $||\psi_n|| < K$, $\lim\limits_{n\to\infty} \psi_n(z) = 0$ *a.e.* $z \in C \implies s-\lim\limits_{n\to\infty} \Phi(\psi_n) = 0$.

Proof. Using notation $\psi(\widehat{T}) = \Phi(\psi)$, we check multiplicativity:

$$\begin{aligned}
\psi_1(\widehat{T})\psi_2(\widehat{T})f &= P_\Theta\psi_1(U)P_\Theta\psi_2(U)f = P_\Theta\psi_1(U)\psi_2(U)f \\
&- P_\Theta\psi_1(U)(I-P_\Theta)\psi_2(U)f = P_\Theta\psi_1(U)\psi_2(U)f \\
&- (I-Q_-)(I-Q_+)\psi_1(U)(Q_+ + Q_-)\psi_2(U)(I-Q_-)(I-Q_+)f \\
&= P_\Theta(\psi_1\psi_2)(U)f = (\psi_1\psi_2)(\widehat{T})f, \quad f \in \mathcal{K}_\Theta.
\end{aligned}$$

Above we have exploited identities $(I-Q_+)\psi_1(U)Q_+ = 0$, $Q_-\psi_2(U)(I-Q_-) = 0$. Now we pass to calculation of $\Phi(z)$:

$$\begin{aligned}
\Phi(z)f &= P_\Theta Uf = (I - \pi_+P_+\pi_+^\dagger)(I - \pi_-P_-\pi_-^\dagger)Uf \\
&= (I - \pi_+P_+\pi_+^\dagger)(Uf - \pi_-P_-z\pi_-^\dagger f) = (I - \pi_+P_+\pi_+^\dagger)Uf \\
&= Uf - \pi_+(z\pi_+^\dagger f)(\infty) = \widehat{T}f, \quad f \in \mathcal{K}_\Theta.
\end{aligned}$$

Properties 2) and 3) follow from the corresponding properties of the functional calculus for normal operators. □

The following assertion answers how the model operators depend on "change of variable".

Proposition 7. *Let* $\pi_{2\pm} = \pi_{1\pm} C_\varphi \eta_\pm$, $\varphi \in CM(G_{1+}, G_{2+})$, $\eta_\pm$, $1/\eta_\pm \in H^\infty(G_{2+})$. *Then*

$$1)\ \widehat{T}_2 = Z\varphi(\widehat{T}_1)Z^{-1}; \qquad 2)\ \widehat{M}_2 = \widehat{M}_1\chi_+(\widehat{T}_1)Z^{-1}; \qquad 3)\ \widehat{N}_2 = Z\chi_-(\widehat{T}_1)\widehat{N}_1,$$

where $Z = P_{2\Theta}|\mathcal{K}_{1\Theta}$, $\chi_+ = \sqrt{\varphi'}/(\eta_+ \circ \varphi)$, $\chi_- = \sqrt{\varphi'}\,(\eta_- \circ \varphi)$.

Proof. First we note that $Z^{-1} = P_{1\Theta}|\mathcal{K}_{2\Theta}$. We shall also use the identities $P_{2\Theta}Q_{1+} = 0$, $Q_{2-}P_{1\Theta} = 0$, which follow easily from Lemma 4(5).
1) By Theorem 6 and Proposition 3, we have

$$\begin{aligned}
&\widehat{T}_2 Z P_{1\Theta} - Z\varphi(\widehat{T}_1)P_{1\Theta} = P_{2\Theta}(U_2P_{2\Theta} - P_{1\Theta}\varphi(U_1))P_{1\Theta} = \\
&P_{2\Theta}(\varphi(U_1)(I - Q_{2+} - Q_{2-}) - (I - Q_{1+} - Q_{1-})\varphi(U_1))P_{1\Theta} = \\
&P_{2\Theta}((Q_{1+} + Q_{1-})\varphi(U_1) - \varphi(U_1)(Q_{2+} + Q_{2-}))P_{1\Theta}.
\end{aligned}$$

Since $(I - Q_{2+})U_2Q_{2+} = 0$, $Q_{1-}\varphi(U_1)(I - Q_{1-}) = 0$, we get

$$\widehat{T}_2 Z P_{1\Theta} - Z\varphi(\widehat{T}_1)P_{1\Theta} = P_{2\Theta}(Q_{1+}\varphi(U_1) - \varphi(U_1)Q_{2-})P_{1\Theta}.$$

Finally, using above mentioned identities we obtain $\widehat{T}_2 Z - Z\varphi(\widehat{T}_1) = 0$.
2) Let $f \in \mathcal{K}_{2\Theta}$. Then

$$\begin{aligned}
\widehat{M}_2 f &= \frac{1}{2\pi i}\int_{C_2} (\pi_{2+}^\dagger f)(z_2)\,dz_2 = \frac{1}{2\pi i}\int_{C_2} \frac{1}{\eta_+(z_2)}(\pi_{1+}^\dagger f)(\varphi^{-1}(z_2))\sqrt{\varphi^{-1}(z_2)'}\,dz_2 \\
&= \frac{1}{2\pi i}\int_{C_1} \frac{\sqrt{\varphi'(z_1)}}{\eta_+(\varphi(z_1))}(\pi_{1+}^\dagger f)(z_1)\,dz_1 = \frac{1}{2\pi i}\int_{C_1} \chi_+(z_1)(\pi_{1+}^\dagger f)(z_1)\,dz_1 \\
&= \frac{1}{2\pi i}\int_{C_1} (\pi_{1+}^\dagger \chi_+(U_1)f)(z_1)\,dz_1 = \frac{1}{2\pi i}\int_{C_1} (\pi_{1+}^\dagger P_{1\Theta}\chi_+(U_1)P_{1\Theta}f)(z_1)\,dz_1 \\
&+ \frac{1}{2\pi i}\int_{C_1} (\pi_{1+}^\dagger((I - P_{1\Theta})\chi_+(U_1)P_{1\Theta}f + \chi_+(U_1)(I - P_{1\Theta})f))(z_1)\,dz_1.
\end{aligned}$$

Taking into account that $Q_{1-}\chi_+(U_1)(I-Q_{1-}) = 0$ and $P_{1\Theta} = (I-Q_{1-})(I-Q_{1+})$, we get

$$(I - P_{1\Theta})\chi_+(U_1)P_{1\Theta} = (Q_{1+} + Q_{1-})\chi_+(U_1)P_{1\Theta} = Q_{1+}\chi_+(U_1)P_{1\Theta}.$$

Whence, by the Cauchy theorem,

$$\int_{C_1} (\pi_{1+}^\dagger (I - P_{1\Theta})\chi_+(U_1)P_{1\Theta}f)(z_1)\,dz_1 = 0.$$

Next, since $Q_{2-}P_{1\Theta} = 0$ and $(I - Q_{1+})\chi_+(U_1)Q_{1+} = 0$, we get

$$\begin{aligned}
\chi_+(U_1)(I - P_{1\Theta})f &= \chi_+(U_1)(Q_{1+} + Q_{1-})P_{2\Theta}f \\
&= \chi_+(U_1)Q_{1+}P_{2\Theta}f = Q_{1+}\chi_+(U_1)Q_{1+}P_{2\Theta}f.
\end{aligned}$$

Then, again by the Cauchy theorem, we have

$$\int_{C_1} (\pi_{1+}^{\dagger}\chi_+(U_1)(I-P_{1\Theta})f)(z_1)\,dz_1 = 0\,.$$

Therefore,

$$\widehat{M}_2 f = \frac{1}{2\pi i}\int_{C_1} (\pi_{1+}^{\dagger}P_{1\Theta}\chi_+(U_1)P_{1\Theta}f)(z_1)\,dz_1 = \widehat{M}_1\chi_+(\widehat{T}_1)Z^{-1}f\,.$$

3) Let $n\in\mathfrak{N}$. Then

$$\begin{aligned}\widehat{N}_2 n &= P_{2\Theta}\pi_{2-}n = P_{2\Theta}\pi_{1-}\chi_-(z_1)n = P_{2\Theta}\chi_-(U_1)\pi_{1-}n\\ &= P_{2\Theta}(P_{1\Theta}\chi_-(U_1)P_{1\Theta} + P_{1\Theta}\chi_-(U_1)(I-P_{1\Theta}) + (I-P_{1\Theta})\chi_-(U_1))\pi_{1-}n\,.\end{aligned}$$

Taking into account that $(I-Q_{1+})\chi_-(U_1)Q_{1+}=0$, we get

$$\begin{aligned}P_{1\Theta}\chi_-(U_1)(I-P_{1\Theta}) &= (I-Q_{1-})(I-Q_{1+})\chi_-(U_1)(Q_{1+}+Q_{1-})\\ &= P_{1\Theta}\chi_-(U_1)Q_{1-}.\end{aligned}$$

Since $Q_{1-}\pi_{1-}n=0$, we get $P_{1\Theta}\chi_-(U_1)(I-P_{1\Theta})\pi_{1-}n=0$.
Further, since $P_{2\Theta}Q_{1+}=0$ and $Q_{1-}\chi_-(U_1)(I-Q_{1-})=0$, we obtain

$$\begin{aligned}P_{2\Theta}(I-P_{1\Theta})\chi_-(U_1) &= P_{2\Theta}(Q_{1+}+Q_{1-})\chi_-(U_1)\\ &= P_{2\Theta}Q_{1-}\chi_-(U_1) = P_{2\Theta}Q_{1-}\chi_-(U_1)Q_{1-}\,.\end{aligned}$$

Again, since $Q_{1-}\pi_{1-}n=0$, we get $P_{2\Theta}(I-P_{1\Theta})\chi_-(U_1)\pi_{1-}n=0$. Therefore,

$$\widehat{N}_2 n = P_{2\Theta}P_{1\Theta}\chi_-(U_1)P_{1\Theta}\pi_{1-}n = Z\chi_-(\widehat{T}_1)\widehat{N}_1 n\,. \qquad \square$$

Applying this proposition for "change of variable" $\pi_\pm=\pi_{0\pm}C_\varphi\psi_\pm$, we get that the triple $(\widehat{T},\,\widehat{M},\,\widehat{N})$ is a linearly similar model for a triple $(T,\,M,\,N)$ of the form $(*)$. So, we have proved the following

Theorem 8. *For any pair $\Pi=(\pi_+,\pi_-)$ satisfying conditions* (i), (ii), (iii) *there exist an operator $W\in[H,\mathcal{K}_\Theta]$ and a triple $(T,\,M,\,N)$ of the form $(*)$ such that $W^{-1}\in[\mathcal{K}_\Theta,H]$ and $\widehat{T}W=WT$, $\widehat{M}W=M$, $\widehat{N}=WN$.*

Inverse is also true.

Theorem 9. *Let (T,M,N) be a triple of operators of the form $(*)$. Then there exist pair $\Pi=(\pi_+,\pi_-)$ satisfying conditions* (i), (ii), (iii) *and operators $W,W_*\in[H,\mathcal{H}]$ such that $WW_*^*=P_\Theta$, $W_*^*W=I$ and $\widehat{T}W=WT$, $\widehat{M}W=M$, $\widehat{N}=WN$; $\widehat{T}_*W_*=W_*T^*$, $\widehat{M}_*W_*=N^*$, $\widehat{N}_*=W_*M^*$.*

Remark. *Here the triple $(\widehat{T}_*,\widehat{M}_*,\widehat{N}_*)$ corresponds to the dual pair $\Pi_*=(\pi_{*+},\pi_{*-})$. Note that the following identities hold*

$$(\widehat{T}f,g)=(f,\widehat{T}_*g)\,,\quad (\widehat{M}f,n)=(f,\widehat{N}_*n)\,,\quad (\widehat{N}n,g)=(n,\widehat{M}_*g)\,,$$

where $f\in\mathcal{K}_\Theta$, $g\in\mathcal{K}_{*\Theta}$, $n\in\mathfrak{N}$.

Proof. Since $(T,\, M,\, N)$ is a triple of the form $(*)$, there exists a simple unitary node $\mathfrak{A}_0$ such that $(T, M, N) = (\varphi(T_0),\, M_0\chi_+(T_0),\, \chi_-(T_0)N_0)$. For $\mathfrak{A}_0$, there exists (see Section 1) a pair $\Pi_0 = (\pi_{0+}, \pi_{0-})$ such that the node $\widehat{\mathfrak{A}}_0$ is unitarly equivalent to the node $\mathfrak{A}_0$. Let W_0 be a unitary operator that realizes this equivalence. We take "change of variable" $\pi_\pm = \pi_{0\pm} C_\varphi \psi_\pm$ and $\pi_{*\pm} = \pi_{*0\pm} C_{\varphi^\sim} 1/\psi_\mp^\sim$. Applying Proposition 7, we have

$$\begin{aligned}(\widehat{T}, \widehat{M}, \widehat{N}) &= (Z\varphi(\widehat{T}_0)Z^{-1},\, \widehat{M}_0\chi_+(\widehat{T}_0)Z^{-1},\, Z\chi_-(\widehat{T}_0)\widehat{N}_0),\\ (\widehat{T}_*, \widehat{M}_*, \widehat{N}_*) &= (Z_*\varphi^\sim(\widehat{T}_{*0})Z_*^{-1},\, \widehat{M}_{*0}\chi_-^\sim(\widehat{T}_{*0})Z_*^{-1},\, Z_*\chi_+^\sim(\widehat{T}_{*0})\widehat{N}_{*0}),\end{aligned}$$

where $Z = P_\Theta|\mathcal{K}_0$, $Z_* = P_\Theta^*|\mathcal{K}_0$, and $\mathcal{K}_0$ is the model subspace for Π_0. Using relations from Section 1, we get

$$(\widehat{T}_*, \widehat{M}_*, \widehat{N}_*) = (Z_*\varphi(\widehat{T}_0)^* Z_*^{-1},\, \widehat{N}_0^*\chi_-(\widehat{T}_0)^* Z_*^{-1},\, Z_*\chi_+(\widehat{T}_0)^*\widehat{M}_0^*).$$

Let $W = P_\Theta V W_0$, $W_* = P_\Theta^* V W_0$, where $V \in [\mathcal{K}_0, \mathcal{H}]$ is embedding: $Vf = f$, $f \in \mathcal{K}_0$. Then it is easy to check that $\widehat{T}W = WT$, $\widehat{M}W = M$, $\widehat{N} = WN$; $\widehat{T}_*W_* = W_*T^*$, $\widehat{M}_*W_* = N^*$, $\widehat{N}_* = W_*M^*$. Next, we have $W_*^* = W_0^*V^*P_\Theta$. Whence, using Theorem 5(3), we get

$$\begin{aligned}&W_*^*W = W_0^*V^*P_0P_\Theta P_0VW_0 = W_0^*V^*P_0VW_0 = W_0^*V^*VW_0 = I,\\ &WW_*^* = P_\Theta VW_0W_0^*V^*P_\Theta = P_\Theta VV^*P_\Theta = P_\Theta P_0 P_\Theta = P_\Theta. \qquad \square\end{aligned}$$

Remark. *The relation* $W_*^*W = I$ *can be rewritten in the form* $(W_*f, Wg) = (f, g)$. *Thus here we have the same type of the duality as in* [4, 14].

Remark. *One can also consider the transfer function* $\Upsilon(z) = M(T-z)^{-1}N$. *It is interesting to note that the transfer function* $\Upsilon(z)$ *uniquely determines the pure part of the characteristic function* $\Theta^+(z)$ *provided the functions* $\chi_\pm$ *are known.*

References

[1] Szökefalvi-Nagy B., Foiaş C., *Harmonic analysis of operators on Hilbert space.* North-Holland, Amsterdam-London, 1970.

[2] Brodskiy M.S., *Unitary operator nodes and their characteristic functions,* Uspehi mat. nauk **33** (1978), no. 4, 141–168.

[3] Abrahamse M.B., Douglas R.G., *Operators on multiply-connected domains.* Proc. Royal Irish Acad., **74** A (1974), 135–141.

[4] Yakubovich D.V. *Linearly similar model of Sz.-Nagy – Foias type in a domain.* Algebra i Analiz, **15** (2003), no.2, 180–227.

[5] Naboko S.N., *Functional model for perturbation theory and its applications to scattering theory.* Trudy Mat. Inst. Steklov **147** (1980), 86–114.

[6] Makarov N.G., Vasyunin V.I., *A model for noncontraction and stability of the continuous spectrum.* Lect. Notes in Math., **864** (1981), 365–412.

[7] Tikhonov A.S., *Functional model and duality of spectral components for operators with continuous spectrum on a curve.* Algebra i Analiz, **14** (2002), no.4, 158–195.

[8] Duren P.L., *Theory of H^p spaces*, Pure Appl. Math., vol. 38, Academic Press, New York–London, 1970.

[9] Pommerenke C., *Univalent functions.* Vandenhoek and Ruprecht, Göttingen, 1975.

[10] Nikolski N.K., Vasyunin V.I., *Elements of spectral theory in terms of the free functional model. Part I: Basic constructions*, Holomorphic spaces (eds. Sh. Axler, J. McCarthy, D. Sarason), MSRI Publications **33** (1998), 211–302.

[11] Adamyan V.M., Arov D.Z., *Unitary couplings of semi-unitary operators.* Mat. Issled., **1** (1966), no.2, 3–64.

[12] Garnett J.B., *Bounded analytic functions*, Pure Appl. Math., vol. 96, Academic Press, New York-London, 1981.

[13] Shirokov N.A., *Properties of modulus for analytic functions smooth up to the boundary.* Dokl. Akad. Nauk SSSR, **269** (1966), no.6, 1320–1323.

[14] Yakubovich D. V., *Dual piecewise analytic bundle shift models of linear operators.* J. Func. Anal., **136** (1996), no.2, 294–330.

Alexey Tikhonov
Mathematical analysis division
Taurida National University
Yaltinskaya str., 4
Simferopol 95007 Crimea
Ukraine
e-mail: tikhonov@club.cris.net

Operator Theory:
Advances and Applications, Vol. 154, 233–238

Jacobi Block Matrices with Constant Matrix Terms

Marcin J. Zygmunt

Abstract. We investigate a solution of the difference equation

$$tU_n^{A,B}(t) = AU_{n+1}^{A,B}(t) + BU_n^{A,B}(t) + AU_{n-1}^{A,B}(t)$$

with the boundary conditions $U_0^{A,B} = I$, $U_{-1}^{A,B} = 0$, where A, B are hermitian matrices. $U_n^{A,B}$ are usually called matrix Chebyshev polynomials of the second kind. The above equation cannot be easily simplified as in scalar case because A and B do not need to commute. However we are able to compute spectrum of the corresponding orthogonality measure which is very important to investigate discrete Schrödinger operator related to $U_n^{A,B}$.

Mathematics Subject Classification (2000). Primary: 47B36 secondary: 39A70, 39B42.

Keywords. Chebyshev polynomials, discrete Schrödinger operator, Jacobi block-matrices, matrix orthogonal polynomials.

Let $\mathbb{C}^{N\times N}$ denote a space of all quadratic $N\times N$ matrices. We will write $A \geq 0$ (or $A > 0$ respectively) if A is a positive definite (or strictly positive respectively) Hermitian matrix. In the following the inequality $A \geq B$ (or $A > B$ respectively) will be equivalent to $A - B \geq 0$ (or $A - B > 0$ respectively) for $A, B \in \mathbb{C}^{N\times N}$.

Denote by $\ell^2(\mathbb{C}^{N\times N})$ a space of sequences $X = (X_0, X_1, \dots)$ of matrices from $\mathbb{C}^{N\times N}$, for which the series $\sum_{n=0}^{\infty} X_n{}^* X_n$ converges.

We introduce an "$N\times N$-matrix" product on $\ell^2(\mathbb{C}^{N\times N})$:

$$\langle\!\langle X|Y\rangle\!\rangle_{\ell^2} = \sum_{n=0}^{\infty} Y_n{}^* X_n \in \mathbb{C}^{N\times N}.$$

We have $\langle\!\langle X|X\rangle\!\rangle_{\ell^2} \geq 0$ in the sense stating before. Then a system

$$E_n = (0, \dots, 0, I, 0, \dots),$$

The author was supported by KBN's grant 2P03A00723 and RTN network: HPRN-CT-2002-00279.

where I denotes the identity of $\mathbb{C}^{N\times N}$, appears only in the nth position, forms an "orthonormal" basis of $\ell^2(\mathbb{C}^{N\times N})$.

In a similar way we can define an L^2-space of square integrable matrix-valued functions. Let Σ be a positive matrix-valued Borel measure, i.e., $\Sigma(\Delta)$ is positive definite ($\Sigma(\Delta)\geq 0$) for all Borel subsets $\Delta\subset\mathbb{R}$. For matrix-valued functions $F(x)$ and $G(x)$ we define an "$N\times N$-matrix" product

$$\langle\!\langle F|G\rangle\!\rangle_\Sigma = \int_{\mathbb{R}} F(x)d\Sigma(x)G(x)^* \in \mathbb{C}^{N\times N}\,.$$

Now the space $L^2(\Sigma)$ consists of all matrix-valued functions $F(x)$ for which $\langle\!\langle F|F\rangle\!\rangle_\Sigma$ is convergent. More details can be found in [AN], [B], [D1-3], [DV] and [Z2].

Let $A, B\in\ell^2(\mathbb{C}^{N\times N})$ be Hermitian matrices and let $J_{A,B}$ be an operator on $\ell^2(\mathbb{C}^{N\times N})$ acting as follows:

$$(J_{A,B}X)_0 = BX_0 + AX_1,\quad (J_{A,B}X)_n = AX_{n-1} + BX_n + AX_{n+1}\,.$$

Hence $J_{A,B} = B\mathcal{I} + A(\mathcal{S}+\mathcal{S}^*)$, where $\mathcal{S}$ denotes the "shift" on $\ell^2(\mathbb{C}^{N\times N})$ and $\mathcal{I}$ – the identity operator. With $J_{A,B}$ there are associated matrix-valued Chebyshev polynomials of the second kind $U_n^{A,B}(x)$, i.e., polynomials satisfying the recurrence formula

$$xU_n^{A,B}(x) = AU_{n+1}^{A,B}(x) + BU_n^{A,B}(x) + AU_{n-1}^{A,B}(x)\,.$$

Denote by M_n the n-th moment of $W^{A,B}$

$$M_n = \int_{\mathbb{R}} x^n dW^{A,B}(x) = \langle\!\langle x^n I|I\rangle\!\rangle_{W^{A,B}}\,.$$

Theorem 1. *Let $W^{A,B}$ be an matrix-valued measure which orthogonalizes polynomials $U_n^{A,B}$, i.e.,*

$$\langle\!\langle U_n^{A,B}|U_m^{A,B}\rangle\!\rangle_{W^{A,B}} = \int_{\mathbb{R}} U_n^{A,B}(x)dW^{A,B}(x)U_m^{A,B}(x)^* = \delta_{n,m}I\,.$$

Then the moments M_n of the measure $W^{A,B}$ are equal to

$$M_n = \frac{1}{2\pi}\int_{-2}^{2}\sqrt{4-x^2}(At+B)^n dt$$

Moreover

$$\mathrm{supp}W^{A,B} = \bigcup_{t\in[-2,2]}\sigma(At+B)\,.$$

Proof. We have

$$\langle\!\langle x^n I|I\rangle\!\rangle_{W^{A,B}} = \langle\!\langle (J_{A,B})^n E_0|E_0\rangle\!\rangle_{\ell^2}\,.$$

Let

$$(B + Ax)^n = \sum_{k=0}^{n} C_{k,n} x^n .$$

Then

$$\begin{aligned}\langle\langle (J_{A,B})^n E_0 | E_0 \rangle\rangle_{\ell^2} &= \langle\langle (B\mathcal{I} + A(\mathcal{S} + \mathcal{S}^*))^n E_0 | E_0 \rangle\rangle_{\ell^2} \\ &= \sum_{k=0}^{n} C_{k,n} \langle\langle (\mathcal{S} + \mathcal{S}^*)^k E_0 | E_0 \rangle\rangle_{\ell^2} .\end{aligned}$$

On the other hand $\mathcal{S}E_n = E_{n+1}$, hence the behavior of $\mathcal{S}$ is the same as of the shift operator S on $\ell^2(\mathbb{N})$. Hence $\langle\langle E_n | E_0 \rangle\rangle_{\ell^2} = \delta_{n,0} I = \langle e_n | e_0 \rangle I$, where $\{e_n\}_{n\in\mathbb{N}}$ so

$$\langle\langle (\mathcal{S} + \mathcal{S}^*)^n E_0 | E_0 \rangle\rangle_{\ell^2} = \langle (S + S^*)^n e_0 | e_0 \rangle I .$$

The operator $S + S^*$ is a well-known discrete Schrödinger operator (related to the classical Chebyshev polynomials of the second kind), which spectrum is equal to $[-2, 2]$ and its spectral measure w is equal to $dw(x) = \frac{1}{2\pi}\sqrt{4 - x^2}\, dx$. Hence

$$\langle (S + S^*)^n e_0 | e_0 \rangle = \int_{-2}^{2} x^n w(x) dx .$$

Thus

$$\begin{aligned} M_n &= \int_{-2}^{2} w(t) \sum_{k=0}^{n} C_{k,n} t^n \, dt \\ &= \int_{-2}^{2} w(t)(At + B)^n dt \end{aligned}$$

which proves the first part of the theorem.

To prove the second part note that we have the equality

$$\int_{\mathbb{R}} p(x) dW^{A,B}(x) = \int_{-2}^{2} w(t) p(At + B) dt \tag{1}$$

for every polynomial $p \in \mathbb{C}[x]$. Let now $\Delta = \bigcup_{t\in[-2,2]} \sigma(At + B)$. Δ is a compact subset of the real line $\mathbb{R}$. By polynomial approximation we have

$$\int_{\mathbb{R}} f(x) dW^{A,B}(x) = \int_{-2}^{2} w(t) f(At + B) dt \tag{2}$$

for every continuous function $f \in C(\Delta)$. Now it is not to hard to see that the support of $W^{A,B}$ is equal to Δ. □

Corollary 1.1. *The support of the measure $W^{A,B}$ is equal to*

$$supp\, W^{A,B} = \bigcup_{t\in[-2,2]} \sigma(At + B) .$$

Moreover if the matrix A has non-zero determinant (i.e., it is invertible), the measure $W^{A,B}$ is absolutely continuous with respect to the Lebesgue measure multiplied by the identity matrix I.

Corollary 1.2. *The spectrum of $J_{A,B}$ is equal to $\sigma(J_{A,B}) = \operatorname{supp} W^{A,B}$ and consists of at most N non-degenerate intervals of the real line $\mathbb{R}$.*

Proof. By Theorem 2.4, Ch. VII, [Ber], the spectrum of the operator $J_{A,B}$ is equal to the support of the measure $W^{A,B}$. The second statement of the corollary holds because of the continuity of spectrum. □

Theorem 2. *Let $U(t)\Lambda(t)U(t)^*$, where $U(t)$ is unitary and $\Lambda(t)$ diagonal matrix, be the spectral decomposition of $At + B$. Then*

$$W^{A,B}(E) = \int_{-2}^{2} w(t)U(t)\chi_E(\Lambda(t))\,U(t)^*dt$$

for every Borel subset $E \subset \mathbb{R}$.

Proof. Let

$$At + B = U(t)\Lambda(t)U(t)^* . \tag{3}$$

Putting (3) into (1) gives

$$\int_{\mathbb{R}} p(x)dW^{A,B}(x) = \int_{-2}^{2} w(t)p(U(t)\Lambda(t)U(t)^*)dt = \int_{-2}^{2} w(t)U(t)p(\Lambda(t))U(t)^*dt$$

for every polynomial $p \in \mathbb{C}[x]$. Now by approximation of the characteristic functions of given Borel subset $E \subset \mathbb{R}$ we get the thesis. □

Example 3. This example shows that the assumption on hermitianity of A cannot be omitted if we want to save the absolutely continuity of the orthogonalizing measure.

Let us consider the polynomials $P_n(x)$ which satisfy the recurrence formula:

$$tP_k(t) = AP_{k+1}(t) + A^*P_{k-1}(t),$$

where

$$A = \begin{pmatrix} 0 & b \\ a & 0 \end{pmatrix},$$

and $a > b > 0$. Let

$$V = V^* = \begin{pmatrix} 0 & 1 \\ 1 & 0 \end{pmatrix}.$$

Define now $\tilde{P}_k(t) = V^kP_k(t)$. Polynomials $\tilde{P}_k$ are still orthonormal with respect to the same matrix-valued measure as polynomials P_k, moreover $\tilde{P}_0 = id$, $\tilde{P}_{-1} = 0$. It can be easily verified that polynomials $\tilde{P}_k$ satisfy the following recurrence formula:

$$tP_{2k}(t) = A_1P_{2k+1}(t) + A_2P_{2k-1}(t)$$
$$tP_{2k+1}(t) = A_2P_{2k+2}(t) + A_1P_{2k}(t)$$

where

$$A_1 = V^{2k} A V^{2k+1} = \begin{pmatrix} a & 0 \\ 0 & b \end{pmatrix}, \qquad A_2 = V^{2k} A V^{2k-1} = \begin{pmatrix} b & 0 \\ 0 & a \end{pmatrix}.$$

Hence

$$\tilde{P}_k = \begin{pmatrix} p_k & 0 \\ 0 & q_k \end{pmatrix}$$

where polynomials p_k, q_k satisfy

$$tp_{2k}(t) = bp_{2k+1}(t) + ap_{2k-1}(t), \quad tp_{2k+1}(t) = ap_{2k+2}(t) + bp_{2k}(t),$$
$$tq_{2k}(t) = aq_{2k+1}(t) + bq_{2k-1}(t), \quad tq_{2k+1}(t) = bq_{2k+2}(t) + aq_{2k}(t).$$

Thus, the orthogonality measure for polynomials $\tilde{P}_k$ (and so for P_k) is diagonal, where entries on the diagonal are scalar measures μ_p and μ_q orthogonalizing polynomials p_k and q_k respectively. It was shown in [Ch] that in the case $a > b$ the support of the measures μ_p and μ_q consists of the union of the intervals $[-(a+b), -(a-b)] \cup [a-b; a+b]$, where both measures are absolutely continuous. Moreover, μ_p has a non-zero atom at 0 with the weight $1 - b^2/a^2$. This shows that the matrix-valued measure orthogonalizing polynomials P_k has a non-zero atom at 0, with the weight equal to

$$\begin{pmatrix} 1 - \left(\frac{b}{a}\right)^2 & 0 \\ 0 & 0 \end{pmatrix}$$

References

[AN] A.I. Aptekarev and E.M. Nikishin, The scattering problem for a discrete Sturm–Liouville operator, *Mat. Sb.*, 121(163):327–358; *Mat. USSR Sd.*, 49:325–355, 1965.

[B] Y.M. Berezanski, *Expansions in eigenfunctions of self-adjoint operators.* AMS, Providence, 1968.

[Ch] T. Chihara, *An Introduction to Orthogonal Polynomials*, Gordon & Breach Sc. Pub., New York, 1979.

[D1] A.J. Duran, A generalization of Favárd's theorem for polynomials satisfying a recurrence relation, *J. Approx. Theory*, 74, 83–109, 1993.

[D2] A.J. Duran, On orthogonal polynomials with respect to a positive definite matrix of measures, *Canad. J. Math.*, 47(1):88–112, 1995.

[D3] A.J. Duran, Ratio asymptotics for orthogonal matrix polynomials, *J. Approx. Theory*, 100(2):304–344, 1999.

[DV] A.J. Duran and W. Van Assche, Orthogonal matrix polynomials and higher-order recurrences, *Linear Algebra Appl.*, 219:261–270, 1995.

[Sz] G. Szegö, *Orthogonal Polynomials*, AMS Coll. Pub., vol.23, AMS, Providence, 1975 (4th edition).

[Z1] M.J. Zygmunt, General Chebyshev polynomials and discrete Schrödinger operators, *J. Phys. A: Math. Gen.*, (34) 48, 2001, 10613–10616.

[Z2] M.J. Zygmunt, Matrix Chebyshev Polynomials and Continued Fractions, *Linear Algebra Appl.*,340(1-3):155–168. 2002.

Marcin J. Zygmunt
University of Wrocław
Institute of Mathematics
pl. Grunwaldzki 2/4
50-384 Wroclaw, Poland
http://www.math.uni.wroc.pl/~zygmunt

and

Institute of Mathematics
Wroclaw University of Technology
ul. Wybrzeze Wyspianskiego 27
50-370 Wroclaw, Poland
e-mail: zygmunt@math.uni.wroc.pl

List of Lecture Titles

V. Adamyan
Non-stationary Perturbation Determinants

M. Baro
Dissipative Schrödinger-Poisson Systems

J. Brasche
Discussion of Singular Perturbation of $-\Delta$ with the Aid of $(-\Delta)^2_\mu$

D. Cichon
Three Term Relations Modulo an Ideal and Ortogonality of Polynomials

P. Cojuhari
Spectral Analysis of Jacobi Matrices

M. Demuth
On the Equilibrium Potential in Scattering Theory

J. Derezinski
Simple Models of Infra-Read Problem in Quantum Field Theory

J. Dombrowski
Jacobi Matrices: Perturbation of Double-weights

P. Exner
Two Strongly Singular Point-Interactions Problems

S-M. Fei
PT-Symmetric Point Interactions

R. Fruboes
Singular Perturbations of Operators – Simple Renormalization Models

J.S. Geronimo
Turning Point Theory for Difference Equations with Applications

A. Gheondea
Invariant Hermitian Kernels: A Unifying Approach

E. Giere
The Wegner Estimate in One Dimension for Non – Overlapping Single Site Potentials

D. Gilbert
Bounds for the Points of Spectral Concentration for the One-dimensional Schrödinger Operators

P. Hislop
Continuity of the Integral Density of States for Some Random Operators on $\mathbb{R}^d$

R. Hryniv
Inverse Spectral Problems for Sturm-Liouville Operators with Singular Potentials

Z. Jablonski
Hyperexpansive Composition Operators

W. Karwowski
Unpenetrable Barriers and Canonical Quantization

R. Killip
Trace Formulae and Spectral Theory

A.V. Kisielev
Some Spectral Properties of the Non-Selfadjoint Friedrichs Model

S. Kondej
Spectrum of the Laplace Operator in $L^2(\mathbb{R}^3)$ with a Singular Perturbation Supported by a Not Flat Surface in $\mathbb{R}^3$

V. Koshmanenko
The Aronszajn-Donoghue Theory for Rank One Perturbations of the $\mathcal{H}_{-2}$-class

P. Kurasov
Rank one Perturbations in Mathematical Physics

M. Mantoiu
Anisotropic Differential Operators: the Essential Spectrum and Localization Properties

R. Minlos
The Lower Branches of the Spectrum of the System like Polaron

M. Moszynski
On the Point Spectrum of Some Jacobi Matrices in l^p Spaces

B. Pavlov
Dirichlet-to-Neumann Map and Design of a quantum switch

A. Osipov
On some Issues Related to the Study of Difference Operators Generated by Infinite Band Matrices

A. Posilicano
Nonlinear Singular Perturbations

R. Purice
Eigenfunction Decay for Embedded Eigenvalues of Perturbed Periodic Schrödinger Hamiltonians

R. del Rio
Coexistence of Spectra in Rank One Perturbation Problems

A. Rybkin
On a New Transformation of the One Dimensional Schrödinger Equation and High Order WKB-type Asymptotics

A. Shkalikov
Limit Spectral Curves for the Orr-Sommerfeld Problem and Quasi-classical Eigenvalue Distribution Along the Curves

P.D. Siafarikas
Linear and Non-linear Partial Difference Equations: Some Existence and Uniqueness Results with Applications

S. Sontz
Euler's Constant in Mathematical Physics

J. Stochel
A pecularity of the Creation Operator

G. Stolz
Fractional Moment Methods for Continuum Anderson-type Models

F.H. Szafraniec
Subnormality – the Missed Cell in the Quantization Procedure

G. Teschl
Oscillation Theory for Jacobi Operators

I.M. Tkachenko
Sum Rules and Exact Relations in Quantal Coulomb Systems

W. Van Assche
Jacobi Matrices, Orthogonal Polynomials and the Continuum Limit of the Toda Lattice

J. Yngvason
Bose-Einstein Condensation in Dilute Gases

L. Zielinski
On The Asymptotic Distribution of Eigenvalues for a Class of Schrödinger Operators

M.J. Zygmunt
Generalized Chebyshev Polynomials and Discrete Schrödinger Operators

MIX
Papier aus verantwortungsvollen Quellen
Paper from responsible sources
FSC® C105338

If you have any concerns about our products,
you can contact us on
ProductSafety@springernature.com

In case Publisher is established outside the EU,
the EU authorized representative is:
Springer Nature Customer Service Center GmbH
Europaplatz 3, 69115 Heidelberg, Germany

Printed by Libri Plureos GmbH
in Hamburg, Germany